ENERGY, RESOURCES AND ENVIRONMENT

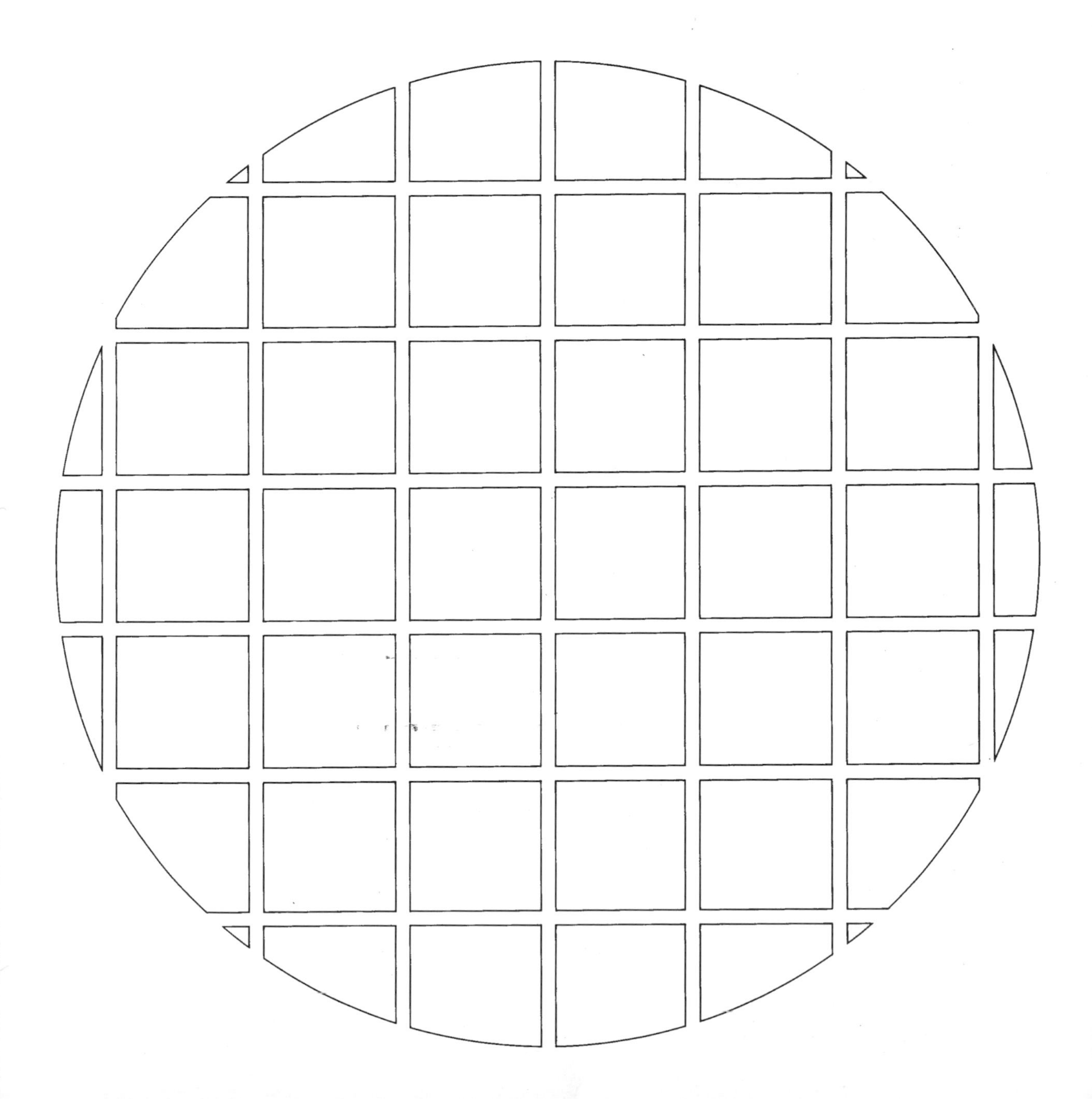

This book is the third in a series published by Hodder and Stoughton
in association with The Open University.

Environment and Society
edited by Philip Sarre and Alan Reddish

Environment, Population and Development
edited by Philip Sarre and John Blunden

Energy, Resources and Environment
edited by John Blunden and Alan Reddish

Global Environmental Issues
edited by Roger Blackmore and Alan Reddish

The final form of the text is the joint responsibility of chapter authors,
book editors and course team commentators.

ENERGY, RESOURCES AND ENVIRONMENT

EDITED BY
JOHN BLUNDEN AND ALAN REDDISH
FOR AN OPEN UNIVERSITY COURSE TEAM

Hodder & Stoughton

A MEMBER OF THE HODDER HEADLINE GROUP

IN ASSOCIATION WITH

This book has been printed on paper made from pulp which was bleached without chlorine; other, environmentally-friendly, oxidation agents were used. The paper mill concerned emits very low levels of effluent, and uses raw material from sources renewable through reforestation.

British Library Cataloguing in Publication Data

A catalogue record for this title is available from The British Library.

ISBN 0–340–66356–1

First published in the United Kingdom 1991. Second edition 1996.

Edited and designed by The Open University.

Index compiled by Sue Robertson.

Typeset by Wearset, Boldon, Tyne & Wear.

Printed in the United Kingdom for Hodder & Stoughton Educational, a division of Hodder Headline Plc., 338 Euston Road, London NW1 3BH, by Butler & Tanner Ltd, Frome, Somerset.

This text forms part of an Open University second level course, U206 *Environment*. If you would like a copy of *Studying with the Open University*, please write to the Central Enquiry Service, PO Box 200, The Open University, Walton Hall, Milton Keynes, MK7 6YZ.

Contents

Introduction

This book is the third in a series of four which present an interdisciplinary and integrated explanation of environmental issues. Throughout it is stressed that environmental issues are complex and that to understand them we must combine scientific evidence and theory, analysis of social processes, knowledge of technological possibilities and awareness of underlying value positions. The series considers a range of environmental issues, some local, many transnational and some global. It aims to widen readers' *awareness* of environmental issues, to increase their ability to *analyse* them and to equip them to *evaluate* policies to influence such issues. It stresses that solutions to particular problems should be complementary with, and ideally contribute to, the solution of international and global problems, including global warming.

The first book identified the basic processes that support the huge variety of life on earth, the impacts of society over millennia and changing beliefs as to how society should relate to environment. The second book examined the relationship between population growth, economic development and environmental change, with particular reference to agriculture and urbanisation. In so doing it began to identify the role of advanced technology in transforming the way society related to environment.

This book makes a new start by considering a few familiar machines in common use in developed countries, and shows that they depend on extensive supply systems for energy and materials. These systems, from natural origin through processing and use to disposal, are the concern of later chapters. Throughout, the book stresses that energy and materials systems are interlinked, not least because mining and mineral processing use large amounts of energy, and that these systems for *supplying* energy and materials should not be thought of in isolation from our *demand* for them, which can change with more efficient use or with our way of life. The book also emphasises the idea that our current notions of the exploitation of finite sources of energy and minerals may not be tenable in the longer term. A change in the way and the rate at which we use these resources may not only be necessary for developed countries and for any future attempts to improve living standards in the developing world, but also may be essential if we are to come to terms successfully with environmental problems, many of which are global.

The first two chapters outline the current systems of the production and consumption of energy and their impact on the environment. The dominance of fossil fuels is a reminder that energy from such sources is being used at a rate which cannot be sustained indefinitely. Moreover, the disparities in levels of consumption between countries is so great that it would be disastrous for the environment if the less developed countries began to use fossil fuels in the same way as the developed countries do. The third chapter considers the alternatives to fossil fuels and how the inevitable transfer to the use of renewable forms of energy may be effected.

The next two chapters explore in detail the impact of current materials systems taking 'a cradle to grave' approach. Thus we begin by examining the effect of visual disturbance, noise and dust around mines and open pits during primary extraction, before moving on to problems of transport spills, chemical pollution from smelting and manufacturing processes, waste dumping, and the disposal of used products. The reduction in the impact of such systems on the environment is also considered as a result of recycling, substitution and the utilisation of 'waste' materials. All these methods have positive advantages in terms of energy conservation and the wise use of other finite resources in a situation, for example, where the rapid depletion of resources might follow a rise in living standards among the peoples of the developing world. Here, the political dimension of the more effective husbanding of such materials is not neglected.

The political responses to the impacts of energy, minerals and manufacturing take centre stage only in the last two chapters. In these, two dimensions of the politicisation process are analysed in terms of waste disposal and public hazard. First, the past and present roles of government in regulating harmful emissions, especially to air and water, are examined. Here, the UK's historical method of negotiating the best practicable means is contrasted with the EU imposition of maximum levels of discharge. Second, the role of public opinion and pressure groups in determining political outcomes is brought to the fore. Frequently public protest arises as a result of a perceived risk of accidents, rather than their reality, or from an aversion to wastes, whether they are unsightly, toxic or radioactive.

One message that certainly emerges from the study of these chapters must be that greater priority needs to be given to strategies for the more efficient use of all forms of waste, hazardous or otherwise, in order to conserve finite resources and reduce pollution.

The fourth book in the series considers two sets of issues: first, the resource and pollution impacts which have recently forced environmental issues on to the global political agenda and, second, the suitability of the concept of sustainable development as a guide to policy at national and international level. The oceans and atmosphere are considered as natural systems and as the scene of changes which have made them global issues and stimuli to international collaboration over ocean management, ozone depletion and the threat of global warming. The concept of sustainable development, which has been so central to the post-Brundtland debate, is critically analysed and shown to be at best an agenda which may bring together the disparate range of environmental issues considered in the series, from habitat and genetic loss, through problems of population growth, agriculture and urbanisation, to the costs and benefits of the technologies of energy and mineral supply.

In thinking out the concept of sustainable development it will be necessary to recall two basic principles from Book One: the biosphere depends on a throughput of energy from the sun and it also depends on the recycling of a finite stock of materials. These are also vital for this book.

Chapter 1 Energy resources

1 Introduction

As you read this chapter, look out for answers to the following key questions.

- What forms of energy exist and how are they converted?
- How are energy and power measured?
- How much energy is currently being used and how much is potentially available?

Raise your eyes from this book and look around you. How much of what you can see is the natural world as it might have been without human intervention, and how much the result of human ingenuity and industry? Even if you are sitting in a garden or the countryside the living things will have been much arranged and organised by human hands. If, as is more likely, you are sitting in your home, the office, a bus or a train, almost everything in sight will be the product of complex manufacturing processes which have transformed the materials of the world into forms we find useful or attractive. Walls and windows, furniture and fabrics, clothes and cutlery, lamps and looking glasses, machines of every kind from cars to refrigerators and television sets – these are all reminders of how far we are from simple survival and how much we depend on the skills of others. (A group of hardy experimenters, used to twentieth century UK conditions, tried to live as though they were in the Iron Age, and they apparently missed above all else good pairs of Wellington boots – so it is not only the obviously advanced products that we take for granted.) A social, industrial and economic organisation of extreme complexity provides this standard of living. Perhaps it is not so surprising that it sometimes shows strain – maybe we should be more astonished that it works as well as it does. The complexity does mean, however, that it is not always at all obvious where environmental problems originate. This book will try to tease out some of the interconnections between the resources used, the services provided and the problems encountered. In the first and fourth chapters we will look at two primary resources, energy and minerals, and then go on in further chapters to the environmental problems created by the ways we use them at present, and some of the choices available for the future.

This first chapter is about energy, which not only makes our present machines turn, but has always been a major influence on the form taken by human society (as described in the first book in this series, *Sarre & Reddish, 1996*, Chapter 2). More fundamentally, the energy of the sun drives the ecosystem (*Sarre & Reddish, 1996*, Chapters 5–7) and, as we shall see, the internal structure of materials implies various forms of energy storage. Most of us have played the 'animal, vegetable or mineral' game to classify different materials; the environmental implications of our production of some of the animals and vegetables by various agricultural processes have been examined in the second book, *Sarre & Blunden, 1996*, Chapters 2–4. The remaining category, mineral, forms the subject of the fourth chapter, but it is not altogether distinct from 'energy'. On the one hand, important energy resources (coal, oil, uranium) are often described as minerals; on the other, mineral extraction and processing require energy, to drive digging and crushing machinery, and to modify chemical structure, e.g. in the smelting of natural ores to produce useful metals.

2 Energy use in three machines

The distinction that is being made between energy and minerals can be clarified, in a preliminary way, by looking at three familiar machines – a car, a refrigerator and a television set. All are complex structures of metals, plastics, glass and more specialised materials (e.g. lead–acid batteries, alumina sparking plugs, quartz–halogen lamps, synthetic rubber tyres; glass wool insulation and chlorofluorocarbon refrigerants; silicon chips and ferrite cores) of wonderfully varied *mineral* origins, and much chemical and manufacturing ingenuity. But they won't do anything (except provide status symbols, and slowly rust away) until supplied with *energy* in a suitable form.

2.1 A car

For a car, the energy source is (at present) almost invariably petrol or diesel fuel, a 'fossil fuel store' of *chemical* energy that burns explosively in the conditions inside the engine to produce *heat*. This is in turn converted into *mechanical* energy in the engine, and hence into movement of the wheels, the vehicle and its occupants, which can all be set in motion from rest and be driven up hills (two basic indicators that energy has been supplied.) Some of the energy produced by the engine is also converted into *electrical* energy and stored in the battery to operate the fuel ignition system, the lights and various electric motors (the starter motor, fuel pump and smaller motors such as in windscreen wipers).

The primary heat-to-mechanical **energy conversion** is characteristic of a **heat engine**. Other examples are the paraffin-fuelled gas turbine of a jet aircraft, or the coal-fired steam turbine of a power station (or indeed the eighteenth-century steam engines that powered the industrial revolution). In a power station the heat engine stage, converting the heat energy of steam into the mechanical energy of steam jets and rotating turbine blades, is followed by a further, mechanical-to-electrical, conversion in the electricity generator attached to the turbine.

2.2 A refrigerator

A refrigerator (of the usual domestic type) has energy supplied as *electricity* from the mains which drives a motor and compressor, an electrical-to-mechanical energy conversion. This compresses the refrigerant, which is a gas such as ammonia or the chlorofluorocarbons (CFCs) which were introduced as they were thought to be more innocuous (until their effect on the ozone layer was observed – see *Blackmore & Reddish, 1996*, Chapter 2). The compression leads to liquefaction and heating of the gas. It is cooled back to the temperature of the room by passing it through an external radiator. It is then allowed to evaporate and expand in pipes surrounding the food storage space where it cools further before returning to the compressor and repeating the cycle. The net effect is that heat is taken *from* the internal cold space *to* the warmer external air. It is an important principle (the famous **Second Law of Thermodynamics**) that this can only be done by supplying energy from some other source to make a 'heat engine in reverse'.

In a refrigerator we are more interested in the cooling effect on the food than the heating effect on the kitchen. But the same principle can be used for heating a building by taking heat from the (already colder) outside air, a nearby river or groundwater, in which case the equipment is called a heat pump.

2.3 *A television set*

A television set also requires a supply of *electrical* energy from the mains, to be converted, among other things, into *light* from the screen and *sound* from the loudspeaker. But this can only form a meaningful picture and message if there is a further supply of electrical energy, to the aerial socket at the back of the set. This can either be, as originally intended, from an aerial picking up energy from *electromagnetic* radio waves sent out by television transmitters, or it can be from a video recorder producing a similar energy pattern from programmes stored on magnetic tape. In both cases this second energy supply is called a **signal** because the amount of energy involved is tiny compared with that supplied from the mains, but it fluctuates in a patterned (coded) way; this ensures that a particular transmitter (or video channel) can be separated out from the others, and the conversion of mains energy into light and sound can be controlled so that the required pictures and speech or music are reconstructed.

Thus a television set is another energy conversion device, to set alongside the machines and heat engines which have contributed so much to the industrial and social fabric of the developed world over the last three centuries. But it is also an **information** device. The feature which is unchanged through all the energy transformations, from the light and sound of the original scene to electrical effects in the television camera, the transmitter, the radio waves, the receiver and back to light and sound, is the distinctive pattern that constitutes information. This sets television alongside earlier information storage methods such as books, pictures and discs, communication methods such as letters and telephones, and the newer access and processing methods of computers. Information transfer too has always been an important factor in the development of society; the great increases in speed and range provided by electronics over the last half-century have already modified work patterns and may gradually transform settlement patterns from those we have come to expect. As we examine the environmental problems of present practice and future options about energy and minerals in the rest of this book, we should also bear in mind the changing possibilities offered by these unprecedented fast, long-distance information transfers.

2.4 *Summary*

The relationship between mineral and energy requirements is illustrated with three familiar machines. All are structurally complex, based on many minerals, and all require energy to operate them: a car uses the chemical energy in petrol to operate a heat engine; a refrigerator uses mains elcctricity to operate a heat engine in reverse; a television set uses mains electricity to produce light and sound, and a further electromagnetic signal determines the programme content. While they are all 'energy conversion' devices, the television set is also an 'information' device.

Activity 1

Before reading any further, pause to reflect on other objects and equipment around you and what you know of their mineral origins, the energy used in making or operating them, and any environmental problems they might cause. Make notes for comparison with ideas that will emerge in later parts of the book.

3 *Energy – what does it mean?*

I have already been using the word 'energy', and implying that it can be converted between various forms, without further explanation, which perhaps has left you feeling rather bemused. The earlier books in this series have similarly made frequent reference to energy of various forms without particularly explaining the terms and units used.

Activity 2

Book One of this series (*Sarre & Reddish, 1996*) made references to coal, nuclear and solar energy, to photosynthesis and energy flow through food webs, to human organisation and its dependence on energy use. Book Two (*Sarre & Blunden, 1996*) referred to energy use in the production of various crops, and their dietary energy value. Make notes on how you think these various usages of 'energy' are related, for reconsideration after Section 4.

Maybe this isn't a problem for you, particularly if you have a background in science or technology. In that case you may be able to skip fairly rapidly to Section 6 of this chapter. But if it all seems obscure and impenetrable I hope Sections 4 and 5 can shed some light on the ideas involved – obviously the material of many textbooks can't all be compressed into two sections of one chapter, but it will be worth catching a glimpse of some general terms and principles to help make sense of current problems. There is no reason to feel surprised or guilty if the ideas seem baffling – 'energy' is an abstract idea of some subtlety, sharper in its scientific definition and richer in its implications than in everyday usage. Much that is now treated as obvious was by no means so two or three centuries ago, and some very able minds have agonised over these concepts. But unifying the terminology and units in different applications continues to be a headache. So take heart, and share in grappling with slippery ideas.

3.1 Different usages

I think there are two kinds of difficulty with the word 'energy'. The first is that it is used along with related words in everyday language in ways which shade into each other. We might describe someone's personality as 'forceful, hard-working, hot-headed, energetic and powerful' and know more or less what we mean by each of these lively adjectives, but we would be hard pressed to disentangle from this exactly how the corresponding five abstract nouns 'force, work, heat, energy and power' were to be distinguished. In scientific language each of these five words is sharply defined and they are by no means interchangeable; we need to watch their use carefully. In Section 4 the main emphasis will be on *energy* in its various forms, but it will be related back to the underlying ideas of *force*, *work* and *heat*. The important distinction from *power* will emerge in Section 5.

The second difficulty is that this sharper scientific use of 'energy' has been adopted by the UK government and picked up by the media and the rest of us, but only up to a point. We used to have a Department of Energy rather than a Ministry of Fuel and Power; the newspapers will write of an 'energy crisis' or of 'energy policy' rather than the specific problems of oil, coal or nuclear power stations. So some of the unifying ideas linking specific **fossil fuels** (coal, gas, oil, which store chemical energy) and nuclear fuels (such as uranium and plutonium) with ways of producing electricity (by heat engines driving generators) are generally recognised, but others, which will emerge below and may be crucial for the future, have often been neglected. So it will be worth trying to get a more general grasp of underlying energy principles so that the continuing debate about future policies (energy efficiency, use of solar or renewable energy versus fossil and nuclear fuels, pollution, the ozone hole, the greenhouse effect and so on) can be better followed.

3.2 Fossil fuel reserves – a first look

Before embarking on this more abstract analysis we can stay briefly with the conventional idea of energy resources by looking at the way 'proved reserves' of oil, gas and coal were distributed around the world in 1994 (Figures 1.1a, 1.2a and 1.3). Since the industrial revolution, national prosperity has been intimately linked with ready access to these fossil fuels. For Britain in the eighteenth century, coal was crucial; it soon became equally important throughout the world, as it still is. For the USA at the end of the nineteenth century, oil also became vital, supporting the development of the automobile, again soon a world issue. Throughout the developed world in the mid-twentieth century, natural gas has been adopted as a cleaner domestic and industrial fuel. So current perceptions about the amounts of these three fuels now available give an instant image of powerful economic and political influences at work at the end of the twentieth century.

Figures 1.1a, 1.2a and 1.3 (which use data from the oil industry) show the map of the world with proportional representations of **proved reserves** of the three fuels. This is a term to be treated warily; it does not represent the total resource judged to be ultimately available, but only that part which can be extracted 'under existing economic and operating conditions'. That is, it depends on the state of exploration and extraction technology, and on world prices, which all the main operators in the energy industries try to keep favourable. The quoted proved reserves of oil and gas have

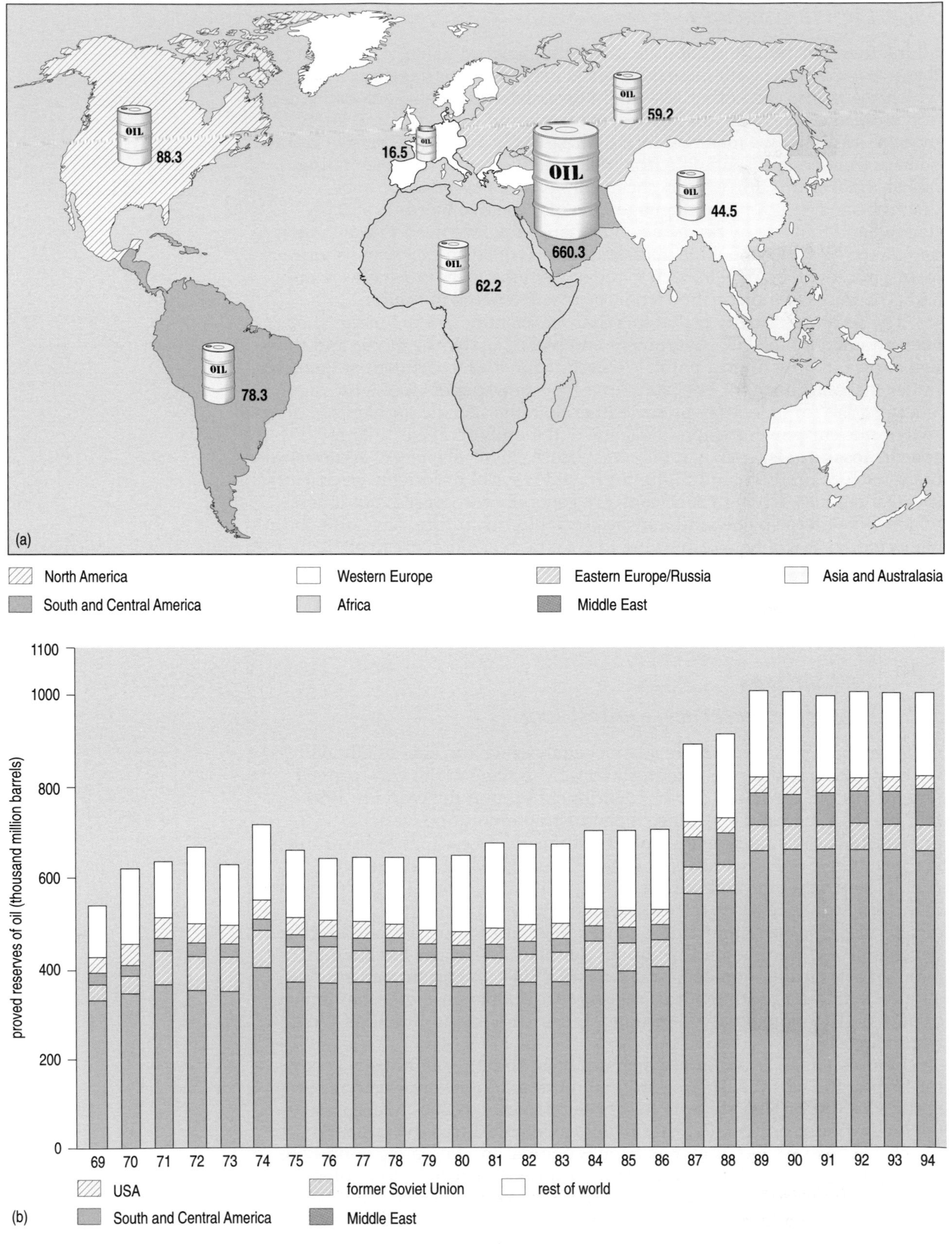

▲ *Figure 1.1 Oil: proved reserves (billion barrels): (a) Global distribution, 1994; (b) Changes since 1969.*

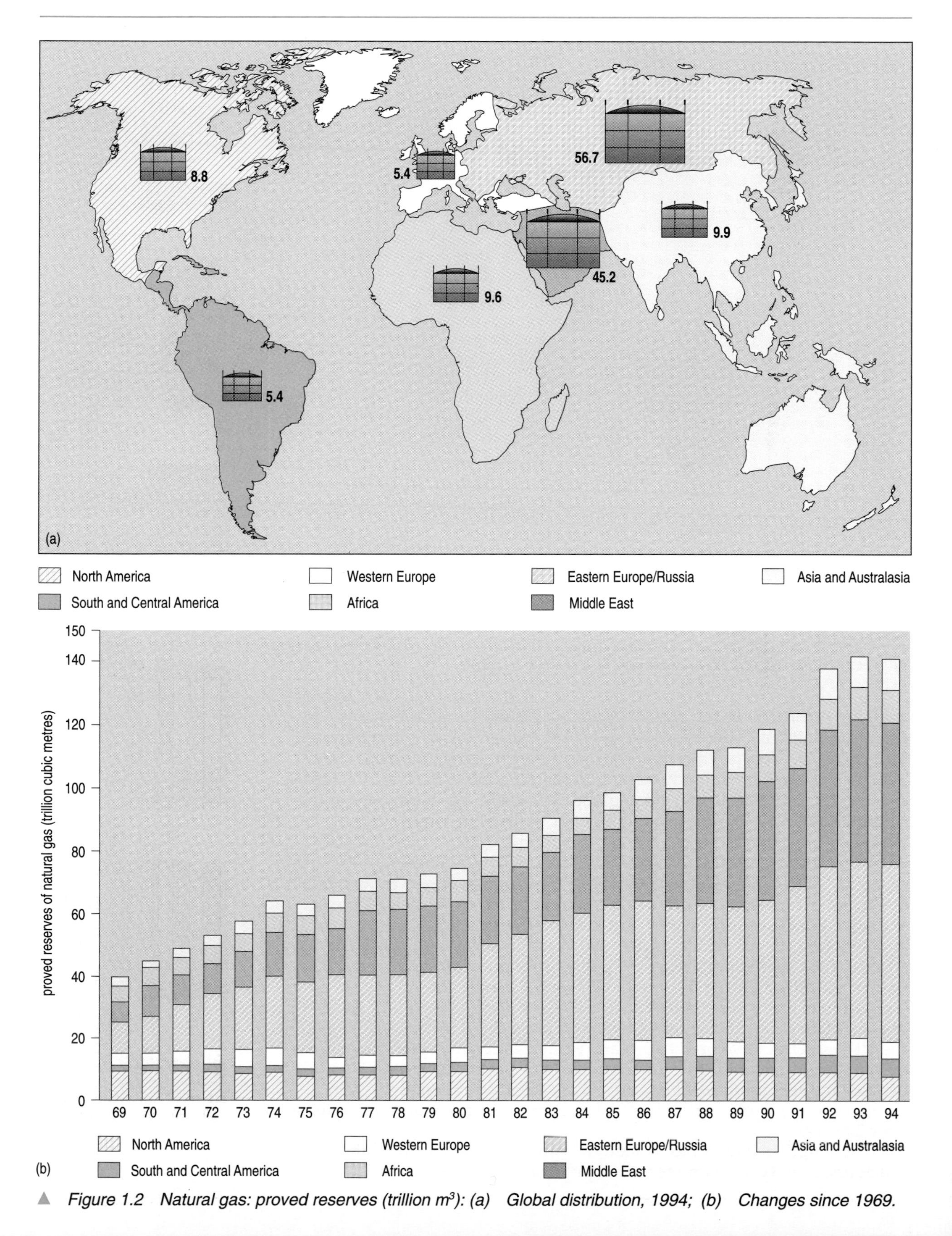

Figure 1.2 Natural gas: proved reserves (trillion m^3): (a) Global distribution, 1994; (b) Changes since 1969.

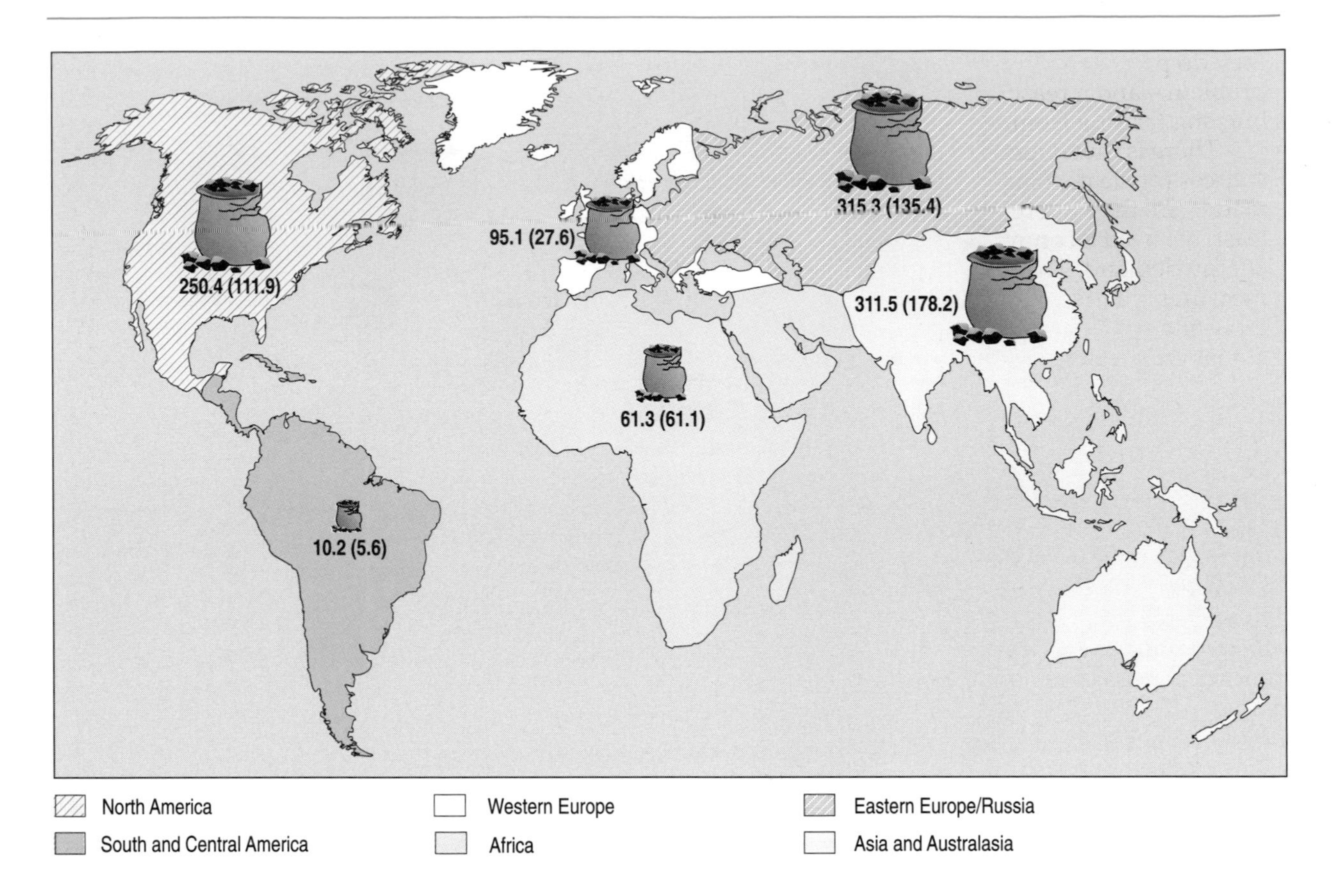

▲ *Figure 1.3 Coal: proved reserves (thousand million tonnes). Map representing global distribution, 1994. The share of anthracite and bituminous coal is shown in brackets.*

actually *increased* over the past 20 years as a result of exploration and changing economic conditions, in spite of the large consumption (Figures 1.1b and 1.2b). This process may well continue for some time as demand increases, though it evidently cannot do so indefinitely. Even so, the most spectacular, oil, is fairly hair-raising; the overwhelming dominance of the Middle East gives an immediate reason for political instability in that region.

What these maps do not show is the relation of these reserves to world demand. This is conveyed by a figure for how long the reserves would last at current extraction rates. This turned out to be for oil (in 1994) 43 years averaged over the whole world, but ranging from a mere 5 years for the UK and 19 for North America to nearly 100 years for several Middle Eastern states, which again emphasises the political volatility of the oil issue. For natural gas, the picture presented in Figure 1.2 is not much more comfortable: Eastern Europe and Russia now outstrip the Middle East. For the world, the 'life' at current extraction rates is about 66 years, but again the UK and USA have under 10 years, while Eastern Europe/Russia have 80 years and several Middle Eastern states have over 100 years. Only for the original industrial fuel, coal (Figure 1.3), is the distribution more even throughout the world, with projected lives at current consumption rates of over 200 years.

These figures have all to be viewed with some scepticism, as there is much uncertainty and commercial or political calculation involved, but they do present an immediate first impression of one part of the energy problem – and a reason why coal-rich countries view any constraint on coal burning (arising from the greenhouse effect, for example) with such alarm.

There is much more comparison between these and other energy sources (nuclear, solar) to be done before a more general assessment can be made. Are they interchangeable? Can the system be made sustainable? Not least, we need a common way of comparing the energy available with energy demand. These three maps, for example, all use different physical measures – barrels of oil, cubic metres of gas, tonnes of coal (all by the thousand million or more). We will reach a common measure, and a more global view of supply and demand, in Section 5.

3.3 *Summary*

Two sources of confusion about the use of the word 'energy' are suggested. There is the overlap in everyday language with words like force and power; the media tend to use it to cover only fuels and power stations.

A first impression of some current energy resource problems is conveyed by figures on the world distribution of oil, gas and coal reserves; these emphasise the need for a more general basis of comparison with other energy resources and with demand.

4 *Forms of energy and energy conversion*

For present purposes it is useful to distinguish five forms of energy – mechanical, electromagnetic, thermal, chemical and nuclear – which will be discussed in turn. The most familiar energy process, burning fuels to produce heat, is chemical-to-thermal energy conversion in these terms. These two forms of energy are formally related to the first two in the list, but it is convenient to continue to keep them separate. Each of the five forms covers a wide range of effects, including some kinds of energy conversion within the form; other kinds of conversion take place between forms.

These many kinds of energy conversion can be summarised in a table (Table 1.1), which we can build up as we proceed. There is no need to remember all the details, but try to get a general picture of the wide range of possible conversions.

There are potentially 25 'boxes' to be filled in as we consider ways of converting from one of the five forms, listed at the left, to the same or a different one of the five listed along the top – though there will be no need to work patiently through each of them in turn (you will be glad to hear). The boxes along the diagonal represent conversions *within* one form, the rest conversions *between* forms. I shall start at the top left with conversions within the mechanical form, which introduces some fundamental ideas. Then as I consider each new form of energy in turn, I can fill in the conversions within it, and also with each preceding form, completing a larger part of the table each time, as you will see.

Table 1.1 Conversion between different forms of energy

To / From	Mechanical	Electromagnetic	Thermal	Chemical	Nuclear
Mechanical					
Electromagnetic					
Thermal					
Chemical					
Nuclear					

4.1 Mechanical energy

Mechanical energy is associated with machines of all kinds; old-fashioned clockwork is a good example. It also arises in everyday experiences with rolling and bouncing balls, in musical instruments, and in less mellifluous producers of noise and turbulence such as pneumatic drills, waterfalls and earthquakes.

A raised weight, a wound-up spring or the stretched elastic in a catapult has **potential energy**. 'Work' has been done against a 'force' to provide this energy: the gravitational force pulls the weight back to Earth and the internal forces inside the spring or elastic try to restore its normal shape. Conversely, anything storing energy, of the many kinds to be considered, is able to do work in this sense, moving something against a force. One weight falling can raise a heavier one by means of a system of pulleys, for example. The heavier weight must move a shorter distance, showing that two factors are involved in work and potential energy: the force and the distance moved. (Try telling a weightlifter that he only does work when he lifts a barbell, not when he is 'just' holding it above his head (Figure 1.4) – so much for precise scientific language!)

This potential energy can be converted into the **kinetic energy** of movement, as when a ball, or a stream of water, picks up speed as it runs down a hill, or a stone is released from a catapult. The kinetic energy of a moving stream of water, air or high-pressure steam can be converted into the *rotational* kinetic energy of a water wheel, windmill or turbine. Conversely, energy must be supplied (from the rotating shaft driven by the engine) to get a stationary car moving, or to raise it to the top of a hill. Similarly, energy must be supplied (from a pump) to move or raise water or (from a fan) to move air.

▲ Figure 1.4
A weightlifter doing no work!

Vibrations and waves

The conversion of potential to kinetic energy needn't take place once and for all: it can be recovered (nearly), as when a ball bounces back into your hand. Or it can be converted regularly back and forth, as when a ball rolls up and down the sides of a bowl, or a weight hanging on a rod or string forms the pendulum of a clock or a child's swing (Figure 1.5). In these cases potential energy at the extreme positions is converted into kinetic at the intermediate position, and back again. These various types of oscillation about a central position are similar to the vibrations that can arise in all kinds of things when they are disturbed from their resting position. The time taken by each vibration can vary over a wide range, from several seconds for large systems (bridges or the water of the sea) to ever smaller fractions of a second in musical instruments producing higher and higher notes. In ultrasonic dog whistles and medical scanners the notes are too high for us to hear, and each vibration can take only a few millionths of a second.

Of course, a swing (or any other vibrating system) left to itself won't go on for ever converting potential and kinetic energy back and forth. The vibrations die down, because of friction between moving parts, air resistance, or similar effects in liquids and solids. It is a basic principle, the law of conservation of energy (or more grandly, the **First Law of Thermodynamics**) that any energy which seems to be missing must be accounted for in some other form. So where does it go? The usual way energy can seem to be lost, or dissipated, is as heat or **thermal energy**, a different form of energy I shall come back to below. This dissipation is the main reason why mechanical energy conversion 'machines' (windmills, musical instruments) have less than perfect **efficiency**, which is a measure of how much of the energy supplied is converted into the form required. The whole of this section is based on the idea that energy is never destroyed, only converted from one form to another, in all kinds of complicated ways. The use of a common unit for measuring energy, which we shall come to in the next section, is likewise only possible because of this principle.

So when you strike a gong, pluck a string or blow a whistle, you convert kinetic energy (of the mallet, your finger or the air stream) into **vibrational energy**, which is alternately potential and kinetic, but dies away as thermal energy unless energy continues to be converted to compensate. Such a conversion can be very orderly – the steady fall of a weight or unwinding of a spring can be used to maintain the regular oscillations of a pendulum in a clock; the steady motion of a violin bow, or a flautist's breath, can produce the regular vibrations, in the string or the air, of a pure note. Or it can be disordered – the hammering in a machine shop, the crash of dropped crockery, producing the confused sounds we call **noise**. In either case there is energy conversion 'within the mechanical form', from steady effects to rapid vibrations of some kind.

The reason you can hear the sounds of musical instruments or breaking crockery is because **sound waves** travel from them through the air to your ear. You may have seen the demonstration of an electric bell ringing in a glass jar from which the air can be pumped away; as it is sucked out the sound dies away, though the hammer can be seen to be striking the bell as before. This emphasises that air has to be present for sound to be heard at a distance, and is an example of a more general energy transfer process.

When a vibration, or indeed any kind of mechanical disturbance, is set up in one place in a gas, liquid or solid, energy can be 'handed on' progressively *through* the material as waves. These are of various kinds depending on the exact conditions. For each kind of wave there is a distinctive speed at which the disturbance travels, though the material at each point only vibrates around a fixed position. We normally experience sound waves in air, but they travel similarly in other gases, and in liquids and solids. You can hear sound under water, or by putting your ear to the ground – the speed of sound is then greater than in air.

There are other types of mechanical wave, each with their own speeds. Most obvious are those we first learned to call waves at the seaside, those on the surface of water. Others arise in the strings and drumskins of musical instruments, and, more disturbingly, in the Earth, carrying away the energy from earthquakes.

Waves, too, die away as their energy is dissipated as thermal energy.

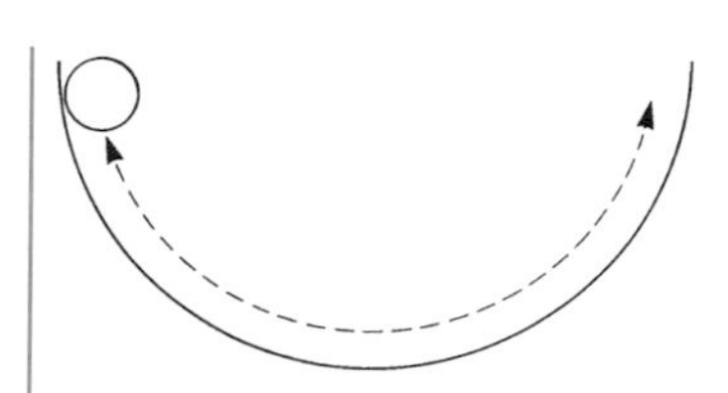

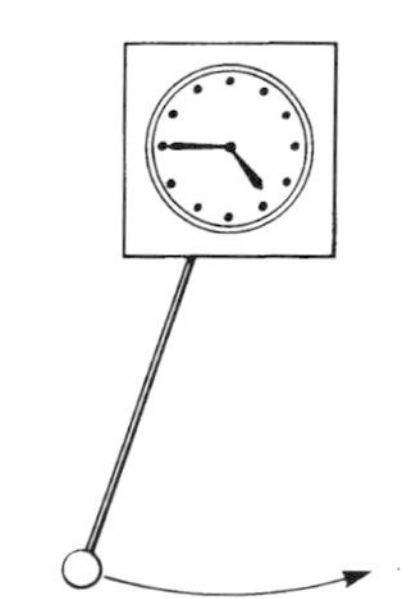

Figure 1.5 Oscillations – ball, pendulum, swing.

Think of the sequence of events when a bucket of water is poured into a water-butt. The original potential energy of the raised water is successively converted into the kinetic energy of the falling water, the turbulent motion (and sound) of the splash, and waves travelling outwards, then back and forth as they bounce off the sides. As the ripples die down all the energy initially supplied is finally turned into thermal energy (shown by a slightly increased water temperature, which in turn falls back to normal as heat is lost to the surroundings).

The conversions 'within the mechanical form' covered in this section can now be incorporated into the table (Table 1.2); dissipation as thermal energy can also be included as a first example of conversion 'to another form'.

Table 1.2

From \ To	Mechanical	Electromagnetic	Thermal	Chemical	Nuclear
Mechanical	potential/kinetic turbine vibration sound waves turbulence		dissipation (friction etc.)		
Electromagnetic					
Thermal					
Chemical					
Nuclear					

4.2 *Electromagnetic energy*

This rather long-winded term 'electromagnetic' is used rather than 'electrical' to emphasise that the electricity we all use every day is only part of a large family of related forms of energy. The familiar sources of electricity are batteries (in portable equipment such as torches and radios, and in cars) and the mains that link our homes and workplaces to power stations. They are used to operate all kinds of equipment – radios and television sets, lights, the motors of vacuum cleaners and washing machines, heaters and cookers.

When we use them to operate radios, microwave cookers and lights we are taking advantage of 'energy conversions within the electromagnetic form'. The ideas of mechanical vibrations and waves described above have electromagnetic equivalents, but with some important differences. The most basic is the question of 'what is doing the vibrating?', which cannot be answered simply. The answer given now for most materials is **electrons**, which are minute constituents of matter, part of its basic structure. They can move fairly freely through conductors such as metals to form an electric current. This can flow steadily in one direction (direct current, d.c.), which is what happens when a wire is connected across a battery. Or it can change direction many times a second: in a wire connected across the alternating current (a.c.) mains, the current changes back and forth 50 times a second –

a frequency of 50 hertz (Hz). But vastly higher frequencies are possible – for example your radio and television will have some way of indicating the frequencies of different transmitters, from kHz (kilohertz, thousands of Hz) to MHz (megahertz, millions of Hz). These imply that each transmitter involves its own oscillator, converting electrical energy from a battery or the low-frequency a.c. mains into currents fluctuating at these high frequencies (much higher than the mechanical vibrations of sound). Furthermore, such currents in the aerial of the transmitter produce **electromagnetic waves** of corresponding frequency which travel away and can be picked up by your receiving aerial. These waves differ profoundly from sound, or other mechanical waves: their speed is vastly greater ('the speed of light', 3×10^8 metres per second, apparently the highest speed which can be reached by any means) and they can travel through vacuum, a fact on which our survival crucially depends.

Why is this? The clue is the reference to the speed of *light*. Fluctuations of currents at radio frequencies, and the corresponding electromagnetic waves, are by no means the end of the story. Electrons are so light and mobile that in various circumstances they can move between or within atoms so as to produce electromagnetic waves of much higher frequency still, all travelling through a vacuum at this same speed. All these waves together form the so-called spectrum of **electromagnetic radiation** – radio frequencies up to about 1000 MHz (10^9 Hz); the microwaves used in cookers, radar and satellite communications (10^9 to 10^{12} Hz); the infra-red radiation from hot objects (10^{12} to 10^{15} Hz); visible light from red to violet in a narrow band near 10^{15} Hz; ultra-violet, x-rays and gamma-rays at even higher frequencies. (These different wavebands are also often described by their *wavelengths*, which decrease as the frequency increases. Thus infra-red to ultra-violet is from some 10^{-4} m to 10^{-7} m, as mentioned in *Blackmore & Reddish, 1996*, Chapter 3.) The sun, our primary source of energy for the Earth, radiates a wide range of electromagnetic waves from infra-red to ultra-violet, which only reach us because they can carry energy across the vacuum of space. (A television set in a bell-jar would not stop working when the air was pumped out; the radio waves would still reach it, and we could see, but not hear, it.)

When we switch on a traditional light bulb, the flow of current makes a tungsten filament white-hot, producing a similar wide range of radiation. A fluorescent lamp is more efficient as a light source because the materials used have been chosen to provide effects limited to narrower bands of frequencies, channelling more of the electrical energy into visible light and less into infra-red ('heat'). An x-ray tube similarly converts d.c. electricity into the even higher frequencies of x-rays. Conversely, a solar cell has materials chosen to convert the high-frequency energy of sunlight into d.c. electricity like that from a battery.

When we use a battery or the mains to drive an electric motor we are moving out of the electromagnetic form to 'conversion to mechanical energy', the kinetic energy of a rotating shaft. This process depends on the subtle relationship between electric currents, magnets and forces that was disentangled, particularly by Michael Faraday, in the first half of the nineteenth century. The process is used in reverse in a **generator**, which is the normal way that electricity is produced in a power station, or in a car, from the rotations of a shaft driven by a steam turbine, or a petrol engine. (Different versions produce d.c. or a.c.) The same principles are also used in some types of loudspeakers and microphones to convert higher frequency electric currents to and from the mechanical vibrations of sound, carefully matching the detailed fluctuations of music, speech or other noise.

Finally, there is one of the commonest uses of electricity: conversion to thermal energy in electric fires and cookers. I could have described this first, as an inevitable process whenever a current flows. Electrons are not completely free to move in metals and other conductors: they encounter resistance, which dissipates electrical energy as heat, much as friction and similar effects dissipate mechanical energy. I kept it to the end for three reasons. First, to emphasise that it is a 'loss' process which limits the efficiency of other types of conversion, as friction does for machines. Second, so that infra-red could be mentioned first, alongside other electromagnetic waves. The fact that it is produced by all hot objects says something about the nature of thermal energy. Third, because in terms of total energy policy it is a wasteful way to use electricity, as we shall see, though it may be convenient in other ways.

We can now fill in these various new conversion processes in the table (Table 1.3).

Table 1.3

To From	Mechanical	Electromagnetic	Thermal	Chemical	Nuclear
Mechanical	X	generator microphone	X		
Electromagnetic	motor loudspeaker	oscillator lamp x-ray solar cell	dissipation (resistance)		
Thermal					
Chemical					
Nuclear					

4.3 *Thermal energy*

This term is used rather than 'heat energy' to acknowledge that there have been some shifts of usage of the word 'heat' over the time that it has been studied more carefully. Three forms of heating have been known about for a long time – that from the sun, long before there were humans to experience it (let alone to describe it as we do now, as the arrival of electromagnetic radiation); that from the burning of fuels such as wood (the release of stored chemical energy), and that dissipated during mechanical work, during thousands of years of human experience. The reverse process, of producing work from heat, has only been available for the two hundred years or so since the industrial revolution, with the invention of the steam engine and other forms of heat engine.

The starting point for more careful consideration of heat is the idea of **temperature**. We can get a direct impression of 'hotness' by touch, and can turn this into a more precise measurement by using a thermometer. The temperature reading obtained in this way must not be confused with the amount of thermal energy that needs to be supplied or removed to *change* the temperature. Everyday experience with cookers shows that it takes longer to boil a pan of water than to get the empty pan red-hot (or even to melt!). The amount of energy needed to produce a particular temperature change depends on the amount, and the nature, of the material being heated.

If two objects at different temperatures are brought into contact, but 'isolated' from the rest of the world, so that they can exchange thermal energy with each other, but not with anything else, the hotter object cools, and the colder warms up until they are both at the same temperature. Think of hot tea in a vacuum flask (the best thermal isolation we can provide). When cold milk is added, the mixture rapidly reaches a uniform intermediate temperature. This is normally described as *heat* transfer from the hot tea to the cold milk, though it could equally be described as a conversion of some of the internal thermal energy of the tea into that of the milk. It is *not* described as a transfer of 'cold' from the milk to the tea! Even cold milk has a thermal energy content – it could be even colder. Heat 'flows' from higher to lower temperatures.

A further normal experience is that (even in a vacuum flask) the uniform temperature of the milky tea gradually falls as heat is transferred to the surroundings. The law of conservation of energy requires that no energy should actually be lost, but as energy is shared with the cooler surroundings the temperature of the tea falls – sometimes described as a reduction in the 'quality' of the thermal energy. One version of the Second Law of Thermodynamics is that there is no *spontaneous* process by which the opposite can happen – energy cannot flow of its own accord from the cool surroundings into the tea to warm it up! Put like that it might seem pretty obvious, but there is nothing in the First Law, that of energy conservation, to say that it can't. The Second Law points out that there is a direction, that of falling temperature and energy 'quality', which is always followed in purely thermal processes.

But these conversions purely 'within the thermal form' are only a small part of the story. We have already seen that unwanted dissipation as thermal energy always occurs to some extent in mechanical and electromagnetic conversions, because of friction and resistance. Such processes are irreversible, and together with the cooling just described, point to a general tendency for energy to degrade from mechanical and electromagnetic forms to thermal energy, which then further degrades to progressively lower temperature. The example of water poured into a water-butt described mechanical energy converting into turbulence, and the ripples dying away into thermal energy. In a comparable way microwave radiation in a microwave cooker is absorbed in foodstuffs (mainly in their water content) and turned into thermal energy.

The sun's radiation is absorbed by the Earth's atmosphere and surface, heating them to a temperature at which they re-radiate just as much energy back into space as they receive (with complications producing the greenhouse effect, as described in *Blackmore & Reddish, 1996*, Chapter 3) – in which case the temperature remains constant. The principle mentioned above, that hot objects give out infra-red radiation is, more correctly, that all objects give out radiation over a wide frequency band depending on their temperature, mainly microwave at low temperatures, reaching further into the infra-red, visible (red-hot and white-hot), and ultra-violet at progressively higher temperatures. The radiation arriving from the sun covers the range of frequencies appropriate to its high temperature; the same *quantity* of energy leaving the Earth is in the much lower range of frequencies appropriate to its lower temperature, i.e. of lower *quality*.

These effects all lead to the view that the internal thermal energy associated with the temperature of an object is a combination of mechanical and electromagnetic fluctuations within the material, differing from the orderly vibrations described previously in their 'noisy', random character. Thus the temperature of a gas can be directly related to the random motion of its **molecules**, which dart about between collisions with each other and the walls of the container at speeds comparable to the speed of sound. In liquids and solids the molecules are not so free to move, but there are also possible electron movements within the material, so that the thermal energy is a combination of random mechanical vibrations and high-frequency currents dependent on the temperature and the particular structure, with much diversity in the details.

While this picture makes it plausible that mechanical or electrical work done on a material can lead to an increase of internal thermal energy, because of the *irreversible* effects of friction or resistance, it doesn't yet explain how a heat engine can perform the converse operation of turning heat into work. The thermodynamic ideas involved were first developed to explain the steam engine, which like other heat engines involves a cycle of compression, heating, expansion and cooling of a gas in a cylinder with a moving piston. The average effect of the random thermal motions of the molecules is to exert pressure, a force balancing that applied from outside to the piston. When the piston moves to compress the gas, *work* is done on the gas, increasing its internal thermal energy (and temperature). Conversely, in expansion, work is done by the gas at the expense of its internal thermal energy. The internal thermal energy can also be increased, of course, by supplying heat from some external source, and conversely decreased by removing heat to an external cooling system. These processes are in principle *reversible*, and in practice nearly so. When the implications are followed through, it turns out that heat supplied from the external source *can* be converted into work, but that even in the most favourable circumstances, only a fraction of the heat supplied can be converted, the remainder being 'rejected' at the lower temperature of the cooling system.

This is a basic thermodynamic constraint on the efficiency of any thermal-to-mechanical energy conversion system, like a car engine or the steam turbine of a power station (the subsequent mechanical-to-electrical conversion in the generator is a separate operation, not limited by these thermodynamic ideas). This means in practice that present types of power station only convert 30–40% of the thermal energy available from the burning of fuel into electricity, and the rest is 'rejected' as heat. This can either be wasted, lost to the atmosphere via cooling towers as is usual in the UK, or it can be used in a combined heat and power (CHP) system, pumped as hot water round buildings in the surrounding area (district heating), as is often done elsewhere. Furthermore, as this inevitable 'thermodynamic price' has been paid to produce high-grade electricity, it is wasteful to degrade it immediately to heat (in electric fires etc.) without using it to do useful work (motors, lights, electronics, etc.) first. Of course, the irreversible effects of friction and resistance mean it will all turn to heat in the end.

The same thermodynamic principles also show that heat can be transferred from a low to a high temperature after all – but only if external work is supplied, driving a heat engine in reverse, to satisfy the Second Law: this is the principle of refrigerators and heat pumps.

Finally, thermodynamics can be extended beyond thermal-mechanical effects in gases to electromagnetic effects in other materials as well. This has wide implications. An example that can be included in our conversion table is that thermal-electrical conversion can take place *directly*, without the moving parts of heat engines and generators, in the 'thermoelectric effect'. If two suitable conducting wires are joined end to end to form a loop, an electric current flows if the junctions are held at different temperatures – a type of generator. Conversely, if a current is driven through the loop by a battery, one of the junctions is cooled – a type of refrigerator. This effect has so far only found small-scale applications, and is subject to similar fundamental efficiency constraints to those of heat engines, but may have more importance in the future.

Again the table can be filled in (Table 1.4).

Table 1.4

To / From	Mechanical	Electromagnetic	Thermal	Chemical	Nuclear
Mechanical	X	X	(dissipation) refrigerator heat pump		
Electromagnetic	X	X	(dissipation) thermoelectric cooling		
Thermal	heat engine	thermoelectric generation	heat transfer cooling		
Chemical					
Nuclear					

When heat is supplied or work is done to a material, it does not always change the temperature (and the internal thermal energy) in the way described above. It may also produce further internal energy changes. One type is called phase change, such as the melting of ice or the boiling of water, in which the material retains its chemical identity (molecules of hydrogen and oxygen, symbol H_2O), but the interaction between the molecules changes, to the progressively greater freedom of the liquid and gaseous states. This can be accommodated by introducing another kind of internal energy, called latent heat, which is the amount of energy needed to produce the phase change, or released when the change is made in the reverse direction, without changing the temperature. Such phase-change effects contribute to the operation of refrigerators and steam engines.

4.4 Chemical energy

A more deep-seated form of internal energy is that bound up in the structure of the molecules themselves. Chemistry recognises about a hundred distinct elements that make up the molecules. Each has a characteristic atomic structure on a minute scale – **atoms** have a diameter of only about one ten thousand millionth of a metre (which can be written more conveniently as 10^{-10} m; see Box 1.1 later in the chapter). The implication of this submicroscopic scale is that normal quantities of material contain unimaginably large numbers of atoms. They are indestructible by normally available processes (exceptions form the subject of the next section). The wide range of different materials arises from the innumerable possible combinations of such elementary atoms into similarly minute molecular structures held together by **chemical bonds**, subject to various rules about the possible pairs of atoms which can bond together. Each type of bond requires the supply of a certain amount of energy to break it (or, conversely, releases a certain amount of energy when it is formed). Chemical reactions between bulk materials involve rearrangements of the atoms in the reacting molecules to form product molecules of different structure; that is, the chemical bonds break and re-form. The detailed sequence of events on the molecular scale may be quite complicated, but the final state is a collection of molecules with a different total energy in all the bonds from that in the initial state. The difference can be positive or negative; that is, the reaction may either give out or absorb energy on balance (and is called exothermic or endothermic respectively).

In our conversion diagram it is scarcely possible to find any conversions wholly 'within the chemical form' because chemical reactions always involve a balancing release or absorption of energy in this way. Often this is thermal energy, i.e. the disordered combination of mechanical and electromagnetic fluctuations described above. Conversions of this kind are studied in **thermochemistry**.

Fuels and photosynthesis – the carbon cycle

Those materials that react with air to give out heat are candidates for consideration as fuels. For example, when coal burns, carbon (C), its main constituent, combines with the oxygen of the air (O_2) to form the strong carbon–oxygen bonds in carbon dioxide (CO_2), with considerable energy release. Hydrocarbon fuels like natural gas (methane, CH_4) already contain carbon–hydrogen bonds which must first be broken during combustion, permitting the formation of carbon–oxygen and hydrogen–oxygen bonds in the products (carbon dioxide and water) – again with net energy release because of the way the bond energies add up.

The process of **photosynthesis** in plants is in a sense the converse of this. Plants are able to take carbon dioxide from the atmosphere and use energy supplied from sunlight to break up its carbon–oxygen bonds so that the constituent atoms can be recombined, along with other elements, to form the enormous range of organic compounds, such as sugars and carbohydrates, needed for biological processes; some oxygen is released back to the atmosphere. Some of these compounds are useful 'energy stores' for the process of respiration, which uses atmospheric oxygen to provide energy release within the organism when required. Like burning, this produces carbon dioxide. Normal decay also returns carbon dioxide to the atmosphere, but organic materials that are buried without access to air can retain some of their stored energy, and are slowly converted into coal, oil or gas – fossil fuels. So the energy transferred through the ecological food webs and that available from fossil fuels both derive from solar energy, but with a difference of millions of years in the storage times.

Chemistry and electromagnetism

Burning and photosynthesis are reactions taking place in opposite directions, forming and breaking up carbon dioxide respectively. There is another important difference between them. The energy release from burning fuels is typical of a thermochemical process, producing disordered thermal energy. In photosynthesis the energy is supplied as electromagnetic radiation in an appropriate frequency range; this is characteristic of **photochemistry**. Again this can involve either the supply or release of energy in electromagnetic form – some reactions require light or other radiation, while other reactions give it out. (This is not to say that materials formed photochemically cannot be broken up thermally – this happens whenever we burn wood.)

A further link with electromagnetic energy is in **electrochemistry** – the branch of chemistry from which the study of electricity developed in the first place. This studies the way electrical currents passed through solutions can break up chemical compounds (electrolysis), or, conversely, chemical energy can be stored in batteries and fuel cells to be converted into (d.c.) electrical energy when required. Primary batteries, such as torch batteries, convert the chemical energy originally stored in their materials, which change as the battery is discharged by supplying current. Fuel cells have provision for removing spent materials and replacing them with fresh. Secondary batteries, like those in cars, can be recharged from a generator to reverse the chemical changes produced during discharge.

In the light of what we now believe about atomic structure and the nature of chemical bonds, these electromagnetic effects are nearer the heart of the matter. The atom is seen as a minute positively charged **nucleus** containing most of the mass (the chemically unchanging part) surrounded by negatively charged electrons occupying most of the volume, which are rearranged in forming and breaking chemical bonds. In a sense chemical energy *is* electromagnetic energy on the atomic scale, though this makes it so inaccessible to normal experience that it remains convenient to think of chemical energy as a distinct form. Much practical chemistry concerns bulk quantities of materials involving myriads of molecular reactions. The average behaviour is much easier to observe than individual molecular events, so thermochemistry continues to be much used. But the direct chemical-to-electromagnetic energy conversions of electrochemistry and photochemistry are not subject to the same thermodynamic constraints as the chemical–thermal–heat engine–electricity conversion route, so in principle batteries and fuel cells can be more efficient than power stations. In practice, more needs to be learned about the use of such primary molecular processes in a controlled and economic way if they are to replace existing methods.

Life processes

All the above ideas are now generally believed to apply to the **biochemistry** of living materials as well as inanimate ones. The complexity, delicacy and virtuosity of life processes are extreme compared with much technical chemistry. For example, metabolism (the conversion of food into useful forms) involves long series of delicately controlled low-energy reactions, dependent on highly specific molecular interactions rather than heating. Genetics involves 'coding' in the complex molecular structure of DNA, using delicate distinctions which are easily damaged – by radioactive emissions, for example. Movement, from basic effects in cell walls to more complex processes in muscles, involves *direct* chemical to mechanical energy conversion, for which we have as yet no technical equivalent. Our vehicles mostly use heat engines, using the unwanted intermediate stage of thermal energy; those that instead use the intermediate stage of electricity are constrained by limitations in battery design, i.e. in electrochemical understanding. Look at birds, and how their foraging, metabolism, flight mechanism, voice production, self-maintenance and replication are so elegantly packaged!

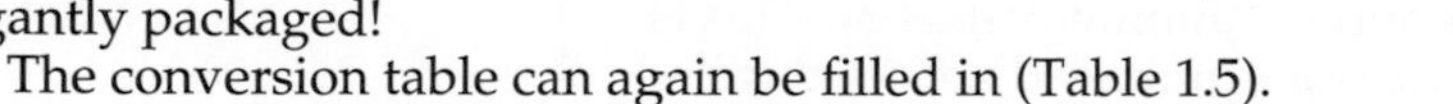

The conversion table can again be filled in (Table 1.5).

Table 1.5

To / From	Mechanical	Electromagnetic	Thermal	Chemical	Nuclear
Mechanical	X	X	X	?	
Electromagnetic	X	X	X	electrolysis battery (charge) photosynthesis	
Thermal	X	X	X	thermochemistry (endo-)	
Chemical	muscle	battery (discharge) fuel cell photochemistry	thermochemistry (exo-)	0	
Nuclear					

4.5 Nuclear energy

Nuclear energy was identified during the present century from the recognition that the 'indestructible' atoms of chemistry are not the end of the story – nothing ever is! The positively charged nucleus of the atom, the chemically unchanging part, is now believed to consist of positively charged **protons** (each balancing an electron's negative charge, but much heavier) and **neutrons** (not electrically charged, but of about the same mass as protons) held together by a distinct type of nuclear attractive force. Much energy must be supplied to prise them apart. This can be compared to the way molecules consist of atoms joined by chemical bonds (i.e. electromagnetic effects of electrons *outside* the nucleus), except that the scale of these effects *within* the nucleus is about a thousand times smaller and the energies required to separate the components are millions of times greater – which is why they were not so readily observed. These large binding energies imply that nuclei are stable, and at first sight this does not seem too promising as an energy source. But, as in the chemical case where energy can be released because some bonds are stronger than others, so in the nuclear case some nuclei are more stable than others, and the change from a less stable to a more stable form can involve a large energy release. There are three distinct ways this can happen.

Radioactivity

Each chemical element is distinguished by the total charge in the atom; electrons outside, protons inside the nucleus. Its nucleus can contain varying numbers of neutrons, changing the nuclear mass, to form different **isotopes** of the same element. They are identified by adding a mass number (the total number of protons and neutrons) to the letter symbol for the element: ^{12}C for the commonest isotope of carbon, ^{14}C for a different one (sometimes written as C-14 etc.). The nature of the nuclear force means that the range of stable combinations of protons and neutrons is limited. There are a few possibilities for each element, but if there are too many or too few neutrons the nucleus sooner or later 'decays' to a more stable combination, spitting out unwanted bits with the appropriate amount of energy. This is the process of **radioactive decay**. (The emissions produced are often called radiation, though this only corresponds to the usage introduced in Section 4.2 for the part of the process called gamma-radiation.) The exact moment of decay of each nucleus is not predictable, but the average decay time over all the atoms in a sample is a well-defined property of each radioactive isotope. The decay time is variable, from small fractions of a second to thousands of millions of years. Natural radioactivity arises from those long-lived isotopes that survive from whatever cosmic process created them in the first place. But many more isotopes which have never previously existed (or would have long ago decayed if they had) can now be made artificially in nuclear reactors. For large nuclei there are no stable forms, so no elements are found naturally with more than 92 protons (uranium is the largest). Larger nuclei, all radioactive, can be produced artificially; plutonium (94 protons, 138–152 neutrons in various isotopes) is a particularly important example.

The energy of the emissions produced by radioactive decay is dissipated, finally as thermal energy, in passing through matter. The gross heating effect of natural radioactivity contributes to the high temperature below the Earth's surface which can be exploited in geothermal energy supply systems. But there are more specific chemical and biochemical damage effects when particular emissions collide with molecules, including those of biological materials. Living things have evolved in the presence of natural radioactivity, so are to some degree adapted to it. Arguments for the safe disposal of artificial radioactivity (nuclear waste) often point to the low levels of emissions from them compared with the natural background, a reasonable argument if the radioactivity can be contained so that it remains external to living things. But radioactive isotopes not known in nature are chemically indistinguishable from their stable counterparts, and may find their way into living organisms and enter the food web. They may then concentrate radioactivity *within* organisms in ways which would not otherwise arise, and the natural background is irrelevant. This means that specific radioactive isotopes particularly related to life processes must be treated as individual poisons, to add to the chemical variety. Weapons tests, nuclear accidents and controlled releases of ^{3}H (tritium), ^{60}Co (cobalt-60), ^{90}Sr (strontium-90), ^{131}I (iodine-131), ^{137}Cs (caesium-137), ^{239}Pu (plutonium-239), for example, have caused particular concern because of the way these isotopes are taken up by the body.

Fission

Some of the heavier elements can decay not only by radioactivity, but also by **fission**, the process which made possible the atomic bomb and all current types of nuclear power station. When an extra neutron is added to the nucleus of some fissile isotopes, notably of uranium (^{235}U) and plutonium (^{239}Pu), the resulting instability blows the nucleus apart into two not quite equal parts (i.e. isotopes, many radioactive, of elements of mass around 120) and other fragments, in particular neutrons, all with a great deal of energy. (These isotopes are therefore known as nuclear 'fuels', but this is not in the sense used above for chemical fuels.) The total energy release is much greater than from radioactive decay. The emitted neutrons can go on to produce further fission, producing the possibility of a chain reaction in a critical mass of the particular isotope which can be made to explode. In the reactors of power stations this is prevented by absorbing some of the emitted neutrons to control the fission process. Much heat is produced, which is used to produce steam to drive turbines and generators in the same way as in fossil fuel stations. The initial fissile isotope in the 'fuel' is gradually used up, to be replaced by (radioactive) fission products and radioactivity created by neutron bombardment, both in the fuel and the surrounding structure (the neutrons also cause physical damage to the material of the structure). The spent fuel can be removed and either stored or reprocessed to separate the various isotopes present. The surrounding structure eventually becomes so radioactive and/or weakened that the station must be shut down, leaving the problem of decommissioning – all controversial issues, as we shall see in Chapter 7 of this book.

Existing so-called burner reactors make use of the fissile isotope of uranium (^{235}U) that is only present in small proportions in natural uranium, with the implication that available supplies will soon be used up if the use of nuclear power continues to expand. It has long been argued that it would be preferable to modify the reactor conditions so that it 'bred' fissile plutonium instead from the main isotope of natural uranium (^{238}U). Such a 'fast breeder' reactor overcomes 'fuel' supply problems, but introduces others: not only technical difficulties in developing a reliable and economic reactor of this type, and continuing waste and decommissioning problems, but also the production of large amounts (many tonnes) of plutonium. This is both highly toxic (in milligrams) and a bomb material (in kilograms), so requires meticulous accounting and security to avoid plutonium being lost to the environment, or falling into the wrong hands.

Fusion

The fission of heavy elements releases energy because the fission fragments, elements in the middle of the mass range, are more stable than heavier ones. But intermediate mass elements are also more stable than lighter ones. So the bringing together, or **fusion**, of light elements can also release energy. In particular the fusion of isotopes of the lightest element, hydrogen, to form helium is believed to be the primary source of energy production in the sun, so we are fundamentally dependent on it. Extremely high temperatures are needed to bring nuclei close enough together to fuse. These have been achieved on Earth so far only in the hydrogen bomb, though much research continues to try to achieve stable fusion that could be used in a power station. The ready availability of hydrogen isotopes in sea water seems to overcome 'fuel' supply problems. The products of fusion are not themselves radioactive, unlike fission products, so the goal of controlled fusion has proved seductive to nuclear scientists and governments. However, even if successful, the neutrons produced in this process will also produce damage and radioactivity in the surrounding structure, making the economics and decommissioning problematic.

It should be noted that in all nuclear power production so far (existing burners, prototype fast breeders and projected fusion) the only energy conversion process proposed is from nuclear to *thermal* energy to electricity, as in conventional fossil fuel stations. This dependence on the unwanted intermediate of heat leads to thermodynamic efficiency constraints and the need for large-scale installations. The study of nuclear forces does not yet seem to suggest any equivalent for the direct chemical–electrical conversion of batteries and chemical–mechanical of muscles.

Thus, as we complete the conversion table, this thermal conversion is the only useful process at present; the others are either damage effects, or methods of studying nuclear processes. As with chemical energy, there are no conversions purely 'within the nuclear form' (Table 1.6).

Table 1.6

To / From	Mechanical	Electromagnetic	Thermal	Chemical	Nuclear
Mechanical	X	X	X	X	(high-energy collision)
Electromagnetic	X	X	X	X	(high-energy collision)
Thermal	X	X	X	X	?(fusion)
Chemical	X	X	X	0	?
Nuclear	?('radiation' damage)	?(radioactivity)	reactor	?('radiation' effects)	0

4.6 *Summary*

Five distinct forms of energy have been discussed (mechanical, electromagnetic, thermal, chemical and nuclear) with the recognition that thermal is a *disordered* version of the first two, and that chemical is an *atomic* scale version of electromagnetic. The other three are at present seen as more fundamentally distinct. Various energy conversion processes within and between forms have been described. These are finally collected together in Table 1.7. This provides a general basis for considering natural and technical conversion processes, in the present and, probably, in the future too. The efficiency of conversion from one form to another in general depends on the extent to which unwanted conversion into heat can be avoided. All other forms can be fully converted *into* thermal energy of progressively lower temperature (degraded); conversion *from* thermal energy to mechanical and electromagnetic forms (work) can never be complete, according to present understanding of the principles of thermodynamics.

Table 1.7

To / From	Mechanical	Electromagnetic	Thermal	Chemical	Nuclear
Mechanical	potential/kinetic turbine vibration sound waves turbulence	generator microphone	friction refrigerator heat pump	?	(high-energy collision)
Electromagnetic	motor loudspeaker	oscillator lamp x-ray solar cell	resistance thermoelecric cooling	electrolysis battery (charge) photosynthesis	high-energy collision
Thermal	heat engine	thermoelectric generation	heat transfer cooling	thermochemistry (endo-)	?(fusion)
Chemical	muscle	battery (discharge) fuel cell photochemistry	thermochemistry (exo-)	0	?
Nuclear	?('radiation' damage)	?(radioactivity)	reactor	?('radiation' effects)	0

Activity 3

Try to classify the various forms of energy identified in Activity 2 into the five forms described in this section.

5 Energy and power

Because energy ideas arose in a number of different fields that were originally studied separately, various ways of measuring energy are still in use and this can be confusing. There is, however, only one internationally recognised unit for **energy**, the joule (symbol J) and all other energy units in this chapter can be expressed in terms of it. All individual *stores* of energy (raised weights, spinning flywheels, charged batteries, chemical bonds, sacks of coal, barrels of oil, loaves of bread, consignments of uranium) have a certain energy content in joules.

One related idea, however, must be kept clearly distinct, and that is **power**, with its own unit, the watt (symbol W). Whenever a 'machine' of some kind (including in this context lamps, living organisms, power stations, the whole energy system of a country or the sun) converts energy from one form to another, or transfers energy from one store to another, it does so at a certain *rate*, in joules per second; this is what is meant by the power in watts.

The important distinction to grasp is that between energy and power. Only the corresponding units, joules and watts, are needed at this point. The *number* of joules involved in different energy storage systems, or watts in different conversion processes, varies over a wide range. Try to follow the calculations about the equivalence of other units, though you will not need to reproduce them. These will enable us to relate local to global effects, using just two scales for energy and power. These will be logarithmic scales, where each interval represents multiplication by 10, in order to cover the wide range of numbers involved. Together these two scales form the key diagram in Section 5.2, Figure 1.6.

5.1 Units of energy

The primary mechanical unit

The joule is a small unit – often described as about the amount of potential energy Newton's famous apple had when it was still on the tree, or the kinetic energy with which it hit his head.

In practice, amounts of energy are often in multiple units, sometimes so large that we run out of the standard metric prefixes. Even the unusual, large ones, such as PJ or EJ (Box 1.1), used for national energy statistics are not always sufficient, as Figure 1.6 in the next section will show.

Box 1.1

1000 joules
$= 1 \times 10^3$ J = 1 kJ (kilojoule)
1 000 000 joules
$= 1 \times 10^6$ J = 1 MJ (megajoule)
and so on:
1×10^9 J = 1 GJ (gigajoule)
1×10^{12} J = 1 TJ (terajoule)
1×10^{15} J = 1 PJ (petajoule)
1×10^{18} J = 1 EJ (exajoule)

This can also be extended downwards to small quantities, using negative powers, as follows:
0.001 joules (thousandth of a joule)
$= 1 \times 10^{-3}$ J = 1 mJ (millijoule)
0.000 001 (millionth of a joule)
$= 1 \times 10^{-6}$ J = 1 μJ (microjoule)
0.000 000 001
$= 1 \times 10^{-9}$ J = 1 nJ (nanojoule)
and so on:
1×10^{-12} J = 1 pJ (picojoule)

Electrical energy

The primary electrical units have also been chosen to give energy in joules.

Standard electrical appliances usually have power 'ratings' quoted in watts (W) or kilowatts (kW): for example, a 60-W electric light bulb, a 650-W microwave cooker, a 2-kW electric fire and so on.

The total amount of electrical energy used depends on the time the appliance is on, so the standard unit of electricity we buy is the kilowatt-hour (kWh), the energy used by a 1-kW appliance in one hour – which is a rather roundabout way of getting back to joules!

Q How long will a 60-W bulb run on one unit, 1 kWh?

A A 1-kW (= 1000 W) appliance would run for 1 hour, a 1-W appliance would run for 1000 hours, so a 60-W bulb would run for 1000/60 = 16.7 hours.

Q How many joules are there in a kilowatt-hour?

A A device operating at a power of 1 kW = 1000 W = 1000 joules per second running for 1 hour (= 60 × 60 seconds) converts
$1000 \times 60 \times 60$ J = 3600 000 J
$= 3.6 \times 10^6$ J
= 3.6 MJ

of electricity into some other form such as thermal energy. So,
1 kWh = 3.6 MJ

Thermal energy

Thermal energy is now normally measured directly in joules, but there used to be heat units based on the properties of water:

- the metric calorie (the energy needed to increase the temperature of 1 g of water by 1 °C), and
- the British Thermal Unit, Btu (similarly, the energy needed to increase the temperature of 1 pound of water by 1 °F).

They are still to be found sometimes:

- the Calorie in dietary requirements – in fact what is meant is 1000 calories, or 1 kilocalorie (1 kcal = 4.19 kJ).
- the therm for gas supplies
  = 10^5 Btu
  = 105 MJ

Q Your electricity and gas bills tell you how much you pay for a unit, 1 kWh, of electricity and a therm of gas (e.g. 6 pence per kWh, 38.5 p per therm). How much are you paying per MJ of energy in the two forms?

A Electricity: 1 kWh = 3.6 MJ costs 6p, so
(6/3.6)p = 1.67p per MJ

Gas: 1 therm = 105 MJ costs 38.5p, so
(38.5/105)p = 0.37p per MJ

Chemical energy

Individual chemical bond energies are typically a few $\times 10^{-19}$ J, a very small amount of energy reflecting the minute size of atoms and molecules. Their rearrangements on the molecular scale when multiplied up to more familiar dimensions correspond to calorific values for fuels measured in kilojoules per gram. For example, the combustion of pure carbon in air would give 33 kJ/g, while real coal of different qualities (which can include some hydrocarbons, as well as impurities) gives values from 15 to 35 kJ/g.

Large-scale national energy statistics often use the unit 'million tonnes of coal equivalent', or mtce. A standard calorific value for coal is taken to be 26.4 kJ/g, which is in the middle of the above range, and the mtce determined accordingly.

Q What is 1 mtce in joules?

A Energy content of 1 million tonnes of coal
= calorific value $\times$ number of grams
= 26.4×10^3 J per gram $\times 10^6$ (grams per tonne) $\times 10^6$
= 26.4×10^{15} J
= 26.4 PJ

The oil industry prefers to express comparative statistics in million tonnes of oil equivalent, or mtoe, which is equal to 41.9 PJ, a larger figure reflecting the higher calorific value of oil (though in the same general range as the mtce).

A conventional practical measure is the standard barrel of oil, which is 0.159 m^3, of approximate energy content 5.7×10^9 J, or 5.7 GJ.

Natural gas is quoted by the cubic metre (m^3), of energy content 38 MJ.

Estimates of the total energy available on Earth from fossil fuels (mainly coal, oil, gas) are obviously technically difficult to make, and are much influenced by commercial and political considerations, as mentioned in Section 3.2 above. It is important to distinguish between proved reserves, which it is considered 'could be extracted in the future under existing economic and operating conditions' and **total resources**, which are much less certain, but have been estimated on geological grounds. The figures quoted in 1994 (which will give a broad impression on the logarithmic chart of Figure 1.6 below) are set out in Table 1.8.

Table 1.8

	Proved reserve		Total resource estimate
oil	1.4×10^{11} tonnes	$= 5.8 \times 10^{21}$ J	1.5×10^{22} J (?)
natural gas	1.4×10^{14} m^3	$= 5.4 \times 10^{21}$ J	1.0×10^{22} J (?)
coal	1.0×10^{12} tonnes	$= 2.6 \times 10^{22}$ J	3.0×10^{23} J (?)

Remember that these apparently small differences represent increases by significant multiples: coal reserve about 5 times oil or gas; coal resource about 12 times higher still.

Nuclear energy

At the atomic scale, nuclear reactions involve energies measured in 10^{-11} J rather than the 10^{-19} J associated with chemical bonds. In the case of ^{235}U fission, 3.2×10^{-11} J per nucleus translates into 8×10^{10} J per gram. Natural uranium contains only 0.7% of ^{235}U, so the available energy from a burner reactor is reduced from this by about 150 times, to about 5×10^{8} J per gram of uranium. This figure is some 2×10^{4} times greater than that from combustion of the same mass of coal. (In a fast breeder reactor, which makes use of all the uranium, the energy per gram is more like the original figure, about 3×10^{6} times greater than for coal.) This spectacular reduction in the tonnage of fuel required provides a primary argument for nuclear power, to be set against its problems.

Estimates of total available energy from uranium are even more commercially and politically sensitive. Although this is drastically oversimplifying to give a general impression, a figure of about 2×10^{6} tonnes has been quoted as a 'reasonably assured resource'. This implies an available energy of about 10^{21} J if used in burners, but over 10^{23} J in breeders.

Fusion of deuterium, a possible hydrogen isotope for use in fusion reactors, releases about 10^{-12} J per nucleus. Allowing for the small proportion of this isotope present in normal hydrogen, this translates into about 12 TJ per cubic metre of sea water. Hence, from the known amount of sea water, there is an apparent world fusion resource of 3×10^{30} J, well off the end of the scale in Figure 1.6.

5.2 *Power ratings*

In Figure 1.6 the first scale shows the energy equivalent of the various units and the amounts stored in various resources. The second shows the power rating of various conversion 'machines'.

Familiar domestic electric appliances have ratings from a few watts to a few kilowatts, and can be marked on the scale. But any energy unit transferred over any time period (e.g. car fuel consumption in gallons per hour, oil production in barrels per year, food intake in calories per day) is equivalent to a power in watts. In this way fuel-burning vehicles can have ratings similarly expressed, up to a hundred megawatts or so for a jet aircraft. Note that the power rating can be the average over a long or short period, and that the vehicle figures are only when they are in use, of course – not averages over a day or a year. A well-insulated house may have a total annual heating requirement expressed as 3000 kWh per year.

Q What is this expressed as an average rating in watts?

A $3000 \text{ kWh} = 3000 \times 3.6 \text{ MJ}$
$= 1.08 \times 10^{10} \text{ J}$

$1 \text{ year} = 60 \times 60 \times 24 \times 365 \text{ seconds}$
$= 3.1536 \times 10^{7} \text{ seconds}$

So, an average power rating
$= 1.08 \times 10^{10} \text{ J}/3.1536 \times 10^{7} \text{ s}$
= about 350 W

But the moment-by-moment heating rate will range from zero in summer to several kilowatts in winter.

This distinction between 'average' and 'instantaneous' power can be confusing. A large power station might have a power rating of 1 GW, which is the electrical power available when it is operating. The power stations of the UK have a total installed capacity of about 60 GW, when they are all operating. The total energy they are called on to supply in a year is about 250 TWh which, by a similar calculation to the above, is about 28 GW *average* power. This difference is a reflection of the varying demand through the day and year. The total energy consumption of all kinds in the UK has an average rate at present of about 285 GW, with the same proviso about variations over the year.

Another distinction which is sometimes made in quoting power ratings is between electrical power, in kW(e) or kW_e, and the thermal power needed to produce it, in kW(th) or kW_{th}. The latter is always greater because of the limited efficiency of power stations.

The total power rating of the world energy supply system (fossil fuel, hydropower and nuclear, with rather uncertain fuelwood) has risen from about 1 TW in 1900 to 10 TW in 1990. This increase by a factor of 10, and the expanding expectations of industrial society, is a major issue in world policy, and explains why 'energy' is on every national agenda. If it were to continue to increase at the same rate (say, to 100 TW in the next century) there would be major resource difficulties.

The power rating of our primary energy resource, radiation reaching the Earth from the sun, is more than 10 000 times larger than current energy demand, about 1×10^{17} W. The rate of energy conversion by photosynthesis throughout the world is less than 1% of this, but still a few hundred terawatts.

The food energy requirements of human beings are usually expressed as about 2000–4000 kcal per day, depending on activity – this can likewise be expressed as a range of metabolic power ratings.

Q What is this range in watts?

A $2000 \text{ kcal per day} = 2000 \times 4.19 \text{ kJ}/60 \times 60 \times 24 \text{ s}$
$= 97 \text{ W}$

The 2000–4000 kcal range is therefore about 100–200 W, or about 150 W on average.

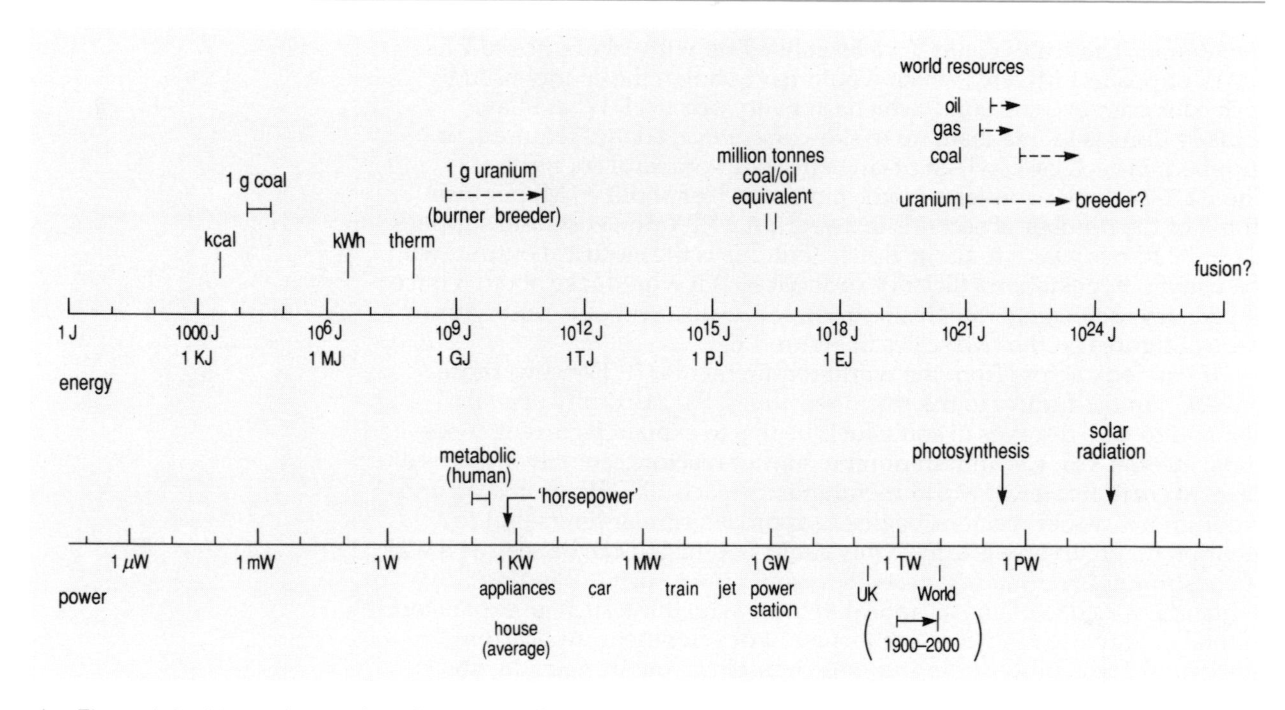

▲ *Figure 1.6 Master log scales of energy and power.*

If this is multiplied by the present world population of 5000 million we reach a primary global human metabolic requirement of 750 GW, doubling, on current projections, by the end of the next century.

The metabolic needs of horses are somewhat greater than those of humans – this is reflected in the old unit for power, the horse-power, which is now standardised at 746 W (not biologically accurate).

In industrial society we all have ready access to many kilowatts of energy from various sources. So we can think of this as access to many '150 W slaves and 746 W horses' – but this is evidently not without limit. If we divide the total world energy use, at a rate of 10 TW, by the present population of 5000 million, we find an average of 2 kW per person throughout the world. But this is unequally distributed, from about 10 kW each in the US, through 4–5 kW in Europe and Japan to less than 1 kW in many less developed countries. If the growing world population follows the development path of industrialised countries, the pressure on existing energy resources will be overwhelming. Energy use on the present US scale by the projected 10 000 million population at the end of the next century would imply a power rating of $10 \text{ kW} \times 10^{10} = 100 \text{ TW}$ for the world energy system. This is the crucial bit of arithmetic for thinking about the energy future.

5.3 *The resource problem*

What this would mean in resource terms can be seen by comparing the two scales in Figure 1.6. In a sense these scales are independent of each other: so much energy in each store, a certain power rating for each converter. But of course they are related by the definition of power as a rate of energy conversion. The scales could have been lined up with 1 W opposite 1 J, 1 MW opposite 1 MJ, etc., which would have shown the energy used by each converter every second, which isn't very practical. What I have actually done is to line them up to show the *annual* energy required, or supplied, in each case, which fits in better with practical energy plans. Thus, a 1-W device converts 1 joule per second, or about 30 MJ per year (think of the number of seconds in a year). A 1-GW power station supplies about 30 PJ per year, i.e. about 1 mtce (but this is the electrical output; with the usual power station efficiency of about 30% it would take about 3 mtce of primary fuel to generate this output). Make sure you can identify these pairs of figures on the two scales in Figure 1.6.

If you look across from the world requirements (10 TW now, more projected in the future) to the world resources, the proximity of *annual* demand to *total* reserves of some fuels begins to explain recurrent crises and concerns. Oil, gas and uranium in burner reactors can only last tens of years at current rates. If world requirements reach 100 TW, even coal, and uranium in breeder reactors, begins to approach similar limits (not to mention other adverse effects). Only controlled fusion can be seen as a way of consuming *terrestrial* resources that would meet such expanding demands, which explains its appeal to those who think such an expansion in total energy use is a necessary feature of development. Apart from questions of feasibility, costs and more immediate environmental effects, such an expansion would in principle increase 'thermal pollution' because the extra energy would disturb the Earth's energy balance, as described below.

At present there is still a large gap, a factor of about 10 000, between human energy production and the power rating – the continuous rate of energy supply – of our 'extra-terrestrial fusion reactor', the sun. So on the one hand, the thermal pollution issue is not serious at present: on the other, the total solar energy resource is more than adequate for our future needs. Even the *annual* energy resource produced by photosynthesis is comparable to the *total* gas and oil reserves. But of course the solar (direct, indirect and biomass) resources are distributed over the Earth, not concentrated as in fossil or nuclear fuels, so their large-scale use would require a different approach to energy policy.

What this broad analysis of the global situation tries to show is the basic interrelationship between three major factors: world population, average energy use per head and the energy resources to be used.

1 The rapid growth of population since industrialisation is in a way a measure of 'success', in biological terms, in utilising resources, including energy. But such growth is bound to be limited sooner or later by resource exhaustion. Current projections of world population rely on less developed countries experiencing the same transition to stability with prosperity as already experienced in the developed ones – but will that require a similar energy consumption? The current projection of stability at a world population of about 10 000 million by the year 2100 already implies major stress on traditional energy resources.

2 Do we need 5–10 kW per head, or can we manage equally well on 1–2 kW, with more care about using energy efficiently?

3 Should we continue to use up terrestrial energy stores, fossil or nuclear, or seek the sustainability and freedom from pollution of a renewed dependence on the continuous energy supply from the sun (and other renewable resources)?

5.4 Summary

Various energy units have been shown to be expressible in terms of a single unit, the joule. The energy content of different energy stores can all be displayed on a single joule scale.

Energy converters are characterised by their power, the *rate* of conversion, in joules per second, or watts. They can all be similarly shown on a single watt scale.

A comparison of individual and world energy consumption with resources provides a broad summary of resource difficulties and policy options.

6 A glance at the detailed statistics

The information shown in Figure 1.6 provides a broad, but sketchy, survey of a wide range of effects. A vast amount of more detailed information is available about various parts of this range, each with its own flavour.

6.1 Global energy balance

The continuous energy supply to the Earth is almost entirely from the sun: some 1.7×10^{17} W reaches the Earth, of which about 35% is reflected back into space by clouds, 20% is absorbed in the atmosphere, and 45% reaches the surface. Two much smaller additions are gravitational effects from the sun and moon interacting with the Earth's rotation to give tides, about 3×10^{12} W, and internal heat from the Earth, 3×10^{13} W.

Solar energy drives the hydrological (water circulation) and carbon cycles and, with the Earth's rotation, the weather system. These provide the range of renewable energy resources: hydropower, biomass, wind and wave power (solar); tidal and geothermal power (non-solar). These will be discussed further in Chapter 3.

The temperature of the Earth adjusts itself so that its low-frequency radiation balances the energy received. In the absence of an atmosphere this temperature would be about –18 °C, but the 'greenhouse effect' of frequency-dependent absorption properties in some low-concentration atmospheric gases raises the surface temperature to an average 15 °C, as further discussed in *Blackmore & Reddish, 1996*, Chapter 3.

Any additional energy supply from fossil fuel or nuclear sources upsets this energy balance to some extent. The total heat added in principle constitutes thermal pollution, though this is not yet globally significant – but there are local effects in cities. If total non-solar energy use continues to increase, this could be a problem in the next century. Solar energy use of course does not affect this balance.

A more immediate concern is the *enhanced* greenhouse effect due to gaseous emissions, of which carbon dioxide from all fossil fuels is a major component, and which leads to predictions of global warming. Some uncertainty remains about the 'natural' mechanisms controlling carbon dioxide in the atmosphere, which varied significantly during the periodic glaciations of the last ice age without human intervention. However, the measured increase in atmospheric carbon dioxide (and other greenhouse gases) is much faster than deduced for earlier times and is generally attributed to fossil fuel burning (and other human activities). Strong international pressure is building up for a reduction in fossil fuel use in developed countries, and the encouragement of less energy intensive economies elsewhere.

6.2 *Human needs*

Except for the 150 W or so of metabolic requirements, human beings do not have any need for energy as such, but only for the services it can provide: space heating in cold climates, cooling in hot ones, hot water, cooking, lighting, food storage, domestic machinery, industrial processes, transport, etc. It is the sum of all these services which account for the 4–10 kW per head of energy use in industrial societies. The way energy consumption is built into society and its economic system is extremely complex, and considerable energy analysis has to be done to tease out the energy contribution to particular products and services. One example is illustrated in Figure 1.7, done some years ago, but still indicative of the many energy inputs involved. The processes shown in the production of a loaf, from farm, mill, bakery to distribution, add up to 20 MJ total energy 'cost'. (Note that the food energy value of a loaf is about 9 MJ, more than twice the energy expended on the farm to produce the wheat, but less than half of all

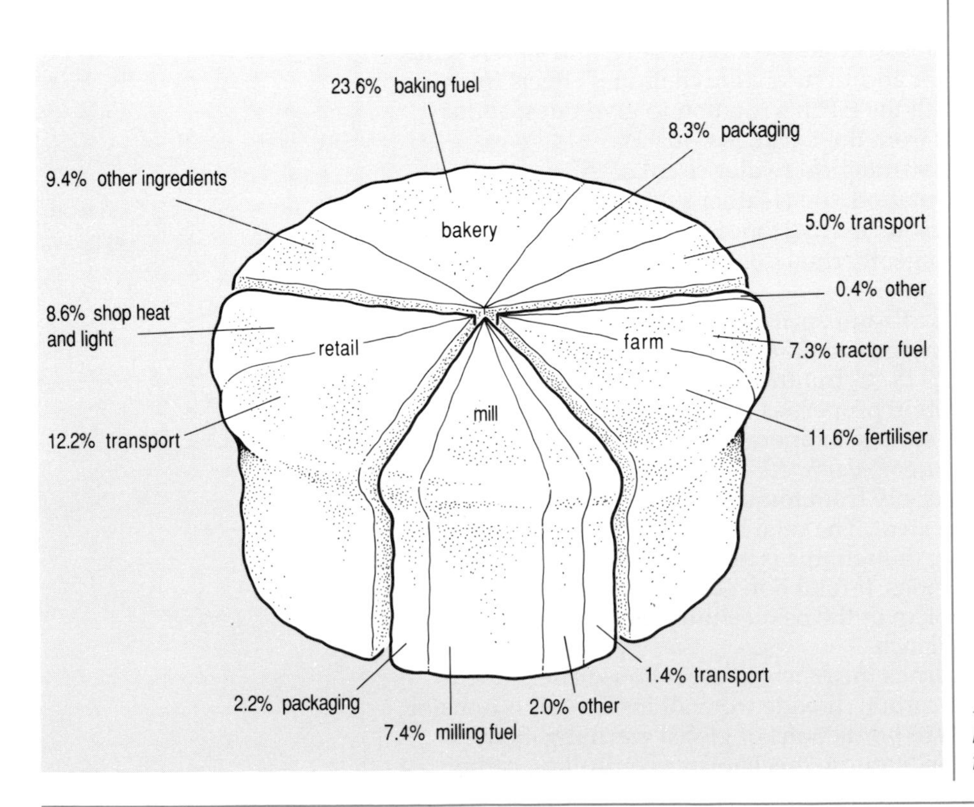

◀ *Figure 1.7*
Detailed breakdown of the fuel cost of a white loaf.

production costs.) All these processes can be carried out with varying energy efficiency – that is, with use of primary energy to a greater or lesser degree.

Day-to-day decisions are made on an economic basis. If energy is cheap there is little incentive to save it. The response to the oil price shocks of the 1970s was a reduction in energy use in the developed countries as a myriad minor efficiency improvements were introduced. Many studies have shown that the services we value can be provided for much less energy, as we shall see in Chapter 3. Whether this happens will depend on the economic and regulatory framework, but perhaps also on individual awareness, as the environmental implications of energy profligacy become clearer.

6.3 *National energy use*

Total world energy use and its changes with time have been mentioned above, but more details of the distribution between different countries, and between different economic sectors, will give a better picture of present practice, with its powerful economic and political influence.

Information on world energy use is speculative before about 1900, but US developments have been traced back to 1850. Figure 1.8 shows the transition from wood to fossil fuels (later in the USA than in older industrial societies), and Figure 1.9 traces the increasing consumption per head to the present 10 kW or more. (Note the logarithmic scale, where the approximate straight-line increase with time implies an underlying exponential, or geometric, growth over this earlier period: that is, a constant percentage increase, about 1% per year. This annual growth increasing along with the total consumption is a 'runaway' process that has, in fact, slowed down in the economic conditions of recent years.)

Figure 1.10 gives one estimate of world consumption from 1900 to 1975, made by the UK Department of Energy, and a revealing scenario from that time for their view of how demand might increase from 1975 to 2050. Figure 1.11 is a more detailed breakdown, produced by the oil industry, for actual consumption in 1969–94. Both graphs show how the rise in total energy consumption was slowed down by the oil price rises of the 1970s.

Q How do the different scales on these two graphs relate to the standard used above, where average energy conversion rate is expressed in watts?

A Figure 1.10 uses units of 10^{20} J yr^{-1} (joules per year). There are 31.6×10^6 seconds in a year, so each unit represents

$$10^{20}/31.6 \times 10^6 \text{ W} = 3.16 \times 10^{12} \text{ W}$$
$$= 3.16 \text{ TW}$$

Thus the graph shows an increase from less than 1 TW in 1900 to about 8 TW in 1975 (and the projection implies about 12 TW in 1988, 30 TW in 2050).

Figure 1.11 uses million tonnes oil equivalent (mtoe) per year, i.e. 41.9 PJ per year.

$$= 41.9 \times 10^{15}/31.6 \times 10^6 \text{ W}$$
$$= 1.33 \times 10^9 \text{ W}$$
$$= 1.33 \text{ GW}$$

So the scale unit of 1000 mtoe represents 1.33 TW, and the value of nearly 8000 mtoe in 1994 represents about 10.4 TW.

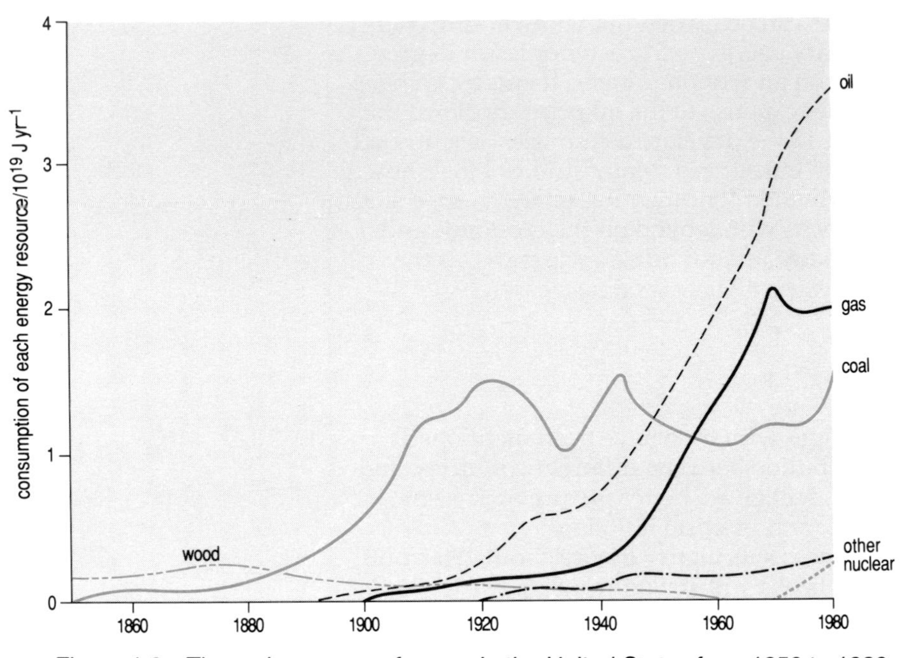

Figure 1.8 The main sources of power in the United States from 1850 to 1980.

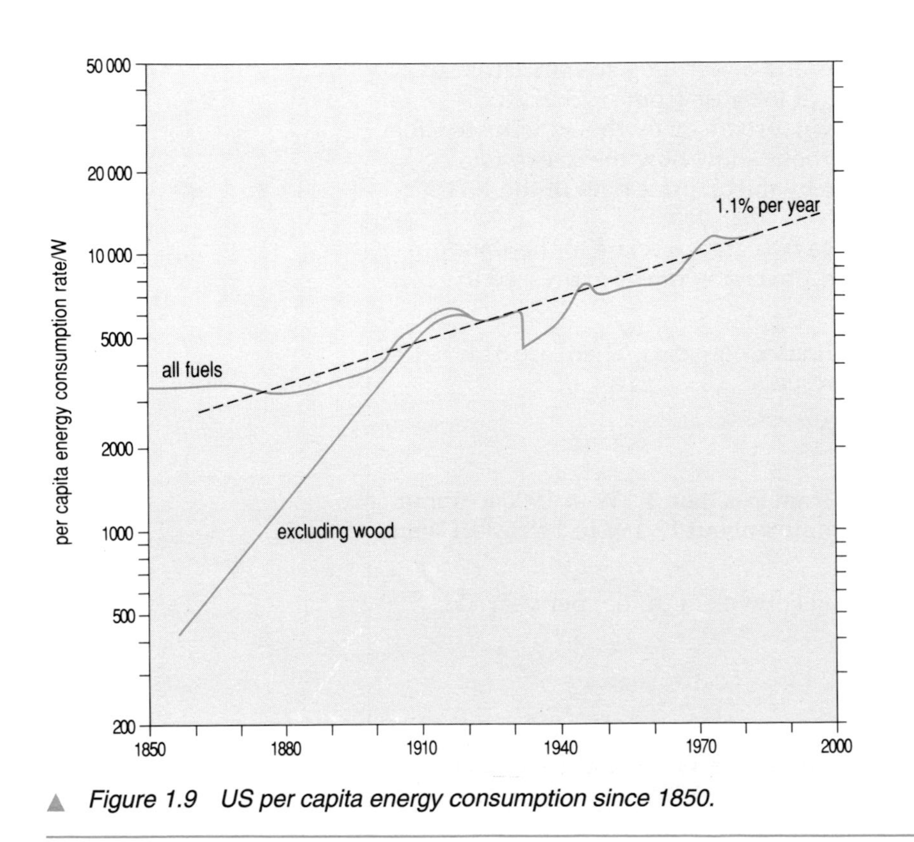

Figure 1.9 US per capita energy consumption since 1850.

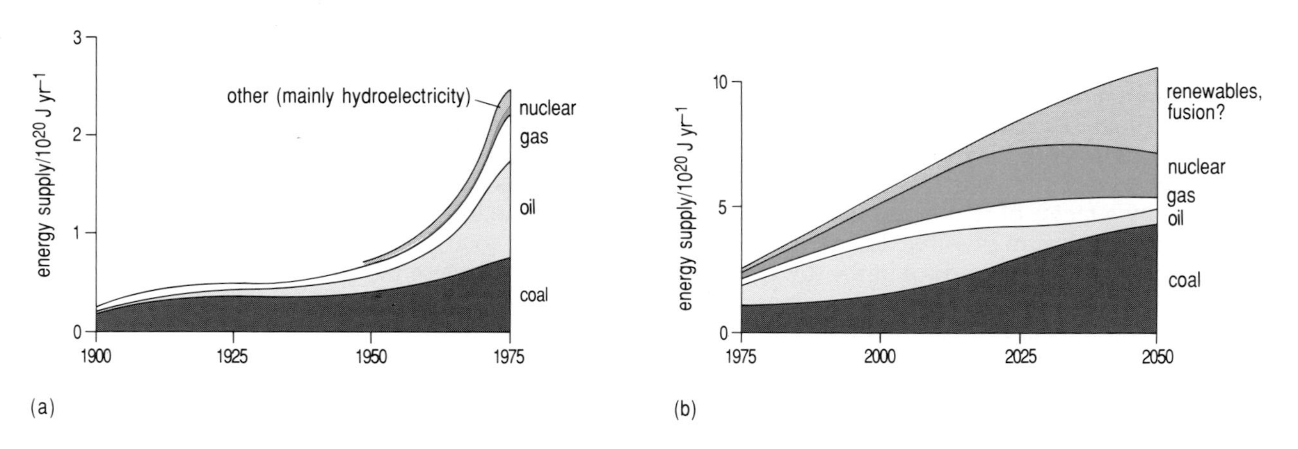

▲ *Figure 1.10 World energy supplies from 1900 to 1975 and the projected demand for the period 1975 to 2050.*

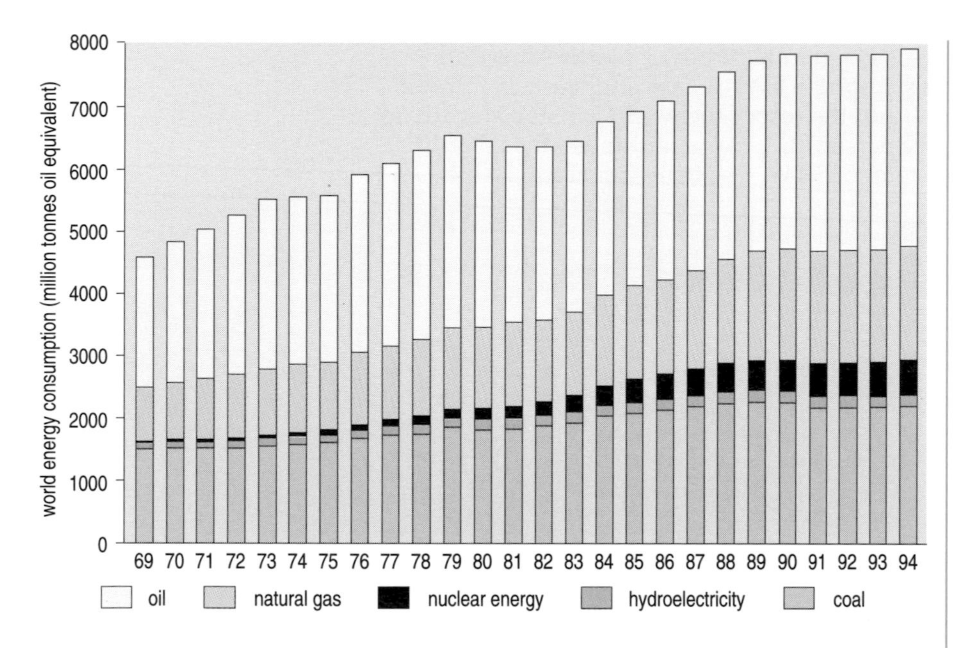

▲ *Figure 1.11 World energy use from 1969 to 1994.*

These figures fill in the details on the rough statement made in Section 5.2 that world energy demand has increased from 1 TW to 10 TW during this century.

At present over 70% of the world's primary energy comes from fossil fuels. Biomass fuels (wood, animal wastes), many of which are not traded and do not appear in commercial statistics, are believed to provide 10–15%. The remainder is accounted for by hydroelectricity, nuclear power and, to a minor extent at present, other forms of renewable energy.

Figure 1.12 shows a comparison made in 1975 between energy consumption, in kilowatts per head, and an economic measure of wealth creation, the gross national product (GNP) per head, for different market economy countries. Such plots were formerly used in the argument that prosperity and energy use were indissolubly linked. While it is obviously true that rich countries use a lot of energy, there are two questions to consider before accepting that this is necessary. First it is clear that there is a wide range of energy uses associated with a particular GNP – compare the US and Switzerland, for example. It has therefore been increasingly recognised over the past two decades that a better measure of economic success is the energy *intensity*, the ratio of per capita energy use to GNP. The dashed lines in Figure 1.12 show two different values for this ratio differing by a factor of two, without any clear implication that the lower value is the minimum possible. Economic theorists will continue to argue how far energy use can reduce in the developed countries, and must rise in the less developed ones, to improve prosperity. But they are now recognising a second difficulty with this approach, the question of whether GNP is a good measure of quality of life, particularly where environmental effects are concerned. The 'sustainable development' objectives of the late 80s have emphasised how GNP increases can be at the expense of environmental damage and resource losses, which are not included in the economic statistics – a debate to which we will return in Chapter 3.

Whatever may be reasoned theoretically, the factual account of regional differences and recent changes is given in Figure 1.13. North American consumption per head remains around 8 kW, eastern and western Europe have converged to about 4 kW, and the rest of the world is rising steadily to 1 kW. What is the real need per person? If 1 kW is multiplied by a 5×10^9 world population this gives a total of 5 TW, less than at present; 10 kW

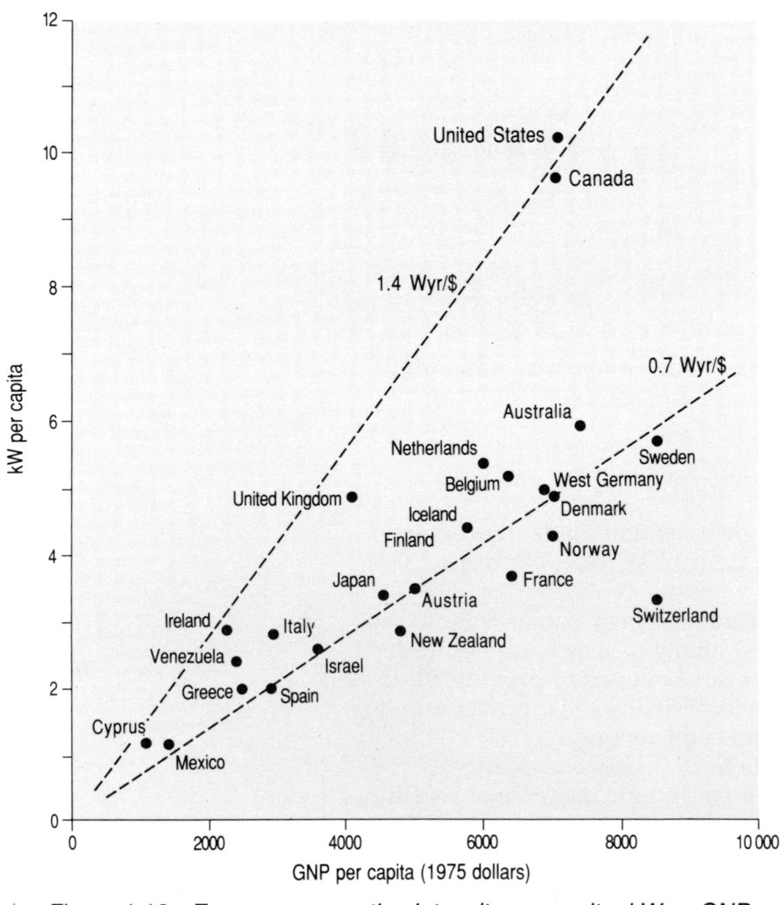

Figure 1.12 Energy consumption intensity per capita, kW vs GNP.

multiplied by 10×10^9 gives a total of 100 TW, which is surely not sustainable by fossil or nuclear methods without serious environmental effects. This emphasises the need for detailed scrutiny of energy use to see how far it can be reduced, and for the introduction of more benign energy sources to meet the demand that remains.

Finally, the way energy use is distributed between different end uses, for the UK economy as an example, is shown in Figure 1.14. This particular analysis was carried out in 1986, since when there have been significant changes in the proportion of *primary* energy supplied by coal and gas, where coal has decreased and gas increased by over a quarter, within a similar total. (The reduction in the amount of *home-produced* coal has been even greater.) However, the point of the diagram is to indicate a representative breakdown of the energy statistics in four different ways (primary, delivered, sector, end use), which could be matched, with local variations, in other industrial countries. The pattern of final *end use* is fairly typical: about 40% of the energy demand is for low-temperature heat, below the boiling point of water, and for space cooling. About 20% is to provide higher temperature heat, 30% for operating vehicles and other machines and only 5–10% for electricity-specific tasks such as lighting, electrolysis and electronic equipment. Human societies perform their various activities at definite times, so there is a daily, weekly and annual pattern in the energy demands of buildings, industry and transport. Many factors affect national energy use: climate, population density, amount and type of industry – Switzerland and Denmark have much less heavy industry than Germany and Sweden, for example. The breakdown for developing countries shows greater energy use for cooking, and less for space heating, but is otherwise similar.

Each national government has its own response to the energy policy required to satisfy this demand efficiently, either by leaving it to the market, or, because large-scale or long-term issues are being neglected, by

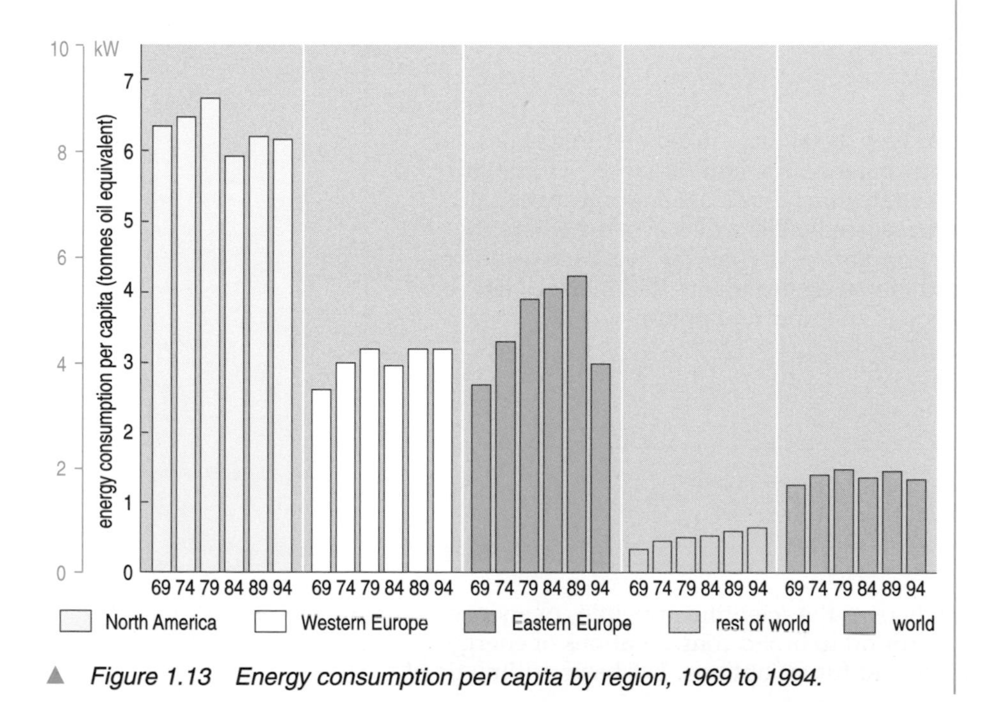

▲ *Figure 1.13 Energy consumption per capita by region, 1969 to 1994.*

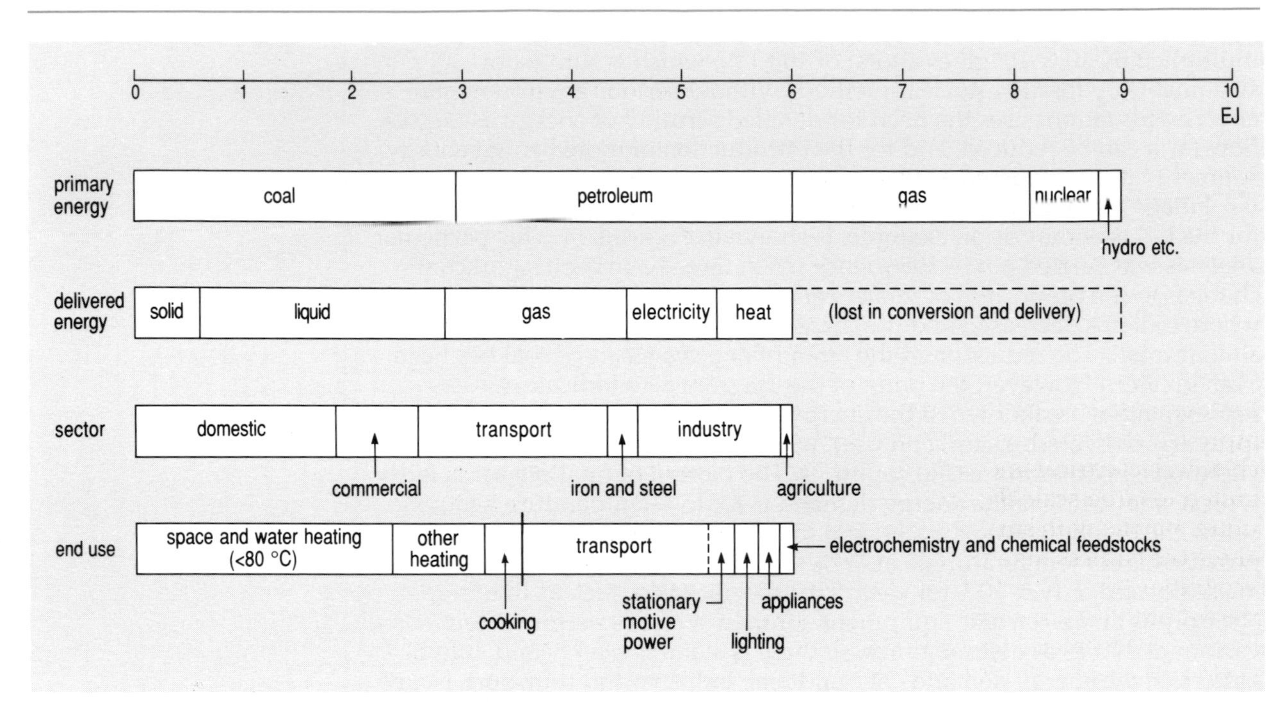

▲ *Figure 1.14 UK energy: end use by function (1986).*

modifying it in some degree through taxation, subsidy, regulation or more centralised planning. Increasingly, world environmental problems are seen to be linked to national energy policies in ways that will require international regimes of agreement to be developed, as we shall see in Chapter 2.

6.4 Summary

Some of the statistics available have been introduced on three scales: the global one of the Earth's energy balance, the individual one of human energy needs, and the intermediate one of national and international provision of energy to meet these needs. The scales overlap where the total provision for a growing population stresses the available resources, and impinges on the local and global environment. Reduction of energy demand by increased efficiency is an important option.

7 Conclusion

This chapter has ranged widely over the scientific principles of energy conversion to provide a background to broad considerations of energy policy, both to present practice and future options. The basic arithmetic of population, energy demand and resources implies growing stresses in the existing energy system, particularly when environmental effects are included: these will be taken up in Chapter 2. There are two related ways forward. The first is to clarify why energy is needed, in order to establish a desirable and equitable provision. The second is to provide what is needed in a benign way. These form the subject of Chapter 3.

Fossil fuels clearly cannot be exploited indefinitely; the long-term choice seems to be between nuclear and solar (i.e. extra-terrestrial nuclear). This is often presented as an ideological conflict between high-tech and low-tech, which is a pity. Present and proposed forms of nuclear energy conversion are undoubtedly a consequence of one path of industrial-scientific development, wedded to large scale and to centralised control. But there is still scope for better application of scientific advance in improved energy conversion and storage principles (chemical–electrical–mechanical–nuclear) without the emphasis being limited to large-scale *thermal* conversion in large power stations. A securer future may lie with smaller scale, more delicate and efficient methods, where the endless ingenuity of human invention is applied with more recognition of global consequences. Meanwhile, can a recognition of present problems and priorities be made politically acceptable, both within the west (particularly the United States) and worldwide? If not, can the world remain so divided without destructive conflict?

References

BLACKMORE, R. & REDDISH, A. (eds) (1996) *Global Environmental Issues*, Hodder & Stoughton/The Open University, London, second edition (Book Four of this series).

SARRE, P. & BLUNDEN, J. (eds) (1996) *Environment, Population and Development*, Hodder & Stoughton/The Open University, London, second edition (Book Two of this series).

SARRE, P. & REDDISH, A. (eds) (1996) *Environment and Society*, Hodder & Stoughton/The Open University, London, second edition (Book One of this series).

Answers to Activities

Activity 1

There can be no general response; it will depend on personal circumstances. You might have considered your central heating system, which may use North Sea gas needing exploration, extraction, pipelines, risks of explosion, leakage (of a 'greenhouse' gas). Or crockery, based on china clay, with the environmental effects of its extraction and the former kilns of the 'Potteries' region. Or domestic refuse, with a considerable energy content in the paper and plastic packaging, which will probably not be recovered if the waste is only buried, also causing a land-fill gas hazard. And so on. The rest of this chapter and book will provide many further examples.

Activities 2 and 3

Coal – chemical energy
Solar – electromagnetic waves; photosynthesis – (photo)chemical
Food webs – (bio)chemical
Human energy use – fire = chemical, thermal, electromagnetic (radiation)
Agriculture – solar – electromagnetic; animals – biochemical
Industry – chemical (fossil fuel), thermal, mechanical, electromagnetic, nuclear
Fertilisers, pesticides – chemical
Machines and transport – mechanical, chemical, electromagnetic

Chapter 2 The environmental effects of present energy policies

1 Introduction

As you read this chapter, look out for answers to the following key questions.

- How can the impacts of energy use be analysed?
- What are the main impacts of using coal?
- What are the main impacts of the other components of the energy system?
- How are these impacts influenced by value systems?

The previous chapter introduced the idea of energy, the natural resources available, and the way these are used at present to provide for human needs. The growth of urban-industrial society over the last three centuries has been strongly dependent on an ever-increasing use of fossil fuels (coal, gas and oil), hydroelectric power and more recently nuclear power. In 'developed' countries they have virtually replaced the traditional wood and other biomass fuels, though these continue to be important in 'less developed' countries. Economic development and an improved quality of life are often believed to depend on a continued extension of this process, but it is clear that energy use worldwide in this way on the scale of the most profligate countries would, at the very least, severely stress the available resources, particularly of oil, gas and uranium. In the market system this is reflected in price fluctuations, with their complicated impact on the rest of the economy – we have already experienced the effects of the sharp rise in oil prices engineered in the 1970s by the Organisation of Petroleum Exporting Countries (OPEC), the cartel dominated by, but not limited to, the Middle East. This caused a reduction in oil use, the development of alternative sources and a price slump in the 1980s.

A corresponding uranium price rise, followed by overproduction and a slump, reflected changing perceptions over the same period about the future role of nuclear power. These basic supply and demand effects on prices are not, however, the main concern of this chapter. Many powerful organisations in the energy industries will continue to manoeuvre, with results that we can hardly expect to predict in detail. Evidently, non-renewable resources can't last for ever, so their prices must rise eventually – but there will be many fluctuations on the way.

What *is* our concern here is the growing recognition over recent decades that our present energy policies also have many undesirable environmental effects (from smog, acid rain and oil spills to Chernobyl and global warming) which are not, at present, reflected in immediate economic costs. We shall come back to the question of whether an 'environmental economics' can be developed which takes better account of these problems, but first we need to be clearer about their nature, and this is considered in Section 2.

Section 3 consists of a more detailed study of the fossil fuel that has been longest used, coal, and the way the environmental effects of its combustion have been seen to widen from a local to a global scale.

The ways that other components of our present energy system (other fossil fuels, transport, nuclear hydroelectric power and biomass) impinge on these global problems are outlined in Section 4, showing a possible progression back to purely local impacts.

Finally the options for technical improvements, and better inclusion of environmental consequences in economic, and other, value systems are introduced in Section 5.

2 A framework for analysis

2.1 Introduction

There is a growing interest in environmental impact assessment, and indeed an increasing requirement that such an assessment be included in planning proposals for new projects. So let us examine now what we mean by the environmental impact of energy use and try to establish a systematic procedure for assessing it.

During the 1980s, the OECD (Organisation for Economic Co-operation and Development, whose members include all the developed countries of Europe, North America, Japan, Australia and New Zealand) carried out the 'Compass' project for the '*comp*arative *ass*essment of the environmental implications of various energy systems'. Its first report in 1983 suggested a framework for analysis which is a suitable starting point. There are four stages:

1 specifying the processes which are the *origins* of environmental effects;
2 listing the *residuals* of these processes which affect the environment;
3 determining the environmental *impact* of these residuals;
4 assessing the *damage* caused by these impacts.

At first sight these four stages may not seem to be very clearly distinguished, and we shall need examples to clarify the way the terms are used. There are some implicit assumptions about the nature of the environment in the approach used, and the language is often revealing; doubts and argument will be found to increase as we progress through the four stages. The procedure is detailed and thorough, as the complexity of the subject demands, but it is also rather tedious and bureaucratic, as such international recommendations tend to be. Some of the implications will be discussed as we proceed, particularly by looking more closely at the word 'environment' itself, and at the more specialised way it is used in systems theory.

2.2 Origins of environmental effects

The OECD approach

We can work our way from the general to the particular in three steps:

1 The fuel cycle – this is the usual term for what is in fact a one-way process turning natural resources in the ground into useful energy and unwanted by-products. In the case of fossil fuels, the time scale for completion of the cycle, turning plant material formed by photosynthesis from carbon dioxide back into coal, gas and oil would be geological, hundreds of millions of years, while that of nuclear fuels would be cosmological, unimaginably longer – so in neither case remotely available on a human time scale. Only biomass fuels can be truly 'cycled', if they are grown at the same rate as they are used.

The OECD (1983) study breaks down this cycle in turn into seven stages, each with its own complications; some stages may occur more than once in the route from energy source to final use:

- exploration: geological studies, prospecting, test drillings, etc.;
- harvesting: a rather quaint term for the variety of mining and drilling techniques used as well as its more normal use for biomass collection (perhaps it sounds more comfortable than 'exploitation'; 'cycle' and 'harvesting' are both 'loaded' words – look out for others);
- processing: crushing, cleaning, grading, extraction, refining, blending;
- transport: by sea, land, air, pipeline, transmission line, of raw and refined materials;
- storage: raw and refined materials, large and small scale; for electricity, in batteries or, indirectly, as pumped storage of raised water, to be converted back to electricity when needed;
- marketing: wholesale and retail, of bulk and consumer products;
- end use: fuels for heat, heat engines for motive power or electricity production, chemical feedstocks; as electricity for heat, power, light, electronics and electrochemistry.

Activity 1

Check your understanding of this breakdown of the fuel cycle by listing the stages as you think they might be involved using uranium ore from Namibia ultimately to power an electric vehicle in England.

2 The second step involves identifying the phases associated with each of the above stages of the fuel cycle (a particular usage of the word 'phase', based on the fact that these activities occur in an approximate time sequence, though they obviously overlap to some extent):

- construction of facilities: mines, oil platforms, refineries, power stations, etc.;
- operation: supply of materials and energy, removal of products and effluents, employment;
- maintenance: scale and frequency;
- decommissioning of plant: restoration of amenity and safety at the end of useful life;
- management of long-lived residuals: spoil heaps, toxic residues, radioactive wastes.

So with 7 stages by 5 phases, potentially 35 phases in all, it's getting complicated. But this is not the end.

3 Finally we come to the specific processes associated with each of the above phases. Here there will be endless technical detail about the particular methods used in each operation, limited only by human ingenuity – mining methods and machines, surface and underground, for coal or uranium; oil and gas drilling and extraction methods; refinery operations to produce particular transport fuels, residual fuel oil, petrochemicals; types of combustion system and nuclear reactor; transport and distribution systems, and so on. No-one can expect to know all the details, which need to be collected for each case considered.

The particular part of the fuel cycle considered in the detailed study of Section 3 will be the combustion of high-sulphur coal, by the methods formerly common for domestic heating, and by those now used in electric power generation. Some other environmentally significant operations in the energy system will be looked at in Section 4.

As all the above processes in the energy system themselves use energy and materials which in turn involve environmental effects in their production, there is some difficulty in knowing where to draw the line – should you include the effluents from the steel works that made the material for the drill of the oil well, or the exhaust from its Managing Director's car, or from the lorries delivering to the supermarket which supplies food for its employees? This will not be taken any further than to say that 'energy analysis' to discover the energy costs of various materials and processes, including energy production itself, has been carried out, with suitable rules for 'cutting off ' the chain of contributions where their effect becomes negligible; that the resulting environmental effects are one stage more complicated; and that putting money values on them is one stage more complicated still – so it is not easy to disentangle 'environmental costs' in detail. But by working back from a particular end-use of energy to its primary source we can at least get a first-order idea of its environmental implications. Such a demand-based analysis is more fruitful in suggesting ways of reducing energy use and/or environmental damage than starting with the energy resource, as the supply industries have a natural financial interest in increasing energy use.

As usual with an international body like the OECD, it is able to produce a thorough analysis of all aspects of the existing energy system which it would be hard to fault – after all these are the Facts, and very complicated they are. As we get into the more uncertain territory of environmental effects, and the more active steps of doing something about them, we may feel the complexity of all the possible options eluding our grasp; this is where 'systems' thinking may help. The first stage is to rephrase this 'origins' analysis in system terms.

A system model

So far we have used 'environment' in the current commonsense way, for the surroundings of human beings that provide various amenities, but can be modified or damaged by our activities. The same word 'environment' is used in a related, but more specialised, way in systems theory. This sets out to provide an orderly way of representing, or modelling, complicated systems of all kinds, whether for analysing existing systems, or for designing new ones. The main difficulty, as we have already seen, is in deciding how much to include in the description, since everything always affects everything else to some extent. The first step is very simple, only putting into a diagram what we often do in our minds without being aware of it. We draw a **system boundary** round the **components** of the system which undoubtedly interact with each other, and call everything outside the boundary the **system environment** (Figure 2.1). This boundary might be a physical one (the walls of a factory or frontier of a country), but more often it will be a conceptual one, only existing in a diagram. The significance of this step will become clearer as we move into the later stages.

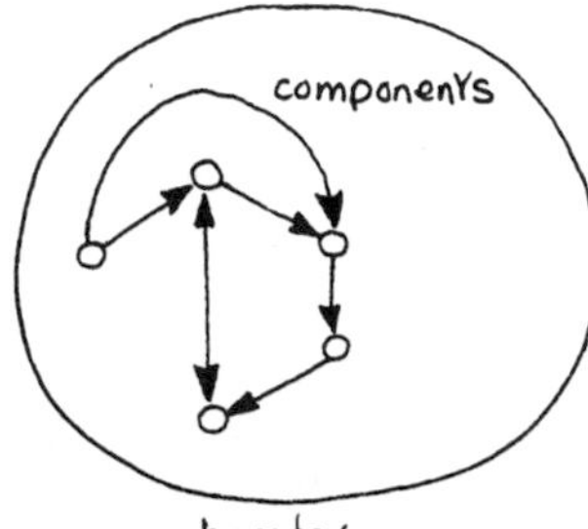

▲ *Figure 2.1*
A system model.

We can take examples, on various scales, from the many operations described above. A small-scale end-use might be a domestic central heating system, a larger installation might be a coal mine, larger still, the UK total energy system – or, going beyond human energy processes, we could consider the world's climate system. These are evidently related in various ways. Some systems are embedded in others to form a hierarchy: a system at one level of description can be a component of a larger one (up to the scale of the universe); and system components can themselves be analysed as smaller systems (down to the atoms, or beyond, of which they are made).

The OECD fuel cycle analysis is an example of this 'embedding' process, and also illustrates different kinds of components and interconnections. The steps 1 to 3 represent progressive breakdown of a particular fuel supply system into more detailed components each time (Figure 2.2). Let us take the final stage first, with one example of a familiar end-use, a domestic central heating system. The components in this case are

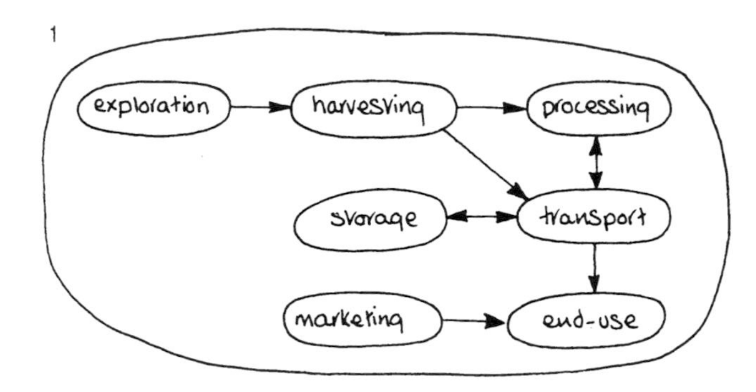

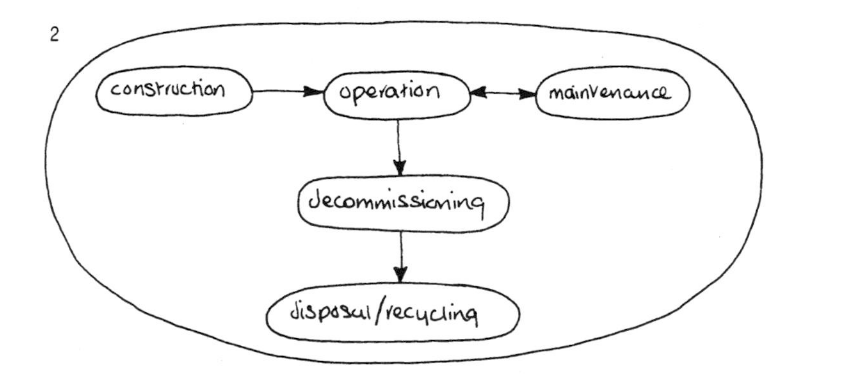

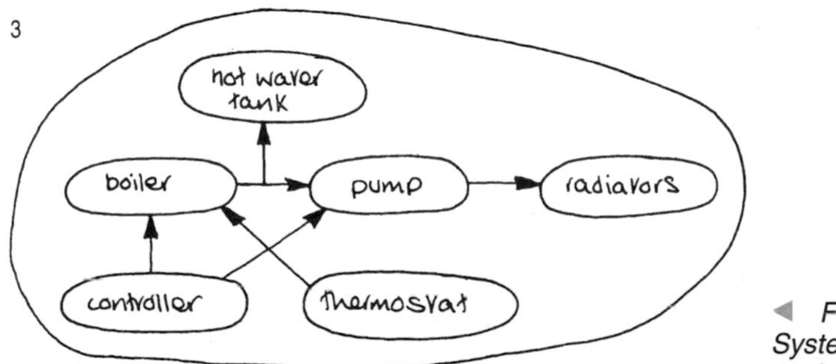

Figure 2.2 System models at each OECD step.

recognisable physical objects (boiler, pump, hot water tank, radiators, thermostat, time switch/controller) and the lines joining them stand for the pipes and wires connecting them together, the first stage of a practical design that could be turned into working drawings. But as we move back to steps 2 and 1, the components, and the connections between them, become more abstract. For example, if step 2 is concerned with the phases in the provision of central heating (in general), it will have as components one covering the construction of central heating equipment (in turn divided into the manufacture of separate parts, and the assembly of complete systems), others with operation (supply of fuel, control, safety), maintenance, decommissioning and disposal (which if not such difficult processes as for some parts of the energy system, nevertheless raise important questions about waste disposal and materials recycling). These system components are no longer physical objects, but each is a complex network of people and equipment; the lines joining them are not single physical connections, but represent all the interactions between them, varying with time. This abstract character is even more pronounced for the fuel cycle system of step 1, in which the final end-use component will include central heating as one of several competing applications, with economic factors determining their relative importance, and similar richness in each of the other components. The interconnections stand for a correspondingly complex pattern of information, energy and material exchange between the components.

At the same time, this cycle for a particular fuel (say, natural gas) is itself only one component of a larger fossil fuel system, in turn part of a larger national, or world, energy supply system, still only part of a total energy system (Figure 2.3), and so on.

You may well feel that drawing a system diagram in this way does not advance much beyond a description in words, and at this stage they are indeed fairly similar. However, in making a particular system diagram we make two kinds of choice: first, we choose how to lump together the underlying complexities as 'components' which we won't analyse further at

(a)

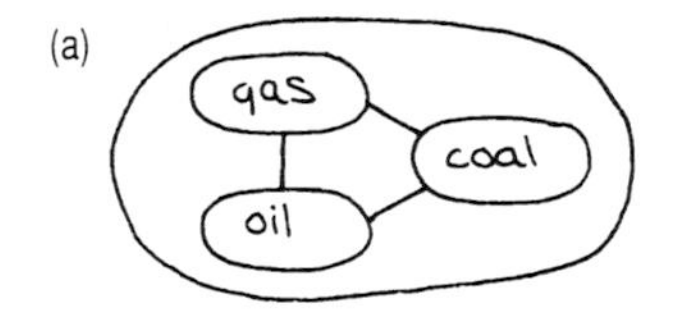

(b)

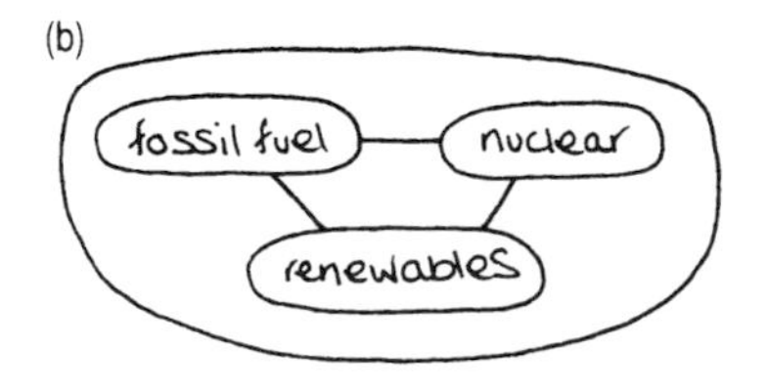

(c)

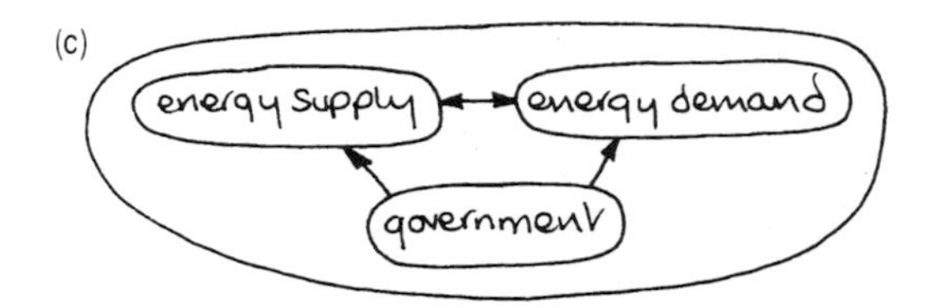

Figure 2.3
Larger systems than the single fuel cycle:
(a) fossil fuel system;
(b) (national) energy supply system;
(c) (national) energy system.

this stage; second, we choose the number of such components which interact together within a boundary to make up the system, with everything outside forming the system environment. It is important to emphasise that these are creative choices, useful to the extent that they provide a clear description of a complex situation. The consideration of effects across the system boundary is the next stage.

2.3 The residuals from energy system processes

The OECD approach

Reverting to the OECD study, with its conventional use of 'environment': each of the above processes involved in the provision and use of energy has environmental by-products – not only the waste materials released *into* the environment which are the most obvious 'residuals', but also, stretching the meaning of the word somewhat, materials removed *from* the environment and structural *changes* to it. The environment may be understood to include not only the geophysical surroundings (air, land, water) and the biological (other living things) but also the socio-economic (employment, markets, welfare, amenity). The assessment process should try to determine the range and quantity of residuals produced over this very wide range.

The OECD study lists the following *Classes of Residuals*:

- Consumption or pre-emption of environmental resources (land, water, biomass, energy, inorganic materials) over and above the primary energy resource which is inevitably consumed at the 'harvesting' stage – a coal mine obviously removes coal, but it has a 'residual' demand for land for spoil heaps, and land pre-empted during the life of the mine for surface buildings, or, for a larger area but shorter time, for open-cast working; for water for coal washing, for biomass like wood for pitprops, for energy to drive the machines and pumps, for inorganic materials like constructional steel.
- Material effluents: solid, liquid, gaseous; inert, chemically active, radioactive.
- Non-material effluents: noise, heat, electromagnetic radiation.
- Other physical transformations: modification of surface or underwater topography (e.g. land filling or dredging), vegetation removal, erection of structures.
- Socio-political influences: changes in employment opportunities, in demands for social services, in population density and distribution.

Some of these last 'residuals' seems to be more like 'impacts' – this is the first sign of confusion, particularly in the 'socio-political' category, which will increase as we continue with the OECD approach, but which will be clarified by taking the systems view.

The residuals involved in the study of coal combustion in Section 3 are straightforward – solid and gaseous material effluents, like ash, dust, soot, sulphur dioxide and nitrogen oxides. Various other residuals will be mentioned in Section 4.

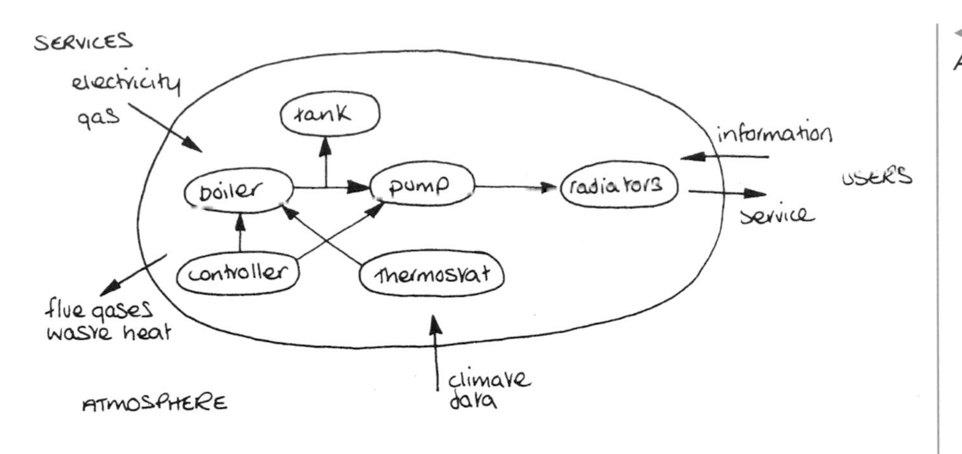

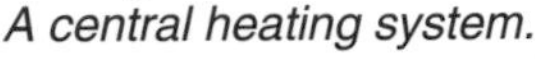
◄ *Figure 2.4*
A central heating system.

Flows across the system boundary

Let us now relate this idea of residuals to our system by looking a little more at the domestic central heating system (Figure 2.4). Its components are undoubtedly connected together to form a system, and affect each other in ways we needn't go into any further, but which obviously had to be carefully considered in the design. A system boundary can naturally be drawn round these components, with an 'environment' outside. But the system is not isolated from this environment; many flows of material, energy and information pass across the boundary: gas, water and electricity supplies in, heat loss and exhaust flue gases out; information about the climate and the wishes of the users, a heating service to the users. These flows across the system boundary largely correspond to the residuals of the OECD method.

As there are these flows across the boundary, just as there are flows between the components within the boundary, how does the 'environment' differ from the system components? Why is it separated by the system boundary?

It can only be that, for the purpose of this system description, the environment is not affected by the flows in and out of it; it can provide whatever is needed and accept whatever is emitted without being changed – or rather, without being significantly changed, so there is some judgement involved about what counts as significant. This is a crucial idea. For example, the climate, the atmosphere, the national gas and electricity supplies are not significantly affected by the operation of one domestic heating system, though they obviously might be by the combined effect of all the country's heating systems. But in this case you're discussing a different system, with a different system boundary – a national energy or climate system, with the many domestic heating systems as separate components within a common boundary.

As well as these physical factors, let us think a bit more about the users of the central heating system. Their demands on the system may well depend on independent factors (their ages, working hours, say) which place them squarely in the system environment. But if the performance of the system affects their behaviour (their temperature preferences, the hours spent at home) then the boundary must be stretched to include them. More significantly, if we include economic factors like their ability to pay for the energy to run the system, the boundary must widen considerably, to embrace rival claims on their budget, employment opportunities, welfare payments and so on.

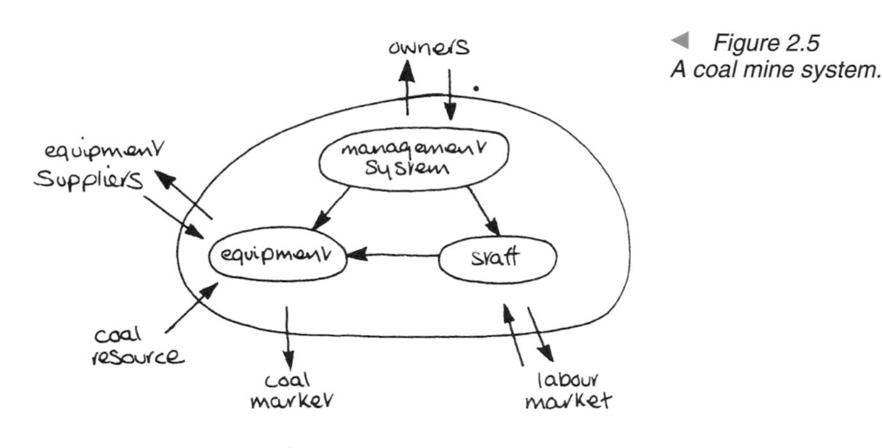

Figure 2.5
A coal mine system.

As a more complex energy system, consider a coal mine, which might be represented by the system diagram of Figure 2.5. This has clearly interacting 'components' inside the boundary – cutting machinery, transport and processing plant, pumps and other facilities, and the staff to run them, with now a significant planning and management system of some kind to be included. But what is outside the boundary as an independent 'environment'? There will no doubt be limited periods when further coal resources, energy, land and water, equipment suppliers, markets for the coal produced, a labour market able to provide or absorb staff, might be thought of as stable and unchanging for day-to-day purposes in running the individual mine. But there will equally obviously be time scales where this is not sensible, and the system boundary has to be enlarged to embrace wider relations between physical resources, people, economics and government. The individual mine can then be seen as one component of a variety of larger systems – political conflicts arise because different groups have different views about what these larger systems should include.

Activity 2

Use the format of Figures 2.1 to 2.5 and your general knowledge to propose your own diagrams for two such larger systems which include the coal mine (e.g. the UK coal industry, and the UK energy system), paying particular attention to the (relatively unchanging) factors which form the system environment, and the flows across the boundary, in each case.

Perhaps it is now clearer that there are two advantages to this component–boundary–environment approach to complex systems.

On the one hand it provides a basis for limiting the scale of a problem so you don't try to think of everything at once – either the endless detail about the way the components work, or the endless possible ramifications of external effects which might be important. There are plenty of cases where for all practical purposes the environment is effectively unchanging – it changes very slowly compared with the system processes of interest, or the effects of the system on it are very local, or very brief. One bonfire doesn't change the climate.

On the other hand we are obliged to remember that there are flows across the boundary and to consider when they would change the environment and what new, larger system boundary has to be drawn to take their effects into account. Perhaps all bonfires can affect the climate.

The 'residuals' are these flows, in and out, across the system boundary. These might be materials extracted or deposited, people employed or laid off, money flows, or more abstract influences like rights or regulations. They will necessarily change the environment in some degree, but the essential idea from the point of view of a particular system is that these changes are negligible: the environment is assumed to go on providing and accepting the flows as required. How to assess the effects when this is no longer true is the next stage.

2.4 *The impacts of these residuals*

The OECD approach

The aim of this stage is to examine the way conditions and processes in the environment are modified by the residuals – the spread of pollutants downwind from a power station is a straightforward example, which we will return to in Section 3. In general, there will be pathways of cause and effect by which residuals alter the environment – transportation through air, water, soil/rock, living things, with physical, chemical or nuclear transformations on the way. Variation with time and space is an important issue – how long does a residual last, how far away from its source can it be detected?

Again, the OECD study suggests various *Classes of Impact* (which have been grouped to indicate some differences of emphasis):

- reduced availability of environmental resources
- altered chemical concentrations: in air, water, soil, biota
- altered circulation patterns: in the atmosphere, water bodies, ground water
- altered temperature, humidity, precipitation
- altered electromagnetic fields
- absorbed radiation dose
- perceived noise

(There is a change of emphasis in these last two from the four previous classes, as these two depend not only on alterations in the distribution of radioactive material and noise, but also on the extent to which humans are affected by them, which depends how the humans are distributed too. It might be better to preserve this distinction.)

- other alterations to, or loss of, habitat

(That is, the attributes of the environment making it suitable for particular species – with a human tendency to concentrate attention on those non-human species that we find more appealing: badgers rather than cockroaches, for example. 'Alterations to habitat' may be 'impact', but 'loss of habitat' sounds more like 'damage'.)

- socio-economic stresses

(This is a bit vague, with the distinctions from 'residuals' and 'damage' again not very clear.)

Changing the system boundary

The 'impacts' of a system are arrived at by extending the system boundary, so that it now contains the previous system as one component, interacting with others via its residuals, the flows across the original boundary (Figure 2.6a). We have a choice as to how to do this, depending on the nature of the particular residuals. If they are gaseous effluents, we will want to consider their interaction with the atmosphere, or other parts of the geosphere (Figure 2.6b); if they are toxic we will need to consider the ecosystem effects (Figure 2.6c); if staff movements, we should consider effects on employment or population distributions; if money flows, effects on the economic system and so on. We are replacing the idea of a generalised impact on 'the' environment by a series of more specific studies of particular larger systems that we judge to be relevant. Each of these systems has its own boundary and flows to its environment which we can examine to see whether we have gone far enough. We might hope during this stage to determine the effects, in time and space, on the system we have chosen (e.g. the atmosphere) of the particular energy process we started with (e.g. a domestic heating system) – its 'sphere of influence', or perhaps better, its 'bubble of influence' (Figure 2.7). This will have a straightforward physical

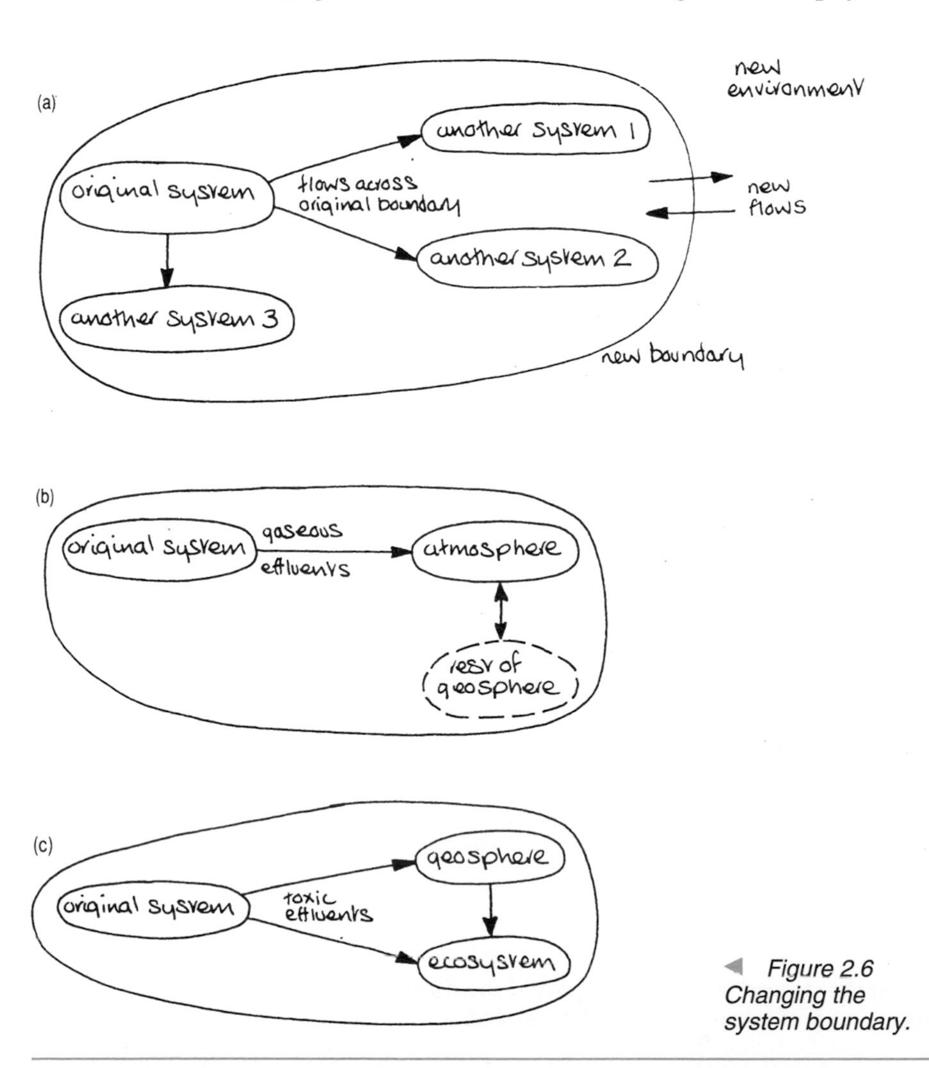

Figure 2.6 Changing the system boundary.

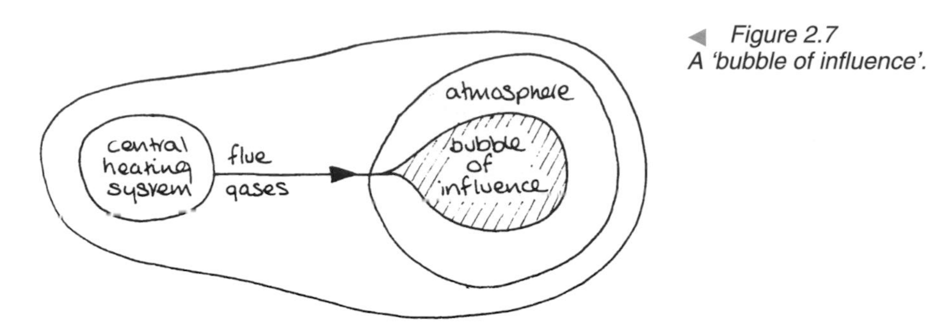

Figure 2.7
A 'bubble of influence'.

meaning for gaseous pollutants in the atmosphere, but the same idea can be extended to other flows and other systems, including more abstract forms of influence.

Evidently, judgement is involved in deciding which larger system is to be looked at. Even if the resulting 'bubble of influence' can be determined (which may not be easy in larger cases), it is still possible that the effect on some other system has been overlooked. This may sound complicated – and indeed it is; the consequences of human actions can be very involved and difficult to predict. But at least this stage, of simply mapping the effects, shouldn't be too controversial, once we have agreed which larger system is being considered. This may be a more political decision than meets the eye: choosing the larger system for study is 'setting the agenda' – we may be looking at the atmosphere and neglecting rivers, or the ecosystem, or employment, welfare, economics, amenity, and so on.

2.5 The damage caused by these impacts

The OECD approach

In considering 'damage' the study now makes human value judgements about the effects of these environmental impacts on us. While the 'origins' and 'residuals' can be defined quite precisely, and the 'impacts' can be traced, even if with difficulty, fairly clearly, the measure and characterisation of the resulting potential harm to human well-being, or to vulnerable parts of the environment, is more difficult and controversial. In Section 3 we shall consider the relatively straightforward example of coal combustion and its solid and gaseous effluents. The resulting impact of smoky, foggy atmospheres was evident, of acid rain somewhat less so, but a *measure* of the resulting damage, in increased respiratory disease, or in ecological effects in lakes and forests, has proved much more difficult to obtain conclusively. Similar difficulties have arisen in tracing the link between nuclear operations and childhood leukaemia, for example.

The OECD study lists the following *Classes of Damage*:

- Direct, to human health and safety – resulting from accidents, sabotage, routine emissions and exposures harmful to humans, either short or long term.
- Direct, to economic goods and services – accidents, sabotage, effluents and other environmental transformations.
- Impairment of environmental goods and services – impaired visibility, dust, ugly structures, annoying noise and odours, reduced diversity.

- Disruption of socio-political conditions and processes – increased centralisation of political power, loss of civil liberties, excessive burdens on social services.

- Psychological distress – physical displacement or unemployment; fears of the above damages, whether or not they materialise (!).

This is a revealing list. As with the three previous ones, you may well feel it progresses from well-defined to increasingly fuzzy issues. There is also a new element, in the attitude it now shows to 'the environment' – one which may be a common human perception, but is surely worth looking at a bit more closely. In the first two items, human welfare and the economic system that provides goods and services seem to be seen as proceeding smoothly in a 'normal' environment; the routine or eccentric behaviour of energy systems may produce environmental impacts, as considered above, that can damage these 'normal' activities. And the last two items, though vaguer, can perhaps be linked with the first two – 'psychological distress' with 'health and safety' (personal), 'socio-political' with 'economic' (social), all providing a human-centred view of our own affairs, and only considering the environment explicitly when it is a potential source of trouble. However, it comes into its own in the middle item, as the source of 'environmental goods and services', which is very much an economist's description of the way we value our surroundings – apparently as a source of pleasant sights, sounds and smells which might be impaired, and, tucked into a corner (ecological?) diversity which might be reduced. Are these just human 'goods' or 'services', to which we might put a price in environmental economics, or do they have other values? What do you think?

Perhaps there are two separate issues. We have been trying to clarify this four-stage environmental assessment procedure by introducing the 'system' view of 'environment', and the choice of progressively wider system boundaries to examine the consequences of our activities. On this view we can hardly see 'environmental damage' as separable from 'normal' activities; it is a potential consequence of any of them. Various responses, using this system approach, are pursued below. But this still leaves the larger ethical question about the attitude of human beings to the world – do changes only matter to the extent that human welfare (which can include our aesthetic responses) is damaged, or are at least some 'environmental impacts' to be seen as unacceptable even if there is no 'damage' to human interests? These are questions which continue to arise throughout this series of books, and to which there will be no final answers; they are left here for continuing discussion.

In the OECD approach, it is argued that if environmental damage can be measured, and if the probability of its occurrence from a particular process can be determined (both considerable 'ifs') then the simplest way of characterising the process is by the 'average or expected harm'. This can be defined as the probability multiplied by the damage, summed over all forms of damage. (This might be seen as providing an illusion of mathematical precision for a very vague process.) Apart from the difficulty of carrying this out, it is not the only possibility. Other relevant attributes might be considered:

(a) the magnitude of the worst-case event, irrespective of probability;

(b) the distribution of damage over time and space – potential damage to future generations or remote communities;

(c) the degree of irreversibility of the damage.

All of these have been cited as arguments against continued and future use of nuclear power, for example, even where estimates of average or expected harm seem favourable.

Responses to 'damage' in the system model

We have seen that by enlarging the system boundary we can interpret the effect of one system on another that we consider to be important as a 'bubble of influence', but to interpret that as 'damage' introduces difficult and controversial value judgements. In some cases the bubble of influence of human activities might not be called 'damage' at all: if the natural energy flows of the wilderness are disturbed to create an elegant garden, a well-run farm or a thriving city, many people would not call that damage – but note the human values embodied in all those adjectives. However, if we think we know what 'damage' is when we see it, we can still seek to assess it or contain it.

The last sentence contains an insidious 'we' that implies the existence of a rational human consensus about such controversial questions. It is only too obvious that this does not exist, either about the nature of the problems, or about the sacrifices of self-interest required to solve them. But if 'we' is understood to stand for some underlying trend of human understanding, which is only turned into action by agonising political processes where conflicts of power and interest are resolved with difficulty (if at all), it may still be a useful shorthand for the rest of this section. But always ask yourself: Who are 'we', making these decisions?

A natural question is 'How big, or long-lasting, is the bubble of influence?'

1 If it is small or brief enough (a puff of smoke) we will ignore it.

2 But suppose it were a speck of plutonium, or any other toxic waste? Obviously there is then another set of questions, about the extent of the overlap (in space and time) between the bubble and the components of the ecosystem (including humans), the flow of materials through the ecosystem, the sensitivity of human or other living 'systems' to toxic inputs. These are usually quite difficult to answer. If we conclude there is a hazard, we will try to put a fence around the rubbish dump where it arises, or find a remote enough place for it. We will then have to deal with all the questions this raises about interference with rights of access and the difficulties of satisfactory containment.

3 Or we will try to disperse effluents, which means making the bubble larger but of lower intensity, so that ecosystem effects are reduced to an acceptable level (again according to someone's value judgement).

4 Or we may conclude that the effects are too serious for the process to be allowed to continue, and we will abandon it.

Finally, and most important, when we have come to terms with the bubble of influence of *one* energy process, by ignoring it, containing it, modifying it, or rejecting it, we have to consider the overlapping effects of *all* energy processes. The difficulties encountered at present arise most of all from the continued growth in the use of processes which might be individually tolerable but are unacceptable when combined. In Section 3 for example we will see how the scale of problems associated with coal combustion has progressively expanded, now perhaps to a global scale, and in Section 4 how other parts of the energy system impinge on these global problems.

2.6 Summary

A procedure for assessing the environmental effects of energy systems has been suggested, involving four stages – identifying their origins in the innumerable processes of the fuel 'cycle', the residuals of these processes, their environmental impacts, and the resulting damage. We have recognised some progressive difficulties of definition and feasibility in carrying out this procedure, particularly over socio-economic and ethical issues, but it provides a useful initial framework. It has been clarified by introducing a 'systems' view of processes on all scales. Interacting components within a system boundary are distinguished from the 'environment', in a new narrower sense, outside it. The effect of flows across the boundary (residuals) can only be examined by studying a larger system with a new boundary. This idea sheds light on 'environmental impact' and 'damage'; there is no longer a single environment to consider, but a range of alternative larger systems affected by different residuals.

Activity 3

Before reading any further, use this framework to make your own notes of the 'environmental damage' you associate with the present use of coal. This will help to define the issues discussed below.

3 Coal: from a local problem to a global one

Let us now look in more detail at one of the simplest and most familiar sources of environmental effects, the burning of coal. The first stage of the assessment procedure is to understand something of the particular process involved.

3.1 Coal composition and combustion

Coal is a rather variable material produced from buried plant remains in swamps by geological effects (temperature and pressure) over very long periods of tens to hundreds of millions of years. Its use as a fuel depends on its mainly carbonaceous composition of modified organic compounds of carbon with oxygen and hydrogen in varying proportions. Initial bacterial breakdown of the plants ceases when they are buried under suitable conditions; the resulting peat is only about half carbon. In the process of coal formation the oxygen content is gradually reduced as the 'rank' of the coal changes, through soft brown lignite to hard black bituminous coal; the final changes involve the reduction of hydrogen content too, to anthracite and finally almost pure graphite (one form of crystalline carbon). The local geological history determines the rank that has been reached by the present time; different coal deposits worldwide represent all of these various stages. (There are of course no significant changes taking place on the human time scale.) The presence of volatile organic compounds (i.e. those driven off by heating) in the lower rank coals makes them easier to burn, but with a smokier flame. The volatile contents can be driven off to form coke, as used in the iron and steel industries, or to provide an artificial smokeless fuel alongside the natural anthracite.

Coal also contains varying amounts of inorganic impurities depending on its source – material that either accumulated along with the plant remains or was introduced during the geological modifications (Table 2.1). Clays and carbonates are the main contributors to the ash that remains after combustion, but small amounts of sulphur (mostly as iron sulphide, pyrite) and salt (sodium chloride) have a great effect on the ability of the ash to corrode boilers and thus on the practical use of particular coals.

Table 2.1 Principal constituents of typical European coals

carbonaceous material	70–90%
water	0–7%
nitrogen	1–2%
sulphur	0.5–2.5%
other inorganic material	5–20%

During combustion the inevitable primary products are the heat released when carbon, hydrogen and their compounds react with the oxygen of the air, and the resulting carbon dioxide and water if the combustion is complete. This is the whole reason for using coal, for domestic or industrial heating or to drive heat engines (steam engines, steam turbines) in factories or power stations. We shall return later to the possible effects of these primary residuals – which would not usually be called pollutants. For the present we shall be more concerned with other products, obtained because the combustion is incomplete, and because of impurities in the coal. They depend on both the particular coal used and on the way it is burned. A domestic open fire produces much less complete combustion than an industrial furnace. There is much combustion engineering knowledge available; however, it will be sufficient here to recognise that in general visible smoke contains particulates, which are unburnt carbon (soot) and ash, contaminated with tarry unburnt hydrocarbons and other impurities, some of which might be carcinogenic or corrosive. The sulphur and nitrogen impurities burn to produce (invisible) sulphur dioxide (SO_2) and various nitrogen oxide gases, collectively known as NO_x, where x stands for varying proportions of oxygen (these can also be formed from the nitrogen in the air when the combustion temperature is high enough).

Let us now examine the impact of these residuals.

3.2 *Air pollution: a local problem*

As the earliest industrial society, Britain had an unenviable reputation for air pollution in its larger cities, which is still vivid for anyone whose memory of them goes back before the 1950s. It didn't require any systematic modelling or elaborate analysis of 'bubbles of influence' to observe the environmental impacts and damage inflicted by coal burning factories, and innumerable domestic open fires. The cities were the dirtiest in the world, with dark skies, acrid smoky air, blackened buildings, and an unhealthy urban population. Charles Dickens in the 1850s wrote in *Bleak House* of a foggy November in London, with 'smoke lowering down from chimney-pots, making a soft black drizzle, with flakes of soot in it as big as full-grown snow-flakes – gone into mourning, one might imagine, for the death of the sun.' He also wrote in *Hard Times* of Coketown, which 'was a town of red brick, or of brick that would have been red if the smoke and ashes had allowed it; but as matters stood it was a town of unnatural black and red like the painted face of a savage. It was a town of machinery and tall chimneys, out of which interminable serpents of smoke trailed themselves for ever and never got uncoiled.' (He acutely went on to use the fog and grime as symbols for the obfuscations of the law and the aridity of a Utilitarian fact-based education system, which perhaps goes some way to explaining why it took another century before the problems were tackled.)

Manchester in 1853 was described by a visiting Frenchman, de Tocqueville, as 'given over to industry's use . . . large palaces of industry . . . keep light and air out of human habitations which they dominate; they envelope them in perpetual fog . . . a sort of black smoke covers the city', while in Sheffield as late as the 1930s 'every square foot of exposed masonry was black and filthy. The sun rarely penetrated a vast umbrella of smoke and soot.' The effects in respiratory disease and stunted growth – of vegetation and humans – were plain to see, though the defiant 'Where there's muck there's brass' was as likely to be mouthed by its victims as its beneficiaries, as they too depended on the employment provided.

Concern grew through the twentieth century, and the turning point for pollution control is usually considered to be the 'Great Smog' of 5–9 December 1952 in London (characteristically, not in one of the industrial cities, where they were also commonplace). Sulphurous soot particles in moist air formed an acrid 'pea-soup' fog that persisted for several days under particular adverse weather conditions. It led to an estimated additional 4000 deaths in the following weeks, due to heart and lung diseases. The Beaver Committee was appointed to examine the problem and two years later estimated that pollution was costing the country (at 1954 prices) over £300 million a year (at last providing the facts to satisfy the Utilitarians?).

The problem, seen as a localised one around areas of dense population and heavy industry, was tackled through the Clean Air Acts of 1956 and 1968. They focused on visible air pollution, their major thrust being to cut soot emission from chimneys and prevent smog formation. The greatest proportion of this was from domestic fires, so local authorities were empowered to declare smoke control zones, within which emission of smoke from houses was banned. Emission from factories and power stations was tackled by removing larger particulates at source, and using taller stacks to dilute and disperse the smaller particulates and gaseous pollutants (SO_2, NO_x). The rules did not set out to stop or control the production of these gaseous pollutants, but only to disperse them over a wider area.

These controls, along with a transfer to cleaner fuels like natural gas and oil, have reduced smoke emissions by more than 85% since 1958. There have been corresponding great improvements in the visible quality of the air in industrial cities (Figure 2.8). There have also been reductions in invisible pollutants like SO_2, mainly from reduced domestic, industrial and transport output. The largest part now comes from power stations, whose output of SO_2 had increased by the early 1980s (Figure 2.9).

▲ *Figure 2.8 Air pollution – before and after the Clean Air Acts: the Tees near Middlesborough (above) in 1924; (below) in 1972.*

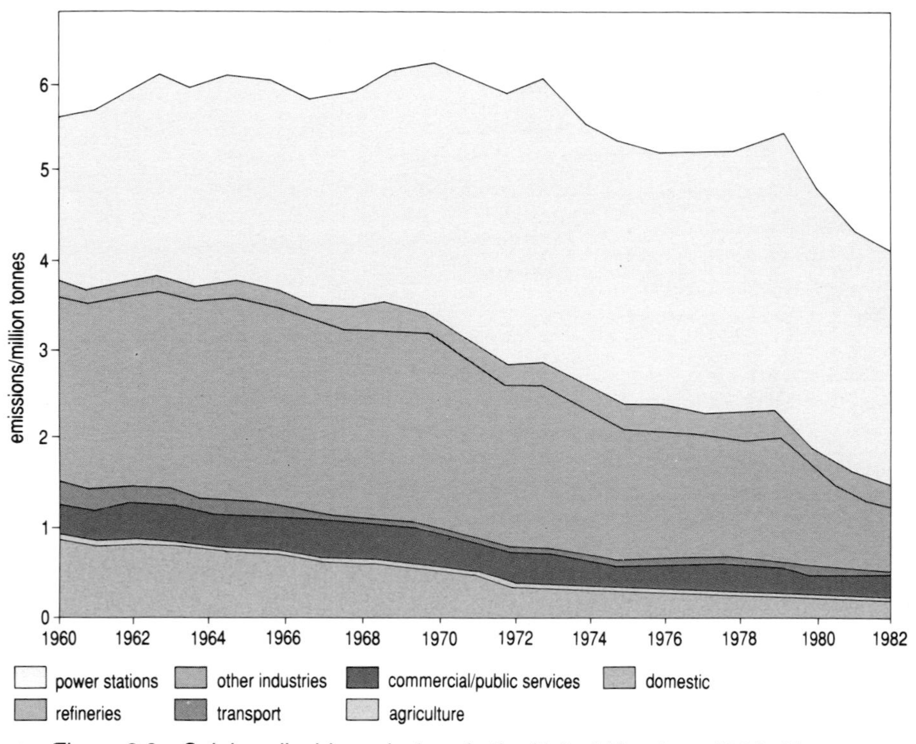

Figure 2.9 Sulphur dioxide emissions in the United Kingdom, 1960–82.

Subsequent reductions in response to EU directives are considered under 'Acid rain' below. This has been associated with a reduction in total coal use, from about 90% of UK energy in 1950 to under 25% in 1994, and a reduction in its use for all purposes other than electricity generation (Figure 2.10). The fall in total coal use has continued, to some 77.5 million tonnes (Mt) in 1994. Even more significant is the fall in home-produced coal, to some 44.7 Mt in that year, reflecting the vast increase in imports for combined economic, technical and political reasons. You may wish to add these figures to Figure 2.10 to emphasise the dramatic scale of these reductions.

The 1990 situation shown in Figure 2.11 indicates the British Coal policy at that time, with large power stations sited near the main coalfields. Since then, major pit closures and privatisation of the coal industry in 1994 leave the future much more uncertain. The privatised electricity industry has also begun to build (more efficient, less polluting) gas-fired power stations – the 'dash to gas' of the 1990s – adding to the pressure on the coal industry. However, the residual contracts for power generation from home-produced coal were guaranteed at 30 Mt a year for three years after privatisation, so, together with the use of imported coal, coal-fired generation is still predominant in the late 1990s.

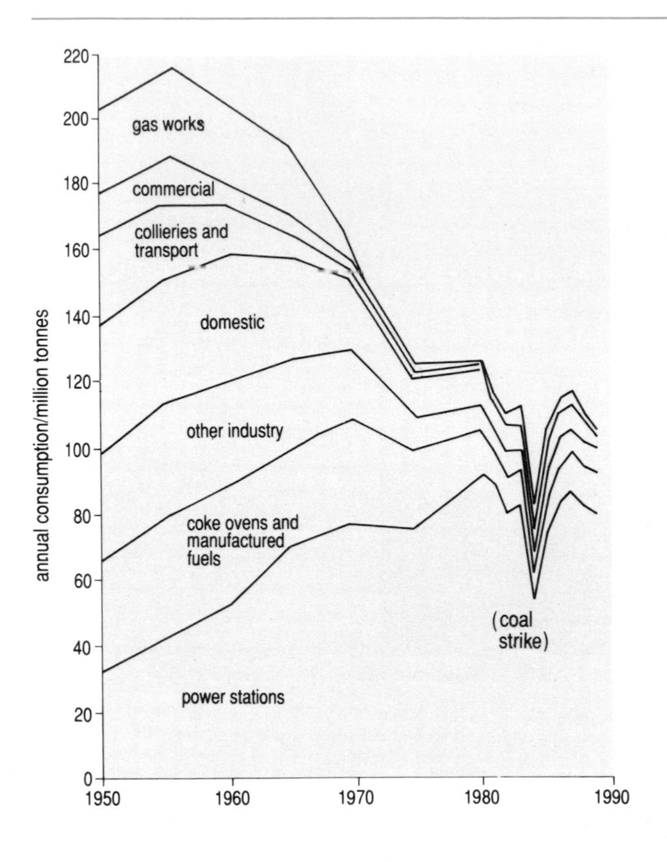

◄ *Figure 2.10 Coal use in the UK 1950–89.*

3.3 *Air pollution: a regional problem*

This national history, of industrial dependence on coal, concern about air quality, clean air legislation and shifts in fuel use from coal, mainly to oil and gas, could be matched with appropriate local modifications in each of the industrial countries of Europe, North America and, increasingly, the rest of the world. The different stages of this process in different countries, and oil price increases, led to a steady increase in total coal use through the 1980s both in Europe and the world as a whole, maintaining a share of total energy use in these areas of just under 20% and 30% respectively.

Following the continuing growth in total energy use, and the recognition that pollutants do not necessarily stay in the country where they were produced, the scene has been shifting during recent decades to larger stages: Europe (as a result of developing European Union controls), the developed world (responsible for 70% of energy use, with agencies like the OECD), and the world as a whole (through various United Nations agencies). Let us concentrate next on the European scale.

European concern

Despite the local improvements in air quality and reduction of SO_2 described above, concern grew in the 1970s about adverse impacts and damage in the UK and in Europe from SO_2 and NO_x produced in various ways (Figure 2.12).

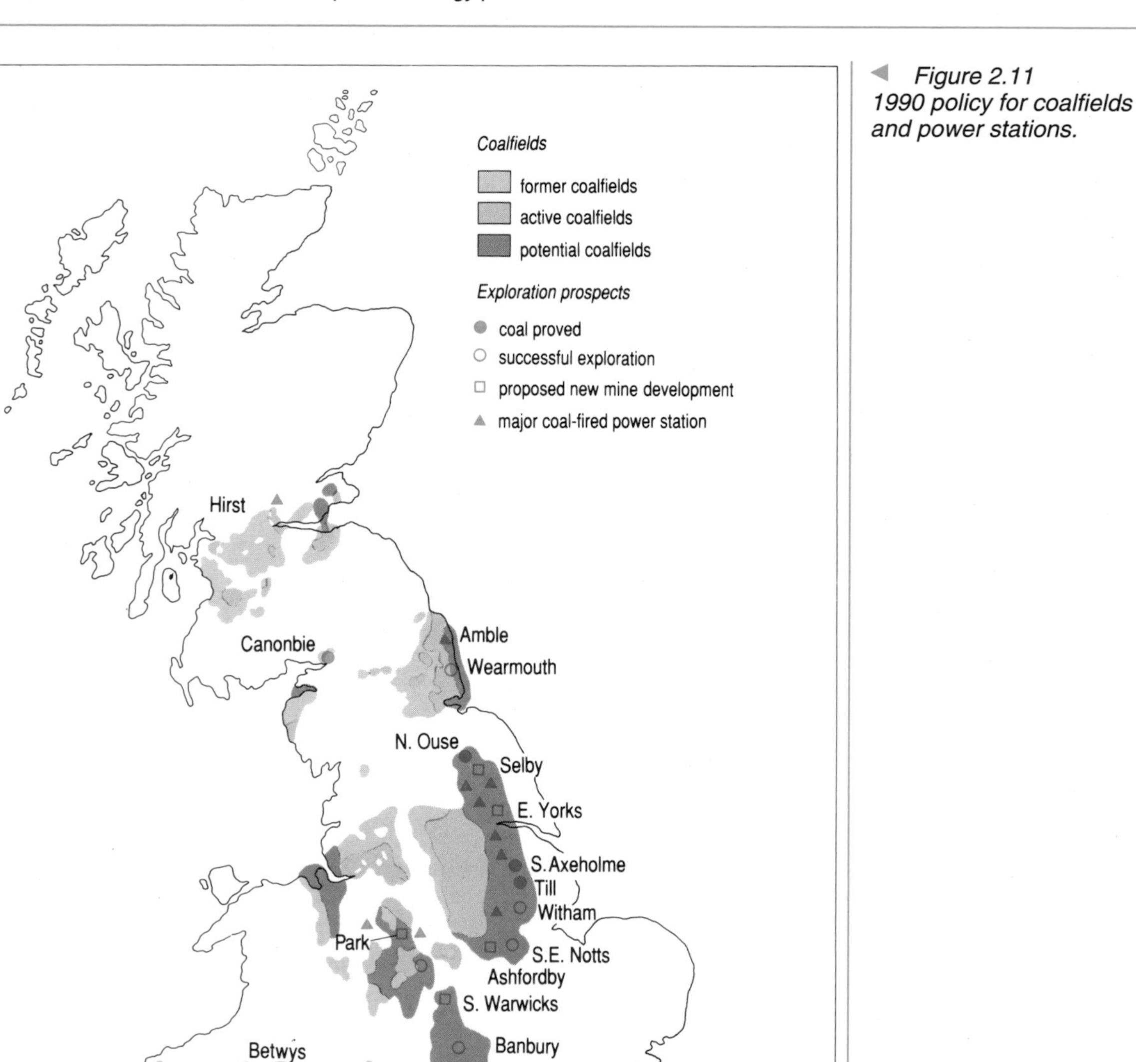

Figure 2.11
1990 policy for coalfields and power stations.

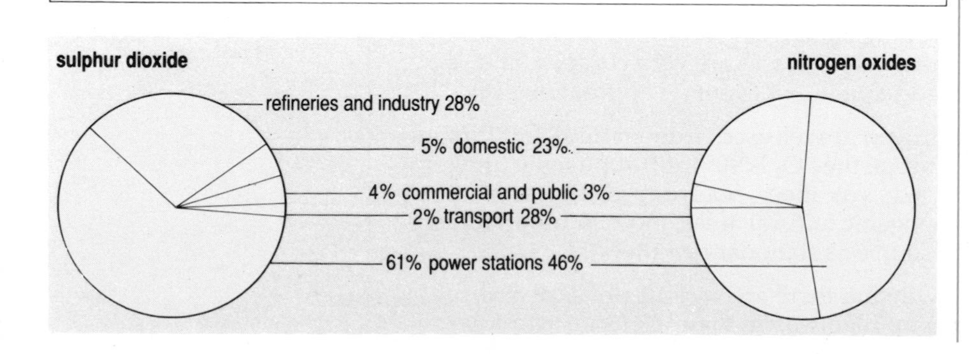

Figure 2.12
Main sources of SO_2 and NO_x in Europe.

Q In these data, what is the main source of each of these effluents?

A In both cases, power stations, which produce more than half of the SO_2 and just under half of the NO_x.

Q What is the next largest source of each?

A This is markedly different in the two cases. For SO_2, it is other industrial processes, including smelting operations like those to be described in Chapter 5. For NO_x it is transport.

Although this section is about the effects of coal combustion, some of them are so interwoven with those of these other sources that they will have to be considered at the same time.

In the case of power stations and other industrial sources, taller stacks had dispersed these effluents and improved conditions locally, but where had the dispersed gases gone? They had been exported to the countryside downwind of the source, carried maybe hundreds of kilometres, and falling as wet or dry acid deposition. This was the 'tall stacks paradox', the inadvertent replacement of the original visible local problem of soot by the invisible long-distance, even transnational, problem of acid precipitation. A study in 1982 by the (UK) Institute of Terrestrial Ecology showed increased rainfall acidity in parts of Scotland (pH up to one unit lower than normal background levels).

An earlier study from the OECD in 1977 had already suggested after a 5-year investigation that the consequences could be more international. The total quantity of SO_2 produced in Europe (about 9 million tonnes in 1973) was stretching the acceptability of dispersal; the UK, with 2.8 million, was the largest single contributor. Furthermore, some countries, with the UK again in the lead, were net SO_2 exporters with more of their emissions falling on other countries than they received from others, while countries downwind of them like Norway and Sweden were net importers, receiving more than they emitted to other countries. This imbalance was gradually shown to exist throughout Europe (and similarly in America), as a consequence of varying industrial activity and pollution control measures combined with prevailing winds.

It is evident that net importers have much more reason for interest in international regulation than net exporters, which no doubt goes some way to explaining controversy about control measures; costs and benefits are not equally shared.

Acid rain: formation

We have seen that SO_2 and NO_x are important residuals of coal combustion, and other processes. Now let us consider their physical transport and chemical transformation in the atmosphere to produce environmental impact from **acid rain**. The chemical processes are very complex and not fully understood, but the broad sequence of events is approximately:

1 Oxides of sulphur and nitrogen are released, from natural and human (anthropogenic) sources. In Europe the SO_2 is 90% anthropogenic (power stations etc.) and 10% natural (e.g. volcanoes, sea spray, rotting vegetation, plankton). NO_x is 50% anthropogenic in rural areas, more in urban regions (with vehicles as important a source as stationary emitters).

2 As they are dispersed downwind there are various physical and chemical effects, shown in a simplified way in Figure 2.13.

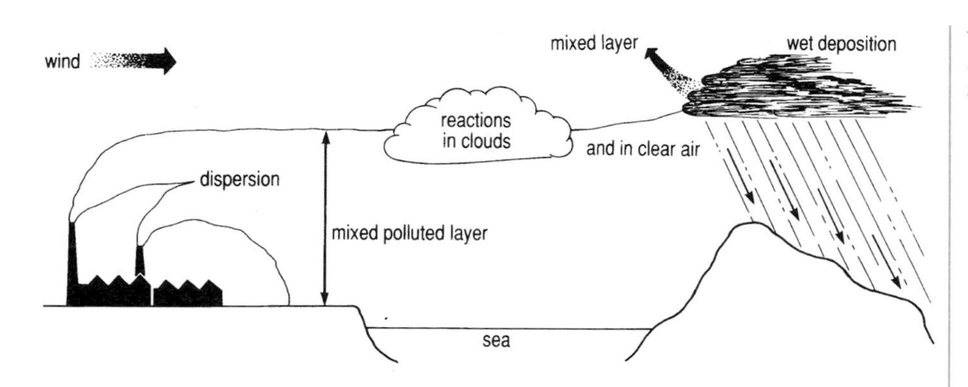

Figure 2.13
Dispersion of pollutants leading to acid rain.

3 Considering SO_2 first: there is some dry deposition, mostly near the source, as adsorption on surfaces or take-up by plants, but this is a comparatively slow process (even for low-level sources), so much SO_2 remains airborne for long distances, particularly in dry weather and if emitted from tall stacks when it can readily reach hundreds of kilometres from the source. While airborne it can be oxidised – that is, have its proportion of oxygen increased, forming the 'higher' oxide SO_3, but this is not straightforward. While it remains a gas the reaction with sources of this extra oxygen is very slow. But it is much faster if the SO_2 is incorporated in clouds or fog as water droplets. The SO_3 produced then reacts with water to form sulphuric acid H_2SO_4, ionised in solution as H^+ and SO_4^{2-}. The dilute acid can finally reach the ground as wet deposition.

4 For this oxidation to happen, there must be an oxidant: a source of extra oxygen. Even in wet reactions, atmospheric oxygen acts only slowly as an oxidant under normal conditions (though its effect can be speeded up by traces of metal ions such as Mn^{2+} and Fe^{3+}). Usually the major oxidants are hydrogen peroxide (H_2O_2) and **ozone** (O_3). These are produced in the atmosphere by photochemical effects: sunlight stimulates their formation from oxygen and water vapour. But these processes are themselves sensitive to the presence of other gases (an effect that is returned to in more detail in *Blackmore & Reddish, 1996*, Chapter 2). In particular, ozone production near ground level is much enhanced by the effect of sunlight on the mixture of NO_x and unburnt and partially burnt hydrocarbons in vehicle exhausts. This is a pollution problem in its own right: the **photochemical smog** of such concern in sunny cities with many motor vehicles, such as Los Angeles. Its importance here is to emphasise the interrelated 'synergistic' character of pollution studies: acid rain production from power station SO_2 is much affected by vehicle NO_x.

5 A further important influence is natural ammonia rising from land or sea and joining in the process; oxidation of SO_2 by ozone is slowed down as acidity increases, so would be self-limiting, but ammonia can neutralise the acid, keeping up the conversion rate.

6 The primary NO_x emissions are transformed in a rather different way. There is little dry deposition of NO, and it is oxidised to NO_2 and HNO_3 as a gas, as neither NO nor NO_2 is very soluble in water. There can be dry deposition of nitrate particles, while incorporation of HNO_3 in clouds and raindrops leads again to wet deposition.

7 Thus, acid rain, containing sulphate SO_4^{2-}, nitrate NO_3^-, ammonium NH_4^+ and hydrogen H^+ ions (in addition to the natural bicarbonate HCO_3^- from CO_2), falls to earth.

Measured increases in rainfall acidity

Direct observations of this impact depend on the **pH** scale of acidity. The pH scale is a logarithmic index of 1 (most acid) to 14 (most basic/alkaline);

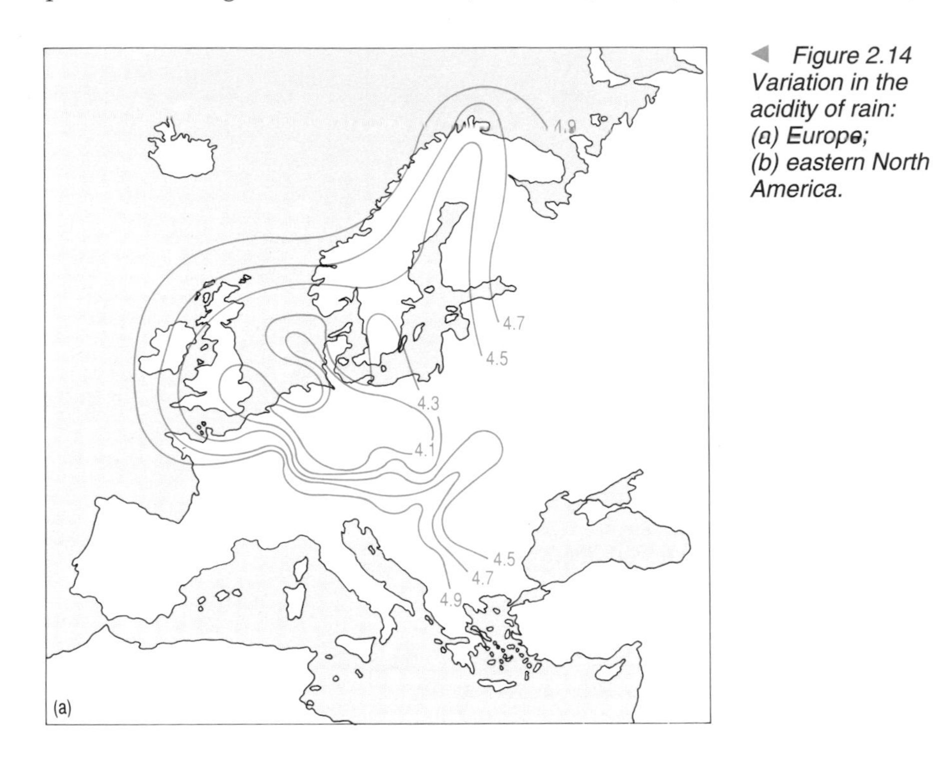

◀ *Figure 2.14 Variation in the acidity of rain: (a) Europe; (b) eastern North America.*

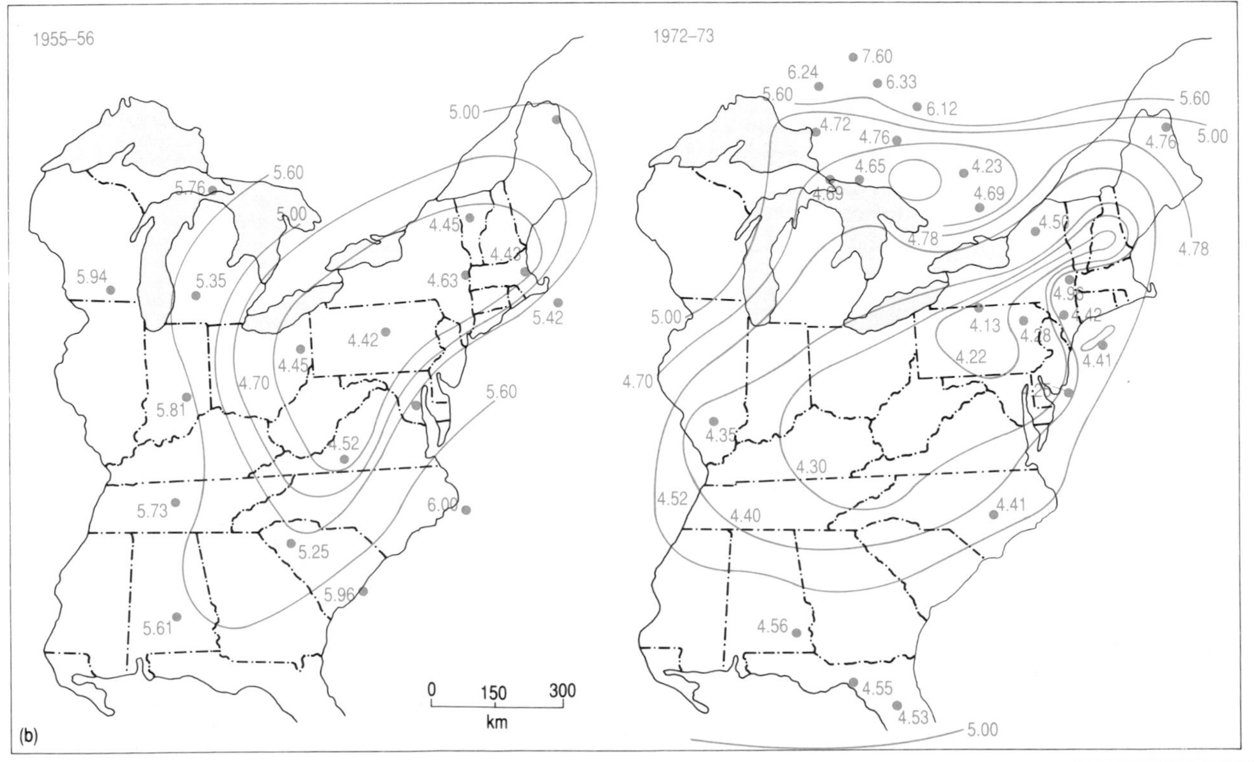

a neutral solution is 7 on the scale. Small numerical changes in pH have a significant effect on many processes, particularly biochemical ones.

'Normal' rainfall is already somewhat acidic, from the carbon dioxide in the atmosphere (dissolved to form carbonic acid) and naturally occurring sulphur compounds (from volcanic eruptions and biological processes). A standard pH figure for 'clean' rain is 5.6, though natural variability can lead to values as low as 4.5 (which confuses argument about anthropogenic change). Systematic comparisons of rainfall acidity show consistent evidence of deepening pH 'holes' below 4.1 (i.e. acidity peaks) over the regions affected by SO_2 and NO_x emissions – see Figure 2.14 for studies of north-west Europe and eastern North America (for two time periods). More widespread global monitoring is now under way.

Damage attributed to acid rain

A further complication is that the acidity of surface waters can be reduced from that of the rainfall by the natural neutralising capacity of the rocks over or through which they flow. Limestone in particular is alkaline, and such areas are less affected, whereas rocks like granite, quartz and sandstone are poor neutralisers. The soils that form and the vegetation grown on them, such as conifer plantations, can increase the acidity of surface water. Nutrient enrichment from sewage, fertilisers and forest fires can reduce it. Factors such as these provide reasons for uncertainty in the secure linkage between emissions, acid rain and ecological effects.

Nevertheless, international concern was first expressed at the UN Conference on the Environment in Stockholm in 1972, when the Swedish Ministry of Agriculture reported massive fish kills which were attributed to acid rain, particularly resulting from UK emissions. These original Scandinavian problems are now seen as part of a wider pattern, including forest decline, particularly in Germany and central Europe (*Waldsterben*, the death of the forest, is shown in Figure 2.15 and Plate 9), lake acidification, and declining fish populations, which have been reported in nearly every European country and related to industrial activity, though there is still not a consensus on their detailed origin. There is also structural damage to buildings – already a familiar observation at earlier times in UK industrial towns.

In the case of forest damage, it has been argued, particularly by the UK's former Central Electricity Generating Board (CEGB), that this results not from acid rain but from the direct toxic effects of ozone in the lower atmosphere, which is derived from photochemical effects on vehicle exhausts, while other workers attach most importance to toxic minerals leached from the soil into surface waters by the changed acidity.

Scientific complexities and political responses

Thus evidence of environmental damage has been emerging in many European countries. While many scientists and politicians, particularly in the countries most affected, attribute this directly or indirectly to acid rain, it is in fact difficult to establish the links beyond doubt. In the UK a Parliamentary Select Committee took evidence in 1984. It echoed these uncertainties, but it proposed action even though the evidence was incomplete:

> While a direct and proportional link between sulphur emissions and environmental damage is not scientifically proven, the case against SO_2 is telling enough and the damage severe enough that action must be taken before all the evidence is in.

Figure 2.15 Damage to forests in Europe.

A similar European Commission study in the same year stated:

> It has not been unequivocally established that environmental impacts are caused by acid pollutant emissions . . . nevertheless circumstantial evidence would suggest that acid emissions and their subsequent chemical transformation and precipitation are at least a partial contributory cause to the observed effects and may be giving rise to as yet unidentified impacts, some of which could be irreversible.

There has been some 20 years of controversy. The UK in particular has been intransigent over control measures, claiming that the inadequate evidence did not justify the increased cost to the electrical industry and the consumer. But action was agreed in November 1988 when an EC Directive to reduce SO_2 and NO_x emissions from large combustion plants was adopted. A 3-stage reduction from 1980 levels of SO_2 from existing plant over 50 MW was accepted, with overall community targets of 23%, 42% and 57% reductions by 1993, 1995 and 2003, and a simultaneous 2-stage reduction in NO_x of 10% by 1993 and 30% by 1998 (and corresponding national targets adjusted for local conditions). The implications for the UK were costs of £2 billion for the installation of expensive flue gas desulphurisation (FGD) in some large coal-burning power stations, as already adopted throughout West Germany, and of £200 million for combustion modification to reduce NO_x. In fact, the timetable is only being met by the use of imported lower sulphur coal and, particularly, the introduction of gas turbines. It should also be noted that FGD requires the provision of large quantities of limestone and disposal of the resulting calcium sulphate – as so often happens, replacing one disposal problem by another.

In March 1990 the final report was issued of the joint £5 million UK–Scandinavian Surface Water Acidification Project (SWAP) funded 5 years previously by the CEGB and National Coal Board (NCB), when claiming research information was still inadequate. It broadly justifies earlier Scandinavian findings, while providing more detailed insight into the complexities outlined above.

These agreements to control emissions from coal-burning plant can be seen as part of the development of an international 'regime' (of the kind explored more generally by *Blackmore & Reddish, 1996*, Chapters 5 and 6). This involves fossil fuels more widely, particularly including transport vehicles, which we have already seen to be significant sources of NO_x and other pollutants (hydrocarbons, particulates and carbon monoxide). Apart from their intrinsic toxicity, they are implicated in the production of photochemical smog and particularly ozone, in turn intrinsically damaging and linked with acid rain production.

An outline chronology of the agreements negotiated by the UN Economic Commission for Europe (UNECE) and by the European Community (EC), now European Union (EU), is shown in Table 2.2, to indicate the interlocking nature of these political processes, and the way 'we' struggle forward when a 'damage' problem has been identified.

A similar pattern of observations, inter-state conflicts, resistance to action and final agreement to (some) emission reductions can be traced in North America. No doubt the coming decades will see similar processes developing in the industrial areas of eastern Europe and the Far East, and increasingly, with industrial development, worldwide.

3.4 Carbon dioxide and the greenhouse effect: a global problem

Thus we are slowly coming to terms with these regional pollution effects caused by the combustion of impure coal (and oil, for transport). They can in principle be removed, though at a price, by better cleaning and combustion methods. At the same time the impact of the primary residual of fossil fuel burning, carbon dioxide, in modifying the 'greenhouse' effect on a global scale is now seriously feared, with its implications for climate change and sea-level rise. This is discussed in more detail in *Blackmore & Reddish, 1996*, Chapter 3. It (almost certainly) cannot be dealt with by cleaning methods and is only modified, not eliminated, by changing to the cleanest fossil fuel, natural gas. The uncertainties and dependence on complex scientific models are even greater, and again it will be prudent to act before effects are certain, or they may by then be irreversible. The scale and nature of international agreement that will be needed to achieve this are unprecedented, and will play a dominant part in the international politics of the coming decades – see *Blackmore & Reddish, 1996*, Chapter 6).

3.5 Summary

The residuals of coal combustion have been shown to produce major environmental impacts and damage on widening scales. The earliest, most local impact was from visible smoke, soot and fog, which are contaminated

Table 2.2 Agreements on pollution controls

	UNECE	EC/EU
1970	Vehicle exhaust emission standards (limits for NO_x, hydrocarbons, carbon monoxide)	Policy to implement UNECE regime
1975		Gas oil directive (max sulphur content)
1979	Geneva Convention on Long-range Transboundary Air Pollution (LRTAP) (general principles)	
1980		Air quality directive (SO_2 and particulates, 1983 target, 1993 compliance; 29 UK local authorities in breach)
1982		Air quality (lead)
1984	An informal Scandinavian agreement to reduce SO_2 emissions – the '30% club'	Industrial plant – 'framework directive'
1985	Helsinki Protocol (ratified 1987, but not by UK) (30% national SO_2 reductions by 1993 from 1980 base)	Air quality directive (NO_x standards, 1987 target, 1994 compliance)
1986		Single European Act (legal and philosophical basis for environmental policy)
1987		Vehicle exhaust emission standards (1985 'Luxembourg' agreement; optional, to be applied by 1988–93, depending on category)
1988	Sofia Protocol (signatories include UK) on NO_x standstill by 1994 from 1987 base (also further informal 12-country '30% club')	Large combustion plant directive (SO_2, NO_x dust standards for new plant; national targets 1993, 1998, 2003 for existing plant; review 1995)
1989		Small cars exhaust emission standards. Amendment to SO_2/dust air quality standards
1991	Protocol for 30% reduction in emissions of VOC (volatile organic compounds)	Stage I limits on car emissions, mandatory 1993
1994		Stage II limits, mandatory 1997. Stage III not yet agreed, will be mandatory 2000)

Source: Skea (1989), updated Open University 1996.

particulate emissions, in urban industrial areas; this was overcome by Clean Air legislation. The second, on the transboundary regional scale, arose from invisible emissions of sulphur and nitrogen oxides, believed to produce acid rain and ecological damage; this is being reduced by flue gas cleaning and improved combustion methods, along with related vehicle exhaust measures. The third is the modification of the global greenhouse effect, still controversial in its detailed implications; this is a problem common to all forms of fossil fuel use, and it seems will only be overcome by a reduced dependence on fossil fuels, and negotiations of an unprecedented nature in order to arrive at an international control regime.

4 Energy and environment – from global to local problems

4.1 A basis for considering the present energy system

We have seen how the environmental problems associated with coal burning have gradually widened from local to global scale issues. In surveying the rest of the present energy system we can order the consideration of the other components by the extent to which they contribute to these global effects.

Q What other components are there in the present energy system?

A Natural gas and oil; nuclear; hydroelectric and biomass; other renewables.

Q Which of these also contribute to 'greenhouse' CO_2 production?

A The other fossil fuels: natural gas and oil.

It seems appropriate therefore to start with consideration of these other fossil fuels, and the transport system that is so firmly dependent on them at present. Nuclear power, though not a direct source of greenhouse gases, has its own environmental problems, and an indirect link with the greenhouse effect needs to be recognised. It should also be recognised that both fossil and nuclear energy release add to the total heat input to the Earth, implying a fundamental 'thermal pollution' limit to the use of either form – though this is rather theoretical at present.

Only the use of solar energy, directly or indirectly, or tidal or geothermal energy avoids all these problems. At present this primarily means hydroelectric power and the use of biomass; these still have local environmental effects, but in principle are free from the global ones.

It should also be recalled for completeness that we have only considered so far the effects of *gaseous* residuals of coal combustion. There are many others, some of which will be covered in Chapter 5: the direct impact of extraction methods, the disposal of (acidified) surplus water, the effects of subsidence and unsightly or hazardous spoil heaps, the radioactive products of coal-burning. To these might be added the level of accidents in a notoriously dangerous industry; constant efforts have been needed to contain them. Any comparison of energy processes worldwide has to consider these, and relate them to the different hazards of off-shore oil and gas operations and of uranium mines – and to the massive transport accident figures (see below).

4.2 Petroleum

There is some variation of usage, but, conventionally, petroleum is the general term for the second broad type of fossil fuel, produced, like coal, by geological processes acting on buried plant remains. The fundamental difference in the end product is that coal is essentially solid carbon with other materials (including some inflammable hydrocarbons), while petroleum is a mixture of hydrocarbons, innumerable compounds of carbon and hydrogen (C_xH_y), many of which are gaseous or liquid, and other materials (particularly oxygen, nitrogen and sulphur compounds). The differences arise from their origins (coal largely from land plants and petroleum from marine sediments), from the detailed geological processes of formation, and particularly from the way gaseous and liquid products can migrate through favourable rock formations to accumulate in 'fields' away from their source. These natural accumulations can be gaseous (natural gas), liquid (crude oil) or solid (asphalt). A complication in this description is that gas fields can be associated with coal as well as with oil formation.

A more detailed classification of petroleum products, which reflects the separation processes used in oil refineries, is based on the carbon number: which is the number of carbon atoms in the hydrocarbon molecule, increasing as the components get denser (Table 2.3).

Table 2.3

Carbon numbers	Term
1–3	gases (methane etc.)
4–10	gasoline ('petrol')
11–13	kerosine ('paraffin')
14–18	diesel fuel
19–25	heavy gas oil
26–40	lubricating oil
>40	waxes

The hydrogen content for each carbon number can vary: as many as 1200 different hydrocarbons have been identified in crude oil, but in general the proportion of hydrogen decreases with increasing carbon number. When hydrocarbons are burned in air the energy-producing reaction with oxygen produces water (H_2O) as well as carbon dioxide (CO_2), so there is less 'greenhouse' CO_2 production per joule of energy released than from coal. This difference is greatest for the lightest hydrocarbon, methane (CH_4), the main component of natural gas, with CO_2 production about 0.6 of that for coal for the same amount of energy. Fuel oil is intermediate between methane and coal in its CO_2 production.

Natural gas

The cleanliness and convenience of natural gas have already led to its largely taking over from coal for domestic heating in Great Britain and in varying degrees elsewhere, and to similar arguments for its use in electricity generation, particularly in 'combined cycle' gas turbines, which are more efficient than, and can now be as economic as, steam turbines. This has the advantage of alleviating, though not removing, 'greenhouse' concerns for the reason just given. But two further problems have to be borne in mind.

The first is that methane is itself a 'greenhouse' gas, 30 times more effective per molecule than CO_2, though at present of so much lower (though rising) concentration in the atmosphere that its total contribution is less. Any leakage of unburnt methane from the natural gas system adds to this concentration, so leakage from an expanded system might outweigh any advantage of the change in fuel use.

The second is that the natural gas resource, which is limited like all non-renewables, is particularly under stress. The figures for 1994 for the time that proved reserves will last at current extraction rates were 10 years for the UK and 66 years for the world, the major reserves being in the Middle East and Russian Federation. Of course the usual economic pressures will produce price changes and new assessments of reserves, but the limits are clearly in view. Synthetic natural gas (SNG) can be made from coal, for which reserves are greater. So, to the extent that economic dependence on fossil fuels will continue at least temporarily in face of greenhouse concerns, a route that would in principle overcome other pollution problems from coal would be large-scale conversion to SNG near the mine, with gas distribution and clean combustion at the points of use. (This of course assumes that the SNG plant is itself designed to be non-polluting and is economic, which is not the case at present. But this is a potential medium-term option for retaining the advantages of natural gas while taking advantage of the much greater coal reserves.)

Oil

All stages of the exploration and extraction process produce residuals, impact and damage broadly similar to those of other mining operations: visual impact, noise and dirt, employment effects, accidents. The extreme conditions of off-shore operations make particular difficulties for the staff, though out of sight of most oil users. A major feature of the present industry is the development of ever-larger crude oil tankers, which have their own impacts and hazards, in order to satisfy the voracious demands of the industrial economies. Refineries, too, are large processing plant with obvious local effects, and which contribute significantly to long-range SO_2.

At the other end of the 'cycle', end-use combustion produces particulates and gaseous effluents, depending on the particular fuel and combustion conditions, which we have already seen being subjected to increasing controls along with those on coal combustion. Internal combustion engines are major contributors to NO_x levels, highly toxic carbon monoxide (and the consequent photochemical smog) and tarry particulates (particularly from diesels) which newer designs seek to reduce:

- Lean-burn engines use a smaller proportion of fuel in the mixture entering the cylinder, and better control of ignition, so that less pollution is produced in the first place.

- Catalytic converters operate on the exhaust gases, effectively removing oxygen from the NO_x and adding it to the hydrocarbons and CO to produce N_2, CO_2 and H_2O only.

The fundamental fossil-fuel effect of the production of the greenhouse gas, CO_2, remains, of course.

Perhaps the most distinctive environmental effect of the oil industry results from spillage at all stages of production and transport to the user, where the unwanted residual is oil itself, because of the direct impact and damage it produces.

Oil is in a sense part of the natural environment, but is normally buried; it enters the surface regions, on land and sea, by natural seepages. The Trinidad asphalts are an example of an enormous seep on land, and there are various oil sands and shales near the surface forming part of the oil resource. Marco Polo in the thirteenth century noted a natural seepage at Zorzania where 'there is a formation of oil which discharges so great a quantity of oils as to furnish loads for many camels.' Over the last 100 years there have been increasing unnatural spillages of oil into ecosystems which have evolved in its absence and are therefore damaged by it.

Oil wells can go out of control, releasing oil on land or in the sea; crude oil tankers have accidents, or release part of their cargo in tank washings, operational discharges or during loading and unloading; storage tanks on shore can leak; refineries discharge their effluents into rivers and the sea; oil or its products can escape from pipelines, tankers or road vehicles.

Off-shore exploration and extraction have increased drilling problems and the risk of contamination; the scale of spillage worldwide is thought to be about 100 000 tonnes per year (100 kilotonnes/year). This is to be compared to a natural seepage into the marine environment of about 600 kt/year, but damage is particularly associated with individual localised incidents: the EKOFISK blowout in 1977 released 30 kt of oil into the North Sea over 7 days; IXTOC-I in the Gulf of Mexico in 1979 released even more – 4 kt/day for 3 months, over 350 kt in all.

Individual tanker accidents likewise attract much attention, such as the *Torrey Canyon* off Cornwall in 1967 (100 kt spilt) and the *Amoco Cadiz* off Brittany in 1979 (220 kt). In 1989 12 accidents involved a total spillage of 162 kt, the two largest being *Kharg-5* off Morocco (76 kt) and the *Exxon Valdez* off Alaska (36 kt) – this last is generally considered to be an environmental disaster because of its particular location, though it spilt the smallest quantity of oil among those mentioned. These spectacular individual events are to be compared with routine discharges of oil from tankers before entering port, which are estimated to involve up to 1 million tonnes per year (1 Mt/year).

One estimate of the total flow of oil into the marine environment is about 5.3 Mt/year, of which nearly half (2.25 Mt) comes from run-off from the land from rivers, drains, municipal and industrial wastes, and a further 1.2 Mt from natural seeps and atmospheric fall-out. Against this background the 0.12 Mt or so from production and refining losses does not seem large, but the 1.76 Mt from all aspects of ocean transportation is significant (Figure 2.16; and a further discussion in *Blackmore & Reddish, 1996*, Chapter 1).

In considering the resulting damage, there is as usual in environmental assessment a distinction to be made between localised events, which are shocking but produce no long-lasting effects, and the more insidious long-term consequences of regular operations. In both cases there can be:

- decrease of aesthetic and economic value due to unsightly spills and oiled beaches (Plate 10);

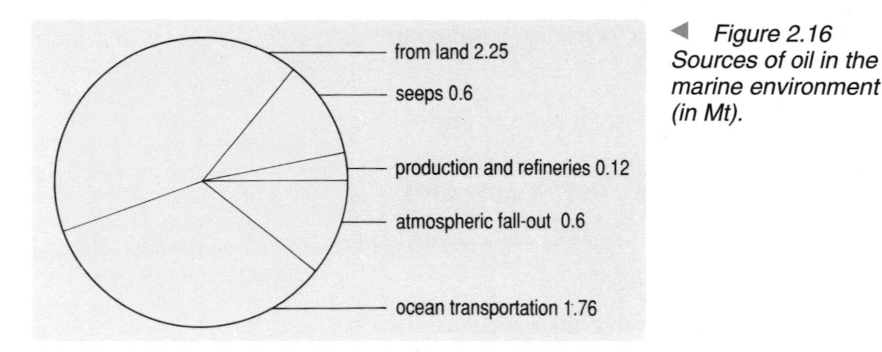

Figure 2.16 Sources of oil in the marine environment (in Mt).

- damage to wildlife such as sea birds and marine animals;
- decrease of fishing resource;
- modification of habitats, delaying or preventing recolonisation by marine organisms;
- contaminated sea-food entering the food web, perhaps reaching humans; and perhaps more permanently;
- modification of the marine ecosystem by elimination of species, with decreasing diversity and productivity.

As with gas, the oil resource is under severe stress. As was noted in Chapter 1, the corresponding figures for the lifetime of proved reserves at 1994 production rates were a mere 5 years for the UK, and 43 years for the world, where the Middle East is even more dominant. Again, economic pressures will modify this picture, but continuing massive oil transportation and political instabilities seem inevitable. As the world transport system is so dependent on oil, it may gradually become more economic to expand synthetic oil production from coal by the various available methods (e.g. the synthesis method as in the South African SASOL process operating since 1955, or the potentially more efficient degradation process). There will be corresponding changes in the environmental effects of production, but not of use. However, the fundamental greenhouse problem will remain, perhaps forcing a more radical reconsideration of transport policy.

4.3 *Transport*

At this stage it may be helpful to change focus from the environmental effects of particular fuels to those of this particular end-use. At present in industrialised countries the transport and oil issues are almost synonymous. Another of the OECD Compass project books, *Environmental Effects of Automotive Transport* (OECD, 1986) noted that OECD member countries use some 70% of the world's energy and 20% of that is for transportation. Of this OECD transport energy over 99% is based on petroleum, of which 70% is on gasoline. About 75% of the transport petroleum is for road transport.

Q What do these figures imply about OECD transport not based on petroleum?

A It uses less than 1% of the energy, presumably as electricity, from coal or nuclear stations, for electric vehicles and trains.

Q How much of the petroleum used for transport is not gasoline, and what is it?

A 29%, presumably kerosine, for jet aircraft, diesel for lorries and trains, heavy fuel oil for ships.

Q What proportion of road transport energy is as diesel fuel?

A We are told 70% of transport energy is gasoline (which is nearly all for non-diesel road transport), implying that the remaining 5% of road transport is diesel. Alternatively, 25% of transport energy is not for road transport, i.e. aircraft, trains, ships, so the remaining 4% of non-gasoline must be for road diesels. Either way, proportion of diesel in road transport energy seems to be 4–5 in 75, i.e. about 6%.

So it is our cars above all else that create the endless demands on (and power of) the oil industry, with all the environmental and political problems outlined above. In thinking of the future it is important to remember that a transport system doesn't have to be based on cars based on gasoline, but it will not be easy to change. We return to this in Chapter 3. Meanwhile we should consider more of the environmental effects of the system we already have.

The Compass study above suggests, as did the previous one (see Section 2.2), a comprehensive and somewhat laborious structure for considering environmental impact. In this case it takes the form of a 6 × 4 matrix, i.e. a diagram with 6 columns representing various aspects of the environment and 4 rows for various parts of the transport system (Figure 2.17). In each of the 24 'cells', produced where rows and columns intersect, there is scope for considering impacts and damage – a daunting task.

The six environment areas chosen (with further subdivisions) are:

- ecosystems: terrestrial, aquatic and human/animal pathways;
- resources: mineral, natural (e.g. land and water), capital/labour;
- physical environment: air, water quality, waste disposal, noise, aesthetic;
- health: occupational, public;
- safety: occupational, public, social (this is to make a distinction between physical hazards and social or economic disruption);
- socio-economic: economic (e.g. employment, costs), institutional (e.g. firms, local authorities, government).

The four transport system areas (like the 'phases' of Section 2.2) are:

- production of vehicles and fuel;
- construction of the infrastructure (i.e. the roads, bridges, railways, airports);
- operation of the components of the system;
- disposal.

Merely to list these topics is to recognise the endless ramifications of our transport system and its impact on (in some places domination of) our environment. This is not a unique scheme of classification, but it clearly

	Ecosystem	*Resources*	*Physical environment*	*Health*	*Safety*	*Socio-economic*
Vehicle and fuel production		mineral shortages, capital for new developments			oil production accidents	
Infrastructure construction	habitat disturbance					
System operation				vehicle exhausts, smog		
Disposal						who pays?

▲ *Figure 2.17 A transport–environment matrix.*

brings out many more factors than the oil production, distribution and combustion effects we have already considered. It can be related to the 'system' ideas of Section 2 by regarding the rows as identifying innumerable primary energy-in-transport systems, which can each be analysed as interlinked components within a system boundary, and the columns as a range of options for extending the system boundary to examine various forms of environmental impact.

Some of the cells have already been filled in: the ones concerned with topics that we have already considered – oil production and distribution, and vehicle exhaust emissions.

Activity 4

Test your understanding of this matrix by filling in at least one topic in each of the remaining cells if you can.

One change has been much hailed as an environmental advance: the availability of lead-free petrol. It is only one small corner of the system, but it removes an added hazard to the oil combustion pollutants that we have already mentioned. Less obviously, the change is essential if catalytic converters are used to control exhaust emissions, because they are poisoned by lead. This has been as important a factor in the change as direct concern about the health hazard of lead itself.

A different problem which stubbornly resists change is the appalling level of casualties in the present system. (This might not be regarded as an 'environmental' problem in one sense, but when viewed as an influence of one technical system on a larger human one it certainly is.)

In Britain transport is responsible for about two-fifths of all accidental deaths; other urban societies are similar or worse. In the one year of 1993, road accidents killed 3820 people, including 1250 pedestrians and 190 cyclists; there were 44 890 serious injuries and over a quarter of a million other casualties. (In contrast, in the whole decade 1982–92, 73 passengers and 94 others were killed in rail accidents.) In the proper concern about the tens of deaths and hundreds of injuries per year in the energy industries, and in the major sea and rail disasters and air accidents that attract understandable attention, we should not lose sight of the daily human cost of this major sector of energy use. Despite the great increase in traffic, road deaths have roughly halved over the last 30 years, though the rates for pedestrians and cyclists, suitably adjusted for numbers and distances, remain higher than in some other European countries, particularly Denmark, Sweden and the Netherlands. There have been considerable changes in life style – loss of freedom of movement except by car contributing to the reductions that have been achieved. The reduction in *total* casualties over the decade 1977–87 was only 10%; in 1989 Paul Channon, then Secretary of State for Transport, set the target of a one-third reduction by 2000, implying a three-fold improvement over the previous decade. The actual reduction from 1989 to 1994 was about 8% in total casualties (though over 30% in fatalities).

4.4 Nuclear power

The fossil fuel industries, and the transport sector in practice if not in principle, are all contributors to greenhouse CO_2, and will have to change if current projections continue to gain credence. Nuclear energy is a (non-renewable) source that does not produce CO_2 directly, a fact to be put in its favour alongside the smaller volumes of fuel that originally seemed to be its advantage. It has its own environmental problems, of course, to which we return below, but it cannot be entirely isolated from a discussion on CO_2 production, for the following reason.

In the present system, any industrial process that uses electrical energy depends on fossil-fuel power stations, which produce CO_2. In a study by Mortimer presented by Friends of the Earth to two public inquiries in 1989, figures were given for the CO_2 emissions associated with the production, or saving, of 1000 MW of electricity by various methods, 'taking into account related mining and fuel producing processes' (Table 2.4).

Mortimer went on to argue that any world increase in nuclear power would lead to the use of lower grades of uranium ore, requiring still more energy, and hence CO_2 production, for extraction and processing. His figures were subsequently disputed by the nuclear industry, and indeed such analysis in terms of energy 'costs' is always open to argument over the details. Nevertheless, the general point is a qualitative one: the acknowledged advantage of nuclear over fossil fuel power generation in terms of *direct* CO_2 production does not escape *indirect* effects, which for a

Table 2.4 CO_2 emission associated with production/saving of 1000 MW of electricity

Method	CO_2 (tonnes per year)
coal	5 912 000
nuclear	230 000
hydro	78 000
wind	54 000
tidal	52 000
loft insulation	24 000
low energy lighting	12 000

Source: Mortimer (1989).

'high-tech' industry are significantly greater than for renewables and efficiency measures.

Further aspects of nuclear policy will be discussed at more length in Chapter 7, but brief general comments are appropriate here, and we will then discuss the radioactive discharges from the Sellafield power station complex in Cumbria for further illustration of the methods of environmental impact assessment described above.

General environmental concerns

The fission process used in current reactor types – Magnox, advanced gas-cooled reactor (AGR) and pressurised water reactor (PWR) – produces large volumes of product isotopes, some highly radioactive. Most environmental criticisms of nuclear power arise from this, and fall into four groups:

1 Abnormal radiation levels from normal operations will cause cell damage, malignant cancers, genetic defects etc. – this possibility was recognised from the outset. It can be argued that the strict exposure levels and checks introduced have minimised this problem, at least in power stations.

2 Reactor operations and the transport of irradiated fuels cannot be guaranteed safe against catastrophic accidents such as those at Windscale in Cumbria in 1957, Three Mile Island in the USA in 1979, and Chernobyl in 1986. While stringent safety precautions may have been successful in nearly all cases (there were more than 400 operational reactors worldwide in 1990), the effects of these relatively few major failures have been so dramatic as to greatly weaken confidence in further expansion. Concern is particularly acute about maintenance in the former USSR.

3 Radioactive waste has to be disposed of safely, which is difficult. An operational 1 GW burner reactor produces some 5 tonnes of radioactive waste a year (apart from uranium and plutonium, which it is present UK policy to recover by reprocessing); for fast breeders, which are nowhere operational and are now abandoned as a project in the UK, it would be less than 1 tonne per year. The activities and decay times mean that the more dangerous wastes must be isolated from the biosphere for at least 1000 years, an engineering challenge of exceptional stringency not yet satisfactorily resolved. Low-level wastes are vented to the atmosphere, discharged into rivers and the sea or buried. The history of such sea disposal at Sellafield, summarised below, provides a vivid example of the environmental assessment method outlined in Section 2.

There is a further problem to be faced in decommissioning reactors at the end of their life, when much of the structure will have become radioactive. The cost of decommissioning was one of the reasons why nuclear power had to be withdrawn from the UK electricity privatisation process in 1989. In the current (1995) renewed proposals for privatisation, only the most recent plant are included; the older stations, with decommissioning more imminent, will be retained under national control. Furthermore, the privatised electricity companies were obliged to provide from their customer charges for a 'non-fossil fuel obligation' (NFFO, see Chapter 3), initially to subsidise the less economic nuclear power (though subsequently used for renewable energy as well). After much technical and moral debate about nuclear power, such economic arguments seem to be currently the most damaging.

4 The large quantities of uranium enriched in its fissile isotope, and even more seriously, of plutonium, if separated by reprocessing, must somehow

be kept secure from theft or other misappropriation for nuclear weapons (the threat of increased nuclear proliferation) by countries or groups that do not at present have them.

Sellafield marine discharges

The history of radioactive discharges into the Irish Sea from the reprocessing plant at Sellafield (formerly known as Windscale) provides a salutary insight into attitudes and environmental assessment methods in a particular case. While there would probably now be few defenders of the detailed policies adopted, particularly in the 1970s, it should be acknowledged that measurements of radioactivity can be carried out at very low levels, and that tracing the movement of particular isotopes has provided a clearer picture of ecosystem effects than used to be available. A great deal of environmental monitoring has been carried out and published, which was certainly not the case for chemical pollutants in the past. Attitudes to pollution standards and the ecological effects of pollutants in general have no doubt been significantly influenced by the monitoring of these radioactive effluents.

They are dwelt on here without any implication that they are *exceptionally* damaging but because the available data fit particularly well into the framework described above in Section 2: the *origin* of the effects is in the reprocessing operations at Windscale/Sellafield; the *residuals* considered are the authorised discharges of low-level radioactive wastes by pipeline into the sea; the *impact* is the resulting level of radioactivity in the marine environment and ecosystem; the *damage* in this case is judged by the dose of radioactivity received by human beings, in various critical groups, as a result. Let us look briefly at each of these in turn.

1 Windscale began as a plant for the production of plutonium for military use in 1950; uranium reactors create plutonium in their fuel rods, and this is extracted in the reprocessing operation. This practice has continued through the civil nuclear programme starting in the 1960s, the plutonium produced being stockpiled for future use, notionally in fast breeder reactors, though a considerable amount has been traded with the USA. The spent fuel also contains about 2% of other radioactive products which are 'waste'. During the initial plant design it was decided that low-level wastes could be disposed of in the sea, where they would be 'diluted and dispersed', like many other industrial effluents. It was argued that radioactivity is a natural property of the sea, so some added level could be negligible, but that discharges should be carefully controlled and monitored. This was at first done within the Atomic Energy Authority, but since 1967 an annual report has been published on *Radioactivity in Surface and Coastal Waters* (from 1977 it has been called *Aquatic Environment Monitoring*) by the Lowestoft Radiobiological Laboratory of the Ministry of Agriculture, Fisheries and Food, who were charged with checking the safety of this operation (as well as the other nuclear installations, and other pollutants). Comparison of their reports over some 25 years provides an interesting reflection of changing perceptions, about nuclear power and the environment generally.

2 The nature and quantities of these particular residuals of the reprocessing operation, the low-level wastes discharged, were reported annually. They rose to a peak in the early 1970s. The Royal Commission on Environmental Protection reported unfavourably in its Sixth Report in 1976, and a general decline in discharge levels began about then. This discharge data gives a clear statement of the specified variation in time of a localised

environmental disturbance – a giant 'blip' of radioactivity injected into the Irish Sea. However misguided the original decision may have been, it is vital to try to learn from it by mapping its consequences, both for public health and more general ecosystem effects.

3 A vast amount of tabulated data makes it possible, in principle, to construct something like the 'bubble of influence' postulated in Section 2 – that is, the impact, in space and time, of these residuals on chosen aspects of the physical environment and ecosystem, though there are so many variables that it is difficult to grasp the whole picture. Figures are given for the specific activity (the amount of radioactivity per unit mass) measured in the laboratory on samples of water, sand and silt, fish, crustaceans and molluscs of many species, taken from many locations. From these the environmental 'impact' has been mapped in some detail, showing how particular radioactive isotopes spread not only through the water and materials of the shore by physical processes, but are concentrated at higher levels in fish, in crustaceans like crabs and most of all in molluscs like winkles.

4 The resulting 'damage' from these complicated 'impacts', as far as the MAFF reports are concerned, is the possible effect on human health (more general ecosystem damage continues to be looked at elsewhere). The principal method used is to identify critical groups of the public whose movements and eating habits bring them into contact with the environmental radioactivity, and to ensure that in the worst case the received dose of radiation falls below some specified 'safe' level; then everyone else must be safer. Obviously there are lots of questions here: who are the critical groups, how do you assess the dose, what is safe? The answers given are complicated and varied over the period considered, and perhaps are still varying.

The broad conclusion was that doses to critical groups never exceeded the prevailing 'safe' level, though they came close to it in the late 1970s. In 1985 the internationally recommended safe level was reduced – this new figure had been exceeded for about a decade, though current levels are well below it and continue to fall.

It is important to remember that the doses quoted in these reports are only estimates from a complicated procedure applied to selected environmental features believed to matter. As reports of leukaemia clusters etc. grow, more direct studies of the population itself are being made; for example, post-mortem measurements of radioactivity in human subjects (not necessarily showing apparent clinical effects) may indicate how good the estimates have been. The results of a statistical analysis of childhood leukaemia published in 1990 and 1993, suggesting a link with occupational exposure of Sellafield workers before becoming fathers, were surprising and controversial (Gardiner, 1990, 1993; Evans, 1990). Disputed at the time, it was concluded in a 1993 High Court action for damages that no link could be proved. Perhaps a clearer picture will eventually emerge.

As a reminder of changed perceptions, a quotation from the first report in the MAFF series noted above (1967) is revealing:

'(This report) shows clearly that the discharge of radioactive wastes in this manner is *completely safe* . . . It is our policy to publish the results of our surveys, *so as to assist in creating an informed climate of public opinion in which the peaceful uses of nuclear energy may be fully exploited* ' (emphasis added) – surely no part of the task of an independent assessment agency!

All the reports suggest a determination to show that the procedures *are* safe rather than to ask *whether* they are.

In fact, the effect of these studies, and others, has been to provide a map of the bubble of influence which has led to a change in practice (one of the general options considered in Section 2) – at least to a reduction in authorised discharges (which however increases the long-term waste disposal problem). Some would argue for a further change: the abandonment of the reprocessing of spent fuel, as has long been the case for civil nuclear power in the US. Others, taking in other environmental and economic concerns as well, would prefer the abandonment of nuclear power completely. Whatever the final conclusion, the studies do indicate the massive effort needed to assess environmental damage from pollutants, and also the need for this to be done as independently as possible from the polluting agency.

4.5 *Waste heat*

After the complexities of the previous sections, this can be dealt with fairly briefly. The first point is that all electric power stations (fossil fuel and nuclear) only convert about a third of the available energy into electricity, so the rest is lost as heat, removed into the environment as a residual. This can produce obvious local effects in the atmosphere or nearby waters used for cooling. Or the system can be changed so that the heat is used for district heating, so is no longer a residual – as has only rarely been done in the UK, but is common in other countries, particularly Denmark and Germany. This replaces further consumption of primary energy for heating, so is an evident improvement in energy efficiency.

The second point is that whatever use is made of primary energy, it all turns into heat in the end, more or less directly. So all the fossil fuel and nuclear energy, formerly locked up in the molecular and nuclear structure of materials, that is released for use adds to the heat input of the Earth, along with the absorbed solar radiation. The Earth is kept in balance by the re-radiation into space of whatever energy it receives from all sources; if this increases, the global temperature (as seen from outside) must increase too. This is a fundamental 'heat pollution' effect (not to be confused with the greenhouse effect caused by particular gases, which modifies the relationship between surface and upper atmosphere temperature, without changing the overall radiation to space). In fact heat pollution is not at present globally significant, with a total world fossil and nuclear power rating of some 10 TW to be compared with a solar input of some 100 PW, ten thousand times greater. There are 'heat islands' associated with big cities, due to higher energy use and changed surface conditions, but a negligible global effect. But it should be remembered that 'success' in meeting expanding energy demand from any non-solar source of supply (such as nuclear fusion) would be in danger of facing this fundamental heat pollution problem. There has been a tenfold increase in energy use this century; if there were a further tenfold increase in the next century, to 100 TW, this effect might begin to be of concern – though such an increase would certainly have other more dramatic effects.

4.6 *Solar energy sources*

All the previous problems (chemical, nuclear, even heat pollution) are avoided if we make use of solar energy, either directly or indirectly via the climate and biomass systems, to meet our needs. Only two methods are at present in widespread use; others of future importance will be considered in Chapter 3.

Hydroelectric power

A few per cent of world energy (some 0.35 TW) is provided by hydroelectric schemes, which depend on favourable topography for dam-building and storage of large volumes of water, which are raised to high level in the hydrological cycle (the solar-powered evaporation of water to form clouds, falling again as rain). It can also be raised by electric pumps, as at Dinorwic in North Wales, to provide effective storage of electricity produced in other ways. Turbines powered by the falling water can be brought in and out of use very quickly, providing flexibility in the electricity grid system. The total resource has been estimated at 2.2 TW worldwide. The large-scale engineering involved has seen its more complete exploitation in Europe and North America than elsewhere, though large schemes now often figure in development plans for the rest of the world. The local environmental effects are considerable, both during construction and in producing a permanent change to the local landscape.

These seem to have been accepted in general in developed countries, though new proposals are now subject to much more detailed planning scrutiny than formerly. The encouragement, or imposition, of large schemes in less developed countries has been much criticised by environmentalists, as often doing more damage to traditional ways of life and land use than is justified by the immediate needs of the local population – perhaps being more useful to an incoming mining company, for example. As always, the kind of scrutiny of environmental effects discussed in this chapter, and the even more difficult balancing of interests, is required in each individual case, implying a developed political framework and equitable planning procedures that are not always present.

Biomass

Wood as a fuel is no longer (at present) a significant part of the energy system in developed countries. But wood products like paper, and other vegetable and animal products, find many other uses, and form a large part of domestic and commercial wastes. In the UK, some 30 million tonnes a year of municipal solid waste is combustible, with over one-third of the energy content of coal. Most of it is buried (landfill), where the decay processes produce methane, hazardous both as an explosive and, as it leaks to the surface, a more potent greenhouse gas than CO_2. If incinerated, preferably in a combined heat and power (CHP) plant for electricity production and district heating, the CO_2 produced is returned as part of the normal carbon cycle (provided new biomass is being produced at the same rate as it is being destroyed), and coal or other fossil fuels with their various problems are replaced. Other European countries and Japan use the energy in wastes in this way – up to 75% in Luxembourg, but only 5% in the UK. There can be pollutant emissions from badly designed incinerators, but the same kind of controls as those for fossil fuel combustion can be imposed, and satisfied by good combustion engineering.

Alternatively, methane can be produced from biomass in a controlled way, collected and itself used as a clean fuel: this is particularly applicable to agricultural wastes and sewage.

In less developed parts of the world, fuelwood (and other biomass like animal dung) remains as a principal source of energy. It does not figure in commercial energy statistics, so its scale of use is difficult to estimate – a figure of 10–15% of total world energy use (i.e. over 1 TW) has been suggested. The collection of fuelwood can form a major part of the day's work. Under stressed conditions of population growth and land shortage it can contribute to deforestation. However, it is inappropriate to see this as a principal cause when set against large-scale forest clearances for commercial logging and ranching, and the general economic and political issue of land ownership. There is a growing interest in **agroforestry** (Plate 11). This practice allows the coproduction of woody and non-woody crops on the same area of land and has economic advantages over monoculture in some areas. Agroforestry also caters for multiple use of land for food, energy, amenity, etc. and has also been shown to be sustainable in some studies. Fundamentally, biomass use of all kinds is sustainable provided it takes place at the rate the biomass can be regrown – obvious, but often overlooked throughout history. At present, this is by no means the case, and progressive reduction in the forested area effectively adds to the greenhouse effect by increasing atmospheric CO_2, apart from its other environmental effects. Conversely, any increase in forested area would reduce atmospheric CO_2 of whatever origin (during the period of change). Strategies for reducing, and preferably reversing, deforestation must form part of the world response to the greenhouse effect.

4.7 Summary

Other aspects of energy policy and their environmental impact have been reviewed, with reference to their global and local effects. Petroleum use can be cleaner than coal, but still contributes to greenhouse CO_2 (and potentially methane); transport is a particularly significant user. Nuclear power does not produce CO_2 directly, though its introduction implies indirect CO_2 production; it presents other environmental problems. All non-renewable sources in principle contribute to heat pollution, though this is not significant at present. Solar energy sources (at present hydroelectric and biomass) are in principle sustainable, with only local environmental effects; changes in biomass production (deforestation or reafforestation) contribute to atmospheric CO_2 changes, in either direction.

5 Environmental economics and other values

If you look back at the 'origins' of environmental effects classified in Section 2, and the numerous details that have followed, it will be clear that there is immense diversity and human ingenuity in the present energy system, and that it is constantly changing. It is also clear that the primary reason for change is economic (new processes are adopted when they are cheaper) and that environmental effects have not usually been included in these considerations as we may now feel they should. How are environmental effects to be taken into account? I would like to suggest five approaches, which are not entirely distinguishable, but which provide a variety of perspectives as an introduction to the discussion in the next book in the series (*Blackmore & Reddish, 1996*). They can be labelled technical, economic, legal, ethical and aesthetic.

5.1 Technical

Fortunately there are some changes where economic interest and environmental concerns coincide, so these should present least difficulty. All the measures broadly described as 'energy efficient' in Chapter 3 (insulation, CHP, efficient appliances) tend to be of this kind, in which case the only question is why they don't happen. A combination of inertia, lack of information and different perceptions of costs between initial outlay and running expenses, or between users and manufacturers, seems to be responsible; enthusiasts and educators can only be patient. Perhaps the most significant factor is that energy policy, at government level, say, has in the past been dominated by supply considerations. It is not in the economic interest of energy suppliers to see energy consumption reduced, and economic statistics can show increased energy consumption as desirable economic 'growth'. This perception has been changing in recent years, as reduction in demand has been seen to be more compatible with sustainable energy use; it is to be hoped that this will gradually be reflected in national policies.

5.2 Economic

This is sometimes crudely characterised as the 'polluter pays' principle, one that could hardly be disputed as desirable, but is not so easy to achieve. The difficulty is that in a free market a transaction between a buyer and seller takes place at one time, using the information then available; the environmental implications (for others as well as the buyer, of course) often only emerge at a later time, when the original transactors may not be recognisable, available or controllable. So the market has to be modified or regulated in some way by government and the law to prevent this, which implies planning, prediction, and policing – not entirely reliable processes, and of varying political acceptability.

However, planning is much used, particularly for large-scale investments, such as the building of power stations. An author who has tackled this particular issue is Hohmeyer (1988), in his *Social Costs of Energy Consumption*, a technical analysis (for West German conditions, but generally applicable to 'developed' economies) of the way 'external' effects (what we have called environmental impact and damage) can be included in the costing when comparing different systems. Cost comparisons of alternative methods (like wind) with existing methods (like nuclear) are then much more favourable. A related idea is the 'least-cost planning' required of US electricity supply utilities by some states, where approval to build a new power station is only given if it can be shown to be cheaper than investing in demand reduction by, for instance, efficiency measures.

A more general study of the way environmental issues can be incorporated in economic decision-making is the 'Pearce Report', by David Pearce and his colleagues for the UK Department of the Environment, and published in 1989 as a *Blueprint for a Green Economy*. This represents a positive move within orthodox 'neo-classical' economics to embrace the ideal of 'sustainable development' – another unarguable, but not so easily defined, doctrine associated with the 1987 World Commission on

Environment and Development (WCED) report *Our Common Future* (the Brundtland Report). Pearce recognises the idea of 'Environmental capital, goods and services', to be considered alongside the human-made varieties of normal economic theory. He specifically rejects the traditional idea that increases in gross national product can be used to measure the improved quality of life, since this can be achieved at the expense of environmental degradation, and proposes new measures which take better account of this.

In a society which is dominated by economics in decision-making, any way of including the environment is to be welcomed, but there are some difficulties, which the earlier examples in this chapter illustrate. They showed how difficult it is to trace the environmental effects of particular processes. Even when the broad links are clear it is not easy to prove who is responsible; often they are not clear, or unforeseen consequences emerge later. More difficult still is to move from environmental effects to environmental costs. It is central to Pearce's ideas that money values can be assigned to environmental factors, which raises difficulties, for instance, in comparing one person's treasured view with another's health or electricity costs, or any of these in different places and times. Finally, there is the mechanism for including these issues in market transactions. These will be much debated, but perhaps it is reasonable to consider charges, tax incentives or subsidies as 'economic' measures (which would have to be imposed on the market) while pure regulation would spill over into the next subsection.

5.3 *Legal*

National laws (like Clean Air Acts) and international 'regimes' of regulation (like the EU pollution directives) may be influenced by economic considerations, but they are surely not controlled by them. Longer term or wider ranging perceptions of human welfare are somehow distilled into these socially imposed constraints; their evolution is discussed in *Blackmore & Reddish, 1996*, Chapters 5 and 6). More controversial is the extent to which other parts of the ecosystem or the natural environment are also 'entitled' to protection, regardless of human welfare. Again attitudes are changing, and there is overlap with the next topic.

5.4 *Ethical*

Many of the world's religions include precepts on the way human beings should treat the natural world; these may as frequently cause conflict as agreement, particularly when different societies impinge on each other. In an increasingly interrelated world we are some way from shared ethical principles. The developed world, for all its residual religious conflicts, has in practice shared some beliefs over the last few centuries derived from a (Utilitarian?) enthusiasm for economic growth, driven by technical innovation and science. The growing environmental movement reveals dissatisfaction with this position, unhappily sometimes coinciding with a rejection of the strength of science – arguably as good an approach to 'truth' as we have, and at best, a creative recognition of the limits of knowledge. A better change of stance would be a greater wariness about innovation, separating it in our minds from improved scientific understanding. Returning to the earlier language of this chapter, we should assume that the residuals of energy, and other technical, processes are guilty until proved innocent, instead of regarding the technically possible as naturally desirable. We will need the best science for this assessment, but must apply it to asking *whether* practices will do more good than harm, not, as so often, demonstrating that they must: humility, not hubris (the Greeks always had a word for it). The implied objectivity of science can also be questioned: 'scientists' usually have employers; whoever sets the agenda of study may control the outcome. However, at best it does provide a systematic way of improving communicable models of the world, provided it is not arrogantly assumed that these instantly license change, which we should know by now often has unexpected consequences.

Mr Harthouse dining at the Bounderbys'.

5.5 *Aesthetic*

We have an alternative method of judging the effects of our activities, our intuitive response: the joy and delight a child shows in a tree or a kitten, the outrage of Blake at a caged bird. It may not be objective, and may be culturally conditioned, but in some way it surely provides a link with our instinct for what is good for us, evolved over millions of years before we had speech, society, laws or economics. Dickens did not have to measure the damage produced by soot and smoke: he knew, a century before society was prepared to act. Such intuitive convictions may not always be right, but we do well to listen to them. Furthermore, aesthetic satisfactions are not only about intuition: there is also a highly complex intellectual process of pattern recognition and formation, in the matching of structure and imagery to underlying perceptions of connections, which artists like Dickens make with lightning concision.

So let us return to *Hard Times*, and its rogues' gallery of the architects of Coketown and its attitudes. It may provide a salutary counterbalance to the rest of the chapter.

Meet the politician and educator Thomas Gradgrind, with his celebrated 'In this life, we want nothing but Facts, Sir; nothing but Facts!':

> A man of realities. A man of facts and calculations . . . With a rule and a pair of scales, and the multiplication table always in his pocket, Sir, ready to weigh and measure any parcel of human nature, and tell you exactly what it comes to.

And his friend Mr Bounderby:

> He was a rich man: banker, merchant, manufacturer and what not. A big loud man, with a stare, and a metallic laugh . . . A man who could never sufficiently vaunt himself a self-made man . . . who was always proclaiming . . . his old ignorance and his old poverty . . . 'Josiah Bounderby of Coketown learnt his letters from the outsides of the shops . . . Tell Josiah Bounderby of Coketown, of your district schools and your model schools, and your training schools, and your whole kettle-of-fish of schools; and Josiah Bounderby of Coketown, tells you plainly, all right, all correct, – he hadn't such advantages – but let us have hard-headed, solid-fisted people – the education that made him won't do for everybody, he knows – such and such his education was, however . . .'

And his schoolmaster M'Choakumchild:

> He and some one hundred and forty other schoomasters had been lately turned at the same time, in the same factory, on the same

> principles, like so many pianoforte legs. He had been put through an immense variety of paces, and had answered volumes of head-breaking questions. Orthography, etymology, syntax, and prosody, biography, astronomy, geography, and general cosmography, the sciences of compound proportion, algebra, land-surveying and levelling, vocal music, and drawing from models, were all at the ends of his ten chilled fingers.

And so on for another page (as it might be, environmental impact assessment, general systems theory, radioactivity and economics), culminating in the flamboyant rhetoric of this account of his teaching:

> He went to work in this preparatory lesson, not unlike Morgiana in the Forty Thieves: looking into all the vessels ranged before him, one after another, to see what they contained. Say, good M'Choakumchild. When from thy boiling store, thou shalt fill each jar brim full by-and-by, dost thou think that thou wilt always kill outright the robber Fancy lurking within – or sometimes only maim him and distort him!

6 Conclusion

Of course we can't quite leave the last word to Dickens' rhetoric, however heart-warming. His opposition of Fact and Fancy is a bit too neat for the complexities of the real world – surely we need reason *and* imagination, not one or the other – and we won't persuade our latter-day Gradgrinds to change by mocking them. We need passion and conviction to stir us into action, but we will also need patient analysis to make a case that will be listened to. The procedures of Section 2, either in the formal checklists of the OECD, or in the more open-ended form of systems theory, provide, it is suggested, a general way to assess environmental effects. Subsequent sections have shown that the Facts are by no means easy to find; there is plenty of scope for Fancy in deciding which larger systems are to be considered in determining impact and damage. There is then some evidence that such an analysis can gain general assent if carried out sufficiently thoroughly, in spite of kicking and screaming on the way (developing pollution control regimes, changes in radioactive waste policies, taxation). We still have some way to go, however, in arriving at value systems that will change the fundamental direction of globally damaging energy policies.

Comments on Activities

Activity 1

All the stages are involved, some more than once. There is no reason why you should know the details, but the following ideas are among those relevant:

- exploration. Someone had to establish, on general geological grounds, that Namibia contained uranium ores; they had to be found, and their grade established so that the economics of mining them could be assessed.
- harvesting. The mining operation itself. In fact, the ore is very low grade, and it only became economic to exploit, in a vast open-pit operation, in 1975, when uranium prices rose sharply. Of course this exploration and mining is conducted by companies in developed countries, with obvious political overtones.
- processing. The ore is processed at a mill using crushing and chemical operations to produce yellowcake, uranium oxide, which is the form in which it is shipped.
- transport. Yellowcake is shipped to UK.
- storage. It may be stockpiled in this form depending on strategic and market considerations.
- processing. It is converted into the forms used as nuclear fuels by British Nuclear Fuels plc (BNFL): metallic uranium for the original Magnox reactors and 'enriched' uranium oxide with an increased proportion of the fissile isotope ^{235}U for the later AGRs and PWRs, which have differences in fuel element design.
- marketing. This is abnormal – there is a powerful world market for uranium in its various forms (even more for the plutonium derived from it after reprocessing), but it is tightly controlled for security reasons, so is very much in the hands of international bodies and government agencies, which make direct agreements with the multinational primary producers.
- transport. Fuel elements are transported (road and rail) to a nuclear power station.
- (1st) end use. The fuel is used in the nuclear power station to produce electricity.
- transport. The radioactive spent fuel is transported in specially robust flasks by rail to reprocessing plant – not part of the essential primary cycle, but seen as important, in particular for plutonium production.
- storage. Radioactive wastes, before and after reprocessing, are stored; this has to be secure for very long periods, e.g. hundreds of years.
- marketing. The electricity is marketed by the production company to the distribution companies. The UK has a partially privatised system with relatively uneconomic nuclear electricity retained under government control and current (1995) plans to privatise some of it too.
- storage. Apart from large-scale pumped storage of electricity as raised water, the transport application requires local electrochemical storage of electricity in batteries which can be carried on vehicles.

- marketing. This is now much more directly aimed at the consumer – marketing of electricity by area boards and of vehicles and batteries by their manufacturers.
- (2nd) end use. For final transport application.

You may know more of the details, though there is no reason to feel guilty if you don't know any of them: were you able to recognise that something like this process, with all its interlocking aspects, had to be involved?

Activity 2

There are no 'correct' answers, but possible suggestions are shown in Figure 2.18.

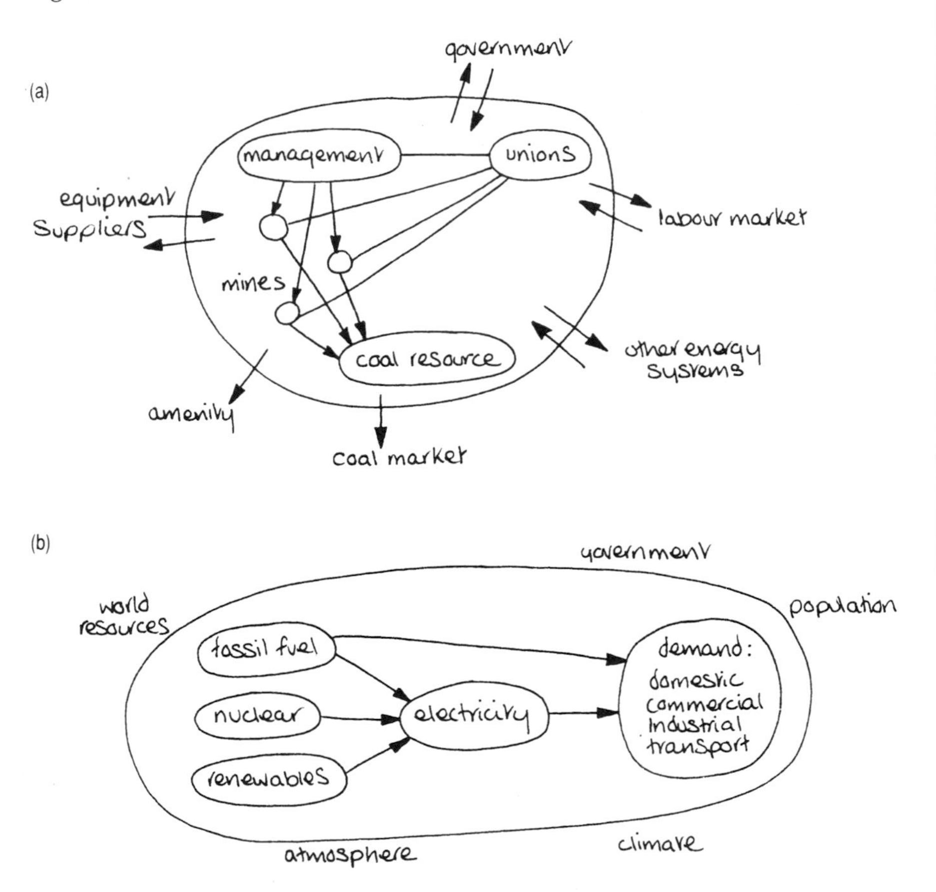

Figure 2.18 Answer to Activity 2: (a) UK coal industry; (b) UK energy system.

Activity 3

Detailed information will be found throughout Section 3. You may have noted

1 visible smoke, soot, etc., from domestic grates and poorly controlled industrial furnaces, now not allowed;

2 acid rain from invisible effluents (SO_2 and NO_x);

3 the enhanced greenhouse effect from CO_2 emission.

Activity 4

Some suggestions are shown below.

	Ecosystem	*Resources*	*Physical environment*	*Health*	*Safety*	*Socio-economic*
Vehicle and fuel production	pollution from mines, oil, factories	mineral shortages, capital for new developments	air, water quality, wastes, aesthetic	occupational and public – toxic materials	oil platform, factory, distribution hazards	employment, geographical shifts
Infrastructure construction	habitat disturbance	materials and capital	dust, run-off noise	pollutants	construction accidents	costs, planning
System operation	toxic pollutants	fuel, maintenance skills	air and water quality, noise, aesthetic	exhausts, smog	accidents	costs to local government – politically sensitive
Disposal	some toxic materials	recycling	junk yards, tyres, batteries, ceramics disposal	toxic materials	accidents	responsibility for cost?

References

BLACKMORE, R. & REDDISH, A. (eds) (1996) *Global Environmental Issues*, Hodder & Stoughton/The Open University, London, second edition (Book Four of this series).

ELSWORTH, S. (1984) *Acid Rain*, Pluto Press, London.

EVANS, H. J. (1990) Leukaemia and radiation, *Nature*, **345**, 16–17.

GARDINER, M. J. (1990) Leukaemia and lymphoma among young people near Sellafield nuclear plant, West Cumbria, *British Medical Journal*, **300**, 423–430.

GARDINER, M. J. (1993) Investigation of childhood leukaemia rates around the Sellafield nuclear plant, *International Statistical Review*, **61**, 231–244.

HOHMEYER, O. (1988) *Social Costs of Energy Consumption*, Springer-Verlag, Berlin.

MILNE, R. (1993) High Court acquits Sellafield, *New Scientist*, **139(1889)**, p.6.

MORTIMER, N. (1989) Proof of evidence, FOE 9, to the Hinkley Point C public enquiry, Friends of the Earth, London.

OECD (1983) *Environmental Effects of Energy Systems – the OECD Compass Project*, OECD, Paris.

OECD (1986) *Environmental Effects of Automotive Transport*, OECD, Paris.

PARK, C. C. (1987) *Acid Rain: rhetoric and reality*, Methuen, London.

PEARCE, D. *et al.* (1989) *Blueprint for a Green Economy*, Earthscan, London.

SKEA, J. (1989) 'Acid Rain and Urban Atmospheric Pollution: Europe', paper presented to the *4th International Energy Conference: Environment Challenge: The Energy Response*, The Royal Institute, December 1989.

WORLD COMMISSION ON ENVIRONMENT AND DEVELOPMENT (1987) *Our Common Future* (The Brundtland Report), Oxford University Press, Oxford.

Chapter 3 Sustainable energy futures

1 Introduction

As you read this chapter, look out for answers to the following key questions.

- What contribution can energy efficiency make to reducing the environmental effects of energy use?
- What are the possibilities for renewable energy supplies?

The first chapter of this book outlined the use of natural energy resources to supply our needs. It placed particular emphasis on *numerical* matters: the existence of a common measure of energy of all kinds, the joule, in which all other energy units could be expressed, and the importance of one other unit, the watt, or joule per second, for the rate of conversion of energy from one form to another. We saw how the numbers of these units involved in energy storage and conversion processes vary over vast ranges as we move from the scale of individual human needs and particular appliances to the world scale of total energy demand and supply. As our future depends, among other things, on our ability to continue somehow providing for our energy needs, it will remain important to be clear about the general scale of these numbers.

However, we have also seen in Chapter 2 that present energy policies have many and complex environmental consequences, which cannot readily be assessed in numerical terms. The values (economic, ethical or aesthetic) to be used in judging between alternative options, and the ways these influence the political process, were introduced briefly there; they will continue to affect our discussion of the future, and are further considered in *Blackmore & Reddish, 1996*, Chapter 6.

A major recent assessment of sustainable development worldwide, the 'Brundtland Report', *Our Common Future* (World Commission on the Environment and Development, 1987) concluded, as far as energy was concerned:

> Energy efficiency can only buy time for the world to develop 'low energy paths' based on renewable sources, which should form the foundation of the global energy structure during the 21st century.

These two topics, energy efficiency and renewables, form the two main strands of this chapter. The first is about ways of reducing energy *demand* while still satisfying human needs. Section 2 outlines the innumerable technical options available; Section 3 indicates some of the political factors involved in bringing them into use. The second is about forms of energy *supply* which are technically feasible and environmentally acceptable. Section 4 compares non-renewable and renewable energy contributions to future supply; Section 5 considers the potential environmental effects of renewables, including a more detailed case study of wind power. Although the discussion in this chapter is biased towards UK policy, this is representative of the issues in all industrialised countries. Furthermore, as these countries use most of the energy at present, creating most of the resulting environmental problems, any reductions in energy use and environmental damage that can be achieved without loss of amenity may help the rest of the world to avoid similar difficulties in their pursuit of an improved quality of life. Section 6 draws the threads of the previous sections together in some provisional conclusions about a sustainable energy future.

Activity 1

Before reading on, reflect on the quantitative and environmental aspects of energy policy raised in Chapters 1 and 2. How has total world energy demand changed during the twentieth century? What does this imply for energy consumption per person in different regions? What is the likely effect on energy demand of a further doubling of world population in the next century, along with demand changes in the 'less developed' countries? What are the implications for the pressure on available fossil fuel, nuclear and solar resources? What are the environmental problems already recognised with fossil fuel and nuclear energy sources? Can these be included in economic considerations?

2 Energy efficiency

Any reduction in the amount of primary energy used to satisfy a particular human need represents an improvement in energy efficiency. It can occur anywhere in the long chain of conversion processes, from fuel extraction and refining, via combined heat and power plant, or improved manufacturing methods, to better insulation of buildings and more efficient domestic appliances. Some of these improvements have been referred to, particularly since the oil crises of the 1970s, as 'energy conservation'. Apart from the pedantic point that, scientifically, energy is *always* conserved, this may include proposals for reductions in the quality of service provided, which may well be resisted – such changes in behaviour should not be confused with reductions of energy use without loss of service. Otherwise the terms can be used interchangeably, but 'efficiency' has perhaps a more optimistic flavour than 'conservation', with its hint of puritanical cheese-paring.

The pursuit of energy efficiency is not new: it reappears with successive economic problems. In the 1920s, Norway and Sweden, with their cold climates and lack of indigenous fuel, had similar building insulation standards to those in the UK from 1990. Since the 1920s many communities in Denmark have used the 'waste' heat from diesel engine power plant in combined heat and power/district heating schemes. In 1934 the US government introduced mandatory thermal insulation requirements on all federally financed homes; they were progressively tightened in the 40s and 50s. When there were coal shortages in the UK in the 1940s it was widely agreed that a ton of coal saved was cheaper, safer and less polluting than a ton of coal mined. In the 1940s Swedish houses were developed with even more insulation and triple-glazing, producing further reductions in energy use.

However, the cheap oil and expectations from civil nuclear power of the 1950s distracted attention from energy efficiency, and it was largely forgotten until the oil crises of the 70s brought it back into consideration with renewed vigour. The vast worldwide reappraisal, often outside official government projects, particularly in countries with a history of such studies (Denmark, Sweden, Germany, Japan, USA) has shown much greater potential for energy efficiency than previously believed. In some applications, cost-effective reductions in energy use by as much as 90% were made. Representative studies from the 1980s in the USA, West Germany and the UK are listed at the end of this chapter. More recently in the UK, official initiatives such as BRECSU's Best Practice Programme, or the Milton Keynes Energy Rating, having been developed, but so far with limited government support for their general introduction.

The renewed emphasis has been on the demand, not for energy as such, but for *energy-related services*: comfortably warm homes, hot water on tap, cold food storage, machines to reduce domestic drudgery, modern telecommunications and health services, comfortable, well-ventilated and pleasantly lit offices, amenities relying on materials like steel, cement, paper and plastics. In the 1990s energy demand in many countries is said to be unexpectedly buoyant. But if the required *services* can be provided using less energy, more cheaply and/or with less environmental impact, then rising energy use may well be seen, globally, as a failure. Unfortunately the separate energy supply organisations tend to regard increasing energy use as a success.

The way the present energy system of the UK is divided up, between different fuels, economic sectors and end uses, was summarised in Figure 1.14 of Chapter 1. It will be useful to remind yourself of the relative proportions of the various components before reading on. This breakdown is representative of that in industrialised countries, though obviously the availability of fuels, climatic and social factors produce many detailed variations.

Q In the UK case, what fraction of the primary energy supplied is lost in conversion and delivery to the user?

A About a third (particularly in electricity generation).

Q What fraction of the delivered energy is used for heating?

A About half.

Q What fraction of delivered energy is used for transport?

A About a third.

2.1 *Space and water heating*

In northern countries like Canada, Scandinavia, Germany and the UK, 25–50% of all energy consumed is used to heat buildings. In the long term it should be possible to eliminate this drain on resources almost entirely.

Some heat input is of course essential to maintain an internal temperature above that outside. In any building, some of this heat is already provided by 'incidental gains' – solar (incoming sunlight), metabolic heat from the occupants, energy uses like cooking, lighting and

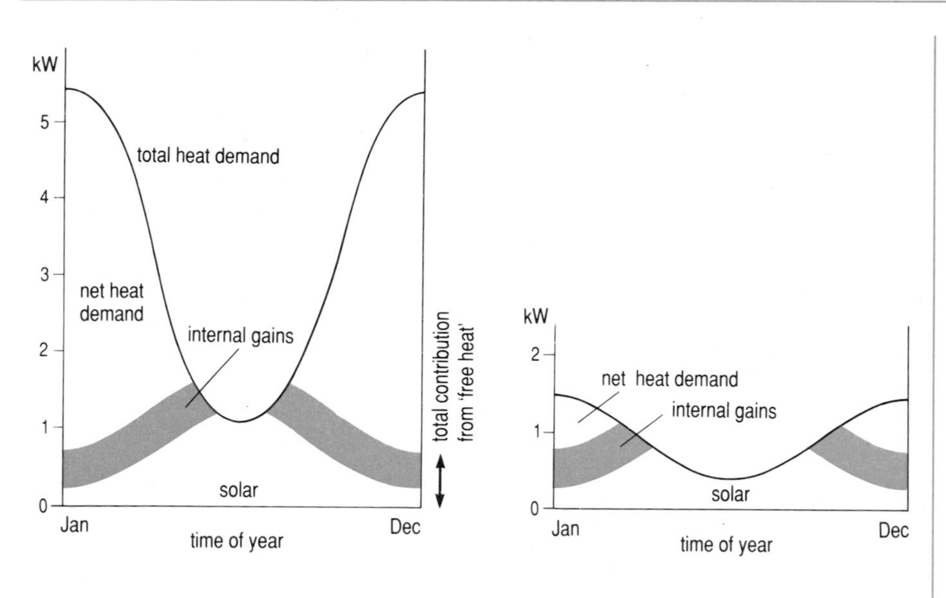

Figure 3.1
Typical variation over the year of the energy demand for space heating in a UK dwelling:

(left) conventional UK dwelling (90 m², semi-detached);

(right) superinsulated dwelling.

electrical appliances. Only the balance has to be provided by the heating system. If the heat losses from the building are greatly reduced and/or the use of solar energy is improved, this 'free heat' can account for most or all of the demand, and the residual requirement from the space heating system can be minimised. Figure 3.1 compares the best available with typical UK practice, showing that large reductions are possible in both the maximum heat required and in that part of the year for which it is needed: the annual heat demand is reduced by 90% or more. The heating system itself can be more efficient in its use of primary energy, as we shall see. The energy demand for water heating can likewise be reduced by more efficient use of hot water, and improved insulation of storage tanks and distribution pipes.

Reduced heat loss

Techniques which have come to be called 'superinsulation' were developed in government research laboratories in Sweden, Canada, Denmark and the northern USA in the late 70s, though their origins are much older. The first such house was built with government support in Regina, Saskatchewan (Canada) in 1977, the second privately in Uppsala (Sweden) in 1978.

These houses have exceptional insulation, draughtproof construction, triple-glazing (or better) and a controlled ventilation system for winter use: usually arranged so that 70–80% of the heat in the outgoing stale air is recovered to preheat the fresh air intake. Constructional techniques have been developed in considerable detail in these countries; a representative design is shown in Figure 3.2. In severely cold climates 95% reductions in space heating costs have been documented; in more moderate climates, up to 100%.

Such advanced design is not yet embodied in the building regulations of the UK, perhaps because of concerns about the detailing and quality control required to avoid condensation or air quality problems and the cost-effectiveness of mechanical ventilation systems. Poorly insulated houses have little or no added insulation in the walls and roof; present insulation standards recommend an equivalent of 150 mm of mineral wool, whereas superinsulated houses have 200–400 mm, with various window, vapour

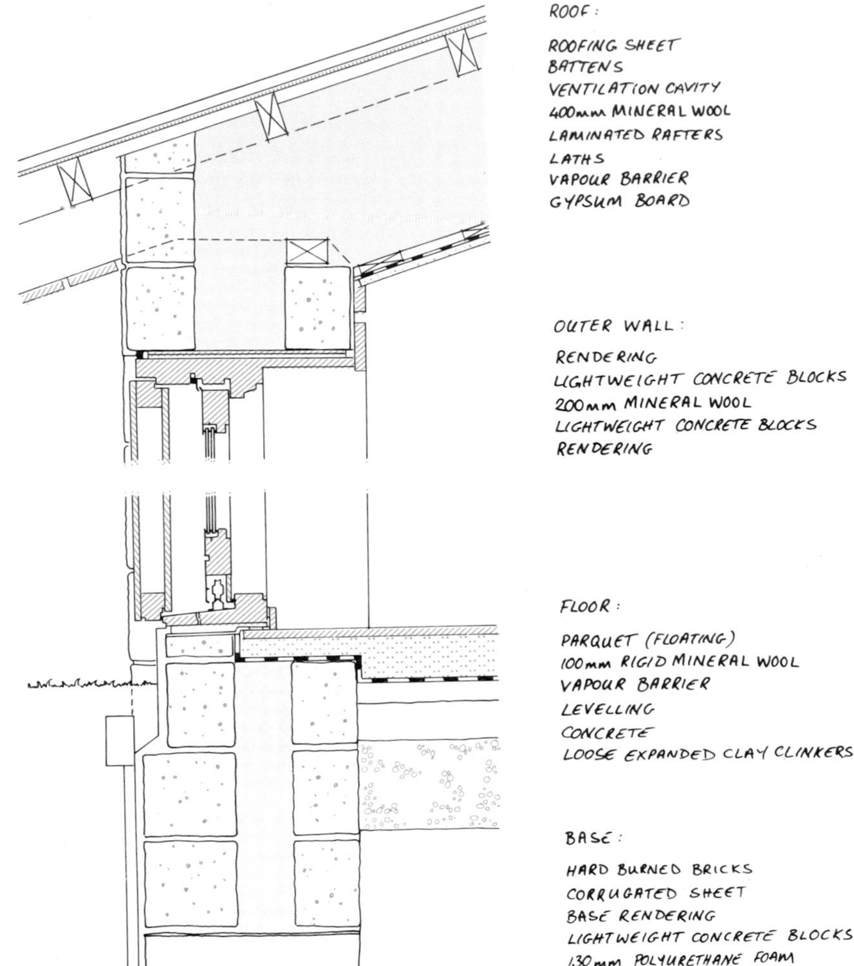

◀ *Figure 3.2* Construction of a Danish superinsulated house.

barrier and air flow details. Differing national standards in the 1990s make new housing in the UK similar in heat loss to that in some other North European countries, but much inferior to that of Denmark and Sweden.

About half the new houses in Saskatchewan province were superinsulated by 1985. By 1989, Scandinavia and North America had hundreds of thousands, mainland Europe several thousand, and the UK about ten superinsulated buildings. The 1989 Swedish building regulations specified insulation that brings much new building into the superinsulated range.

It was also found in Canada and the USA that during major renovation of existing buildings, it was cheaper to superinsulate them than to provide the fuel to heat them. Suitable techniques were developed for different forms of construction such as timber-frame and masonry.

Passive solar design

This refers to techniques developed since the 70s to make the best use of sunlight without summer overheating – an idea which is at least 50, if not 2000, years old, but has been refined with modern materials. Extra south-facing glazing, suitably heavy construction to store heat, orientation to favour heat capture in the winter and rejection in the summer is used in some 20% of new US housing. In less sunny regions like the UK more normal areas of glazing, but preferentially orientated to the winter sun, with rooms requiring heating concentrated on the sunny side, have proved successful when associated with high insulation and airtightness (Plate 13).

Earth-sheltered design

A more controversial development since 1973 has been towards partly underground, 'earth-sheltered' buildings, of which about 20 000 had been built in North America by the 1990s. They exploit the fact that only 3 m into the earth, the temperature variation is just a few degrees, when outside it might be from +30 °C to –10 °C or below. Usually they have concrete walls and a concrete or timber roof, the insulation is less than that of above-ground superinsulated houses, and the floor-slab insulation and ventilation system are similar.

Earth-sheltered buildings have been found to reduce seasonal temperature variation, make efficient use of free heat, reduce the size of the space heating system and reduce or eliminate the need for summer cooling. There are a few earth-sheltered dwellings in the UK, and more are proposed, though with some planning difficulties. One in Monmouth is superinsulated and has negligible space heating requirements (Plate 14).

However, there is more to be learned about the manipulation of underground heat flow; a leading US energy efficiency centre, the Rocky Mountain Research Center, has been experimenting since 1981 on designs in which an insulation/waterproofing 'umbrella' around the house *encloses* a mass of earth large enough to provide inter-seasonal storage, so as to keep the interior warm in winter and cool in summer.

In the long term, a combination of such techniques could enable space heating energy use to be eliminated, even in cold climates.

More efficient heating systems; combined heat and power

Conventional central heating systems, oil- and gas-fired alike, have been radically improved in efficiency during the 80s. In a typical UK domestic system, savings of 30–40% are not unrealistic through using a condensing boiler, thermostatic radiator valves and a room thermostat, pumped hot water and, in larger houses, separate zones with their own controls.

Worthwhile savings can be made by improving systems in this way. However, burning high-quality fuel to produce low-grade heat is thermodynamically wasteful. The savings from improved heating controls are dwarfed if a much more fundamental step is taken: namely, if the fuel which would normally be burnt to provide low-grade heat in the building is instead used for combined heat and power production.

Industrial nations like the USA, UK and France discard as much energy in the form of waste heat from power stations to rivers and up cooling towers as natural gas provides to their whole economies. With combined heat and power and district heating (CHP/DH), energy that would otherwise be thrown away is piped as hot water through insulated heat mains to individual buildings; their heating systems replace the cooling towers of the power station. Modern designs mean that losses are no more than about 10%, comparable to those in electricity distribution. (Note that district heating is sometimes provided by 'heat-only' boilers independent of electricity generation; waste heat from electricity generation can be used for other, say industrial, purposes, so the two ideas 'DH' and 'CHP' are in principle distinct, though often associated with each other.)

The temperature of the supply water of the DH system needs to be higher than the cold river or sea water used in conventional power station cooling, and this involves a small reduction in electrical efficiency. However, the extra fuel required to restore the amount of electricity produced provides low-grade heat to the DH system users at a lower *energy* cost than any other fossil fuel space and water heating system. There is an installation cost for the large insulated hot-water mains, which is larger than that of gas mains. Even so, the low cost of the heat available often makes this the cheapest way to heat towns and cities and provide their hot water.

About 50% of all buildings in Denmark and Finland are heated by DH, mainly from CHP plant. Denmark aims to increase this to 55% by the year 2000; it now only builds CHP stations. Cheaper plastic heat mains and connection systems have been developed to improve the economics, even for highly insulated or dispersed buildings which were previously thought unsuitable.

Recently, 'micro-CHP' plant has become available, based on internal combustion engines, usually burning natural gas or liquid petroleum gas (LPG), with outputs as low as 15 kW electricity and 25 kW heat. These make CHP feasible for small groups of 5–10 houses, individual office blocks, hospitals, etc. They also ease the development of extended DH schemes. Most countries with CHP/DH began with small schemes of 10–1000 dwellings, which slowly merged into town- and city-wide systems. At first, these embryonic schemes used fossil-fuelled heat-only boilers, which produced more expensive heat than CHP plant. Micro-CHP would improve the economics of this early stage.

However, except for Denmark, Finland, Sweden, Germany and Austria, other industrial countries, including the UK, have hardly begun to tap the potential for CHP/DH. The reasons for the UK situation are complex – perhaps the combination of a natural gas industry set up to compete with other energy industries, seeing DH as a rival; the lack of *local* government involvement in energy supply, unlike Germany or Scandinavia; and, until the 80s, low central heating ownership and thermal comfort standards.

Finally, the possibility should be mentioned of cheap and reliable CHP equipment being developed on a small enough scale for the individual house (perhaps based on a different type of heat engine called a Stirling engine). Such a system would take away the need for the heat mains of a DH system, and with suitable forms of grid connection to maintain continuity of electricity supply, begins to point to a more dispersed form of electricity generation. This is not yet available, but developments over the next decade should be watched.

2.2 *Space cooling*

In Japan, the southern USA and many developing countries, space cooling is more important for thermal comfort than space heating. Large amounts of electricity are used – thus, the USA uses 250 GW on hot summer afternoons (equal to the output from 250 large power stations) for this purpose alone, and demand is growing at 5 GW/year. Even in the UK, this is one of the fastest growing uses of electricity.

Countries where space cooling has always been essential for comfort have shown that major efficiency improvements are possible (which is not to say that they are always adopted). Electricity use for this purpose can be almost eliminated in cool and temperate climates, and greatly reduced even in subtropical regions. This depends on improved building construction and orientation to keep out summer heat, energy-efficient lighting and electronic equipment producing less waste heat to be removed, and improved efficiency in the refrigeration equipment itself.

These techniques enable new non-domestic buildings such as schools, shops, offices and hospitals to be kept cool with 90–95% less electricity than in 1980. In cooler climates like north-western Europe, refrigeration plant is not needed at all if natural or mechanical ventilation of suitably heavily constructed buildings is used at night to pre-cool the structure so that it can absorb the daytime gains from the sun, occupants and equipment. In the southern USA, similar techniques have been shown to reduce refrigeration requirements much more cheaply than the cost of the extra electricity. In the humid tropics, one carefully designed building, the headquarters of the Petroleum Corporation of Jamaica, uses less electricity for cooling than would be used for normal air-conditioning in the UK, at no extra cost.

2.3 *Electrical appliances*

Many industrialised countries use 30–40% of their electricity, 10–15% of total energy, to run electrical appliances: refrigerators and freezers, clothes and dish washing machines, television and other electronics, and so on. If everyone used the most efficient commercially available designs this consumption could be reduced by over 80% (some prototypes could achieve over 90%) without affecting the service provided.

Consider as one example the domestic refrigerator. A rough measure of energy performance is the annual energy consumption divided by storage volume: typical mass-produced designs sold in the UK achieve 2–2.4 kWh/litre. Developments in the USA, Germany and Scandinavia show this can be reduced by several, perhaps ten, times: the best German models achieve 0.5 kWh/litre; Danish and US prototypes claim 0.4–0.2 kWh/litre or less.

These radical improvements arise from attention to all stages of the use of electrical energy to transfer heat from the cold food storage space to the warmer surroundings. Poor insulation and loose door seals; an inefficient motor driving an inefficient compressor, with waste heat re-entering the cool space; an internal fan consuming energy and supplying heat; an undersized heat exchanger, reducing the thermodynamic performance – these have all been tackled to achieve the improvements described.

In climates like those of Canada, the northern USA and northern Europe where the external average temperature is only 8–10 °C, separating the condenser and mounting it outside, or using a 'through-the-wall heat pipe', can halve the electricity consumption again, or avoid the need to run the compressor in cold weather. In the USA, re-examination of the traditional inter-seasonal storage of ice made in the winter, but now using modern insulation materials, has shown that the electricity consumption of large refrigeration systems can be almost eliminated.

Decisions about appliance purchase are of course much influenced by costs, but it is not always easy to see their implications. Energy-efficient designs cost little more than standard models, and make major savings elsewhere in the energy system. Thus a particularly efficient 200 litre refrigerator uses 80 kWh/yr and is estimated to cost only about £3 more to manufacture than a standard model using about 350 kWh/yr. To the user this implies a saving of 270 'units' of electricity a year, paying for itself in a year or so depending on the exact retail price charged for the refrigerator and the tariff charged by the electricity company. But it can also be viewed as 'generating' electricity at 270 kWh/yr – what has been termed the generation of 'negawatts', the power that is *not* required as a result of improved efficiency.

Q What is this rate in W?

A There are $24 \times 365 = 8760$ hours in a year, so annual saving, 270×10^3 Wh/yr, is at a rate of $270 \times 10^3/8.76 \times 10^3$ W = about 30 W.

Q If a country has ten million such appliances, what is the power saved?

A $10 \times 10^6 \times 30$ W $= 3 \times 10^8$ W = 300 MW which is equal to one moderate-sized power station.

How does the extra appliance cost compare with the costs of building power stations? These are usually expressed in pence per kilowatt-hour, p/kWh, over the life of the station, with much scope for ingenious 'discounted cash flow' accounting to include initial capital costs, continuous fuel costs and future decommissioning costs in a single figure. These are much debated by proponents of rival generation methods, as we shall see below, but competitive figures are always in the range of a few, say 2 to 5, pence per kilowatt-hour for the various options seriously considered. If our '30 negawatt' refrigerator is assumed to have a 12-year life, a similar calculation can be performed.

Q If discounting is at first neglected, what is the apparent cost of the saved energy?

A At an extra cost of £3 = 300p to save $12 \times 270 = 3240$ kWh, this represents 300/3240 = less than 0.1p/kWh saved.

Of course, this is overstating the case, as the discounting of future savings hasn't been done by an appropriate accounting formula. With the kind of discount rate used in power station calculations, the figure is more like 0.2p/kWh, still considerably less than the cost of operating and fuelling existing stations, let alone building new ones. This is the basis of the 'least-cost accounting' required of some US utility companies. They are required to compare the cost of introducing efficiency measures with that of building new plant, and choose the cheaper option – often the former. This approach is not yet common elsewhere, however.

2.4 *Lighting*

In the UK some 17–18% of electricity is used for lighting; in developing countries the proportion is possibly higher, as this is often the first, and most valued, use of electricity. Methods available at present could reduce this by over 80%; many commercial buildings in the US, and a few in Europe, have achieved over 90% reductions from current standard levels.

During the 80s compact fluorescent lamps became available which can replace the small incandescent lamps still normally used in the home. There have been advances in the design of the 'ballast' used to control the current: the newer electronic types are smaller, lighter, more energy efficient and give a better performance – instant starting without hum or flicker – and the phosphors used now give a light quality resembling the 'warm' incandescent colouring. Replacing incandescent by compact fluorescent lamps reduces *total* domestic lighting costs because the higher initial cost is compensated by a much longer life and about 5 times lower electricity use.

Q If a compact fluorescent lamp costing £12 has a rating of 11 W and a life of 8000 hours, and is equivalent to a 60 W incandescent lamp costing 50p with a life of 1000 hours, how do their total costs compare if electricity costs 6p per unit (kWh)?

A Over its life, compact uses 11 × 8000 Wh = 88 kWh costing 88 × 6 = 528p.
Total cost (for 8000 h service) = £12 + 5.28 = £17.28.
Incandescent over its life uses 60 × 1000 Wh = 60 kWh costing 60 × 6 = 360p.
Total cost (for 1000 h service) = £0.5 + 3.6 = £4.1.
So for 8000 h (neglecting labour of replacement) total cost = £32.8, almost twice that of the compact.

This is again an over-simple calculation for accounting purposes, as it neglects the discounting of future costs. If these are included as in power station calculations, it can be shown that the effective cost saving offered by this appliance change is around 1p/kWh, again comparing favourably with the costs of building power stations. However, these benefits are not widely perceived because of the higher capital investment required. When seen as part of the least-cost planning of the total energy system, they have led some US utilities to give these efficient lamps away rather than build new plant. (The cost is of course absorbed in the price charged for electricity.)

Larger incandescent lamps, like the 500 W types mostly found in industry, can be replaced, usually by discharge lamps, with 75–90% energy reduction at low or negative cost.

Fluorescent tubes provide a little domestic, most public and commercial, and some industrial and agricultural lighting. They have been used since the 1930s, but surprising advances have been made since 1973, so that most systems installed even in the late 70s are technically obsolete. The best lamps in the 1990s, with good colour rendering and high-frequency electronic ballast, give about 2.5 times more light per watt of electricity than standard mid-70s types; US research suggests that twice as much light output again may be achievable by the year 2000.

An integrated energy-efficient lighting system, as well as using the best lamps, makes greater use of daylight, improved 'luminaires' (fittings which waste less light and direct it where it is wanted), task lighting where appropriate, occupancy sensing or on-off schedules (which users can override), and, most simply, ceiling and wall finishes which absorb less light.

Some of these measures applied separately can save 50%; taken together they can save over 90% of the electricity used while maintaining the same lighting standard. Most energy-efficient lighting schemes have an extra capital cost of less than £4–5/GJ (1.5p/kW) of electricity saved; in some buildings designed for energy efficiency from the start, or in carefully planned renovations, they have cost *less* to install than conventional higher consumption schemes. In such cases the energy saving and environmental gain cost less than nothing. In general they cost much less than the operating cost of existing power stations.

2.5 Stationary motive power

Industrial motors consume 40% of the electricity generated in the USA, or 12–15% of its total primary energy. In western Europe the proportions are similar – in the UK about 30%, including the similar motor drives used in commercial buildings and agriculture.

It was shown in the late 70s by researchers at Imperial College in the UK that some 60% of this energy was lost in drives and controls. Electronic variable-speed drives and so on could eliminate most of the losses, reducing electricity consumption by up to 20%. The same principles can be applied to the smaller electric motors used in building ventilation systems and domestic electrical appliances.

2.6 Transport

This is in many ways the most problematic sector of energy use. 'The unrelenting growth of transport has become possibly the greatest environmental threat facing the UK, and one of the greatest obstacles to sustainable development' (Royal Commission on Environmental Pollution, 1994). The share of energy delivered that is used for transport rose, in spite of the oil price rises – in the UK, from 17% in 1960 to 25% in 1994. Over 80% of this is used in road transport (passengers and freight) and 88% of all passenger transport is in private cars, with their current dependence on oil, the least available fossil fuel, and familiar problems of accidents, pollution, and congestion to weigh against their undoubted attractions for many of us.

Energy efficiency, to reduce demand, can be considered in the same general way as for the other sectors, first looking at technical improvements in existing methods, then considering less energy-intensive ways of providing the same service.

The conventional measure of energy efficiency for cars in the UK is miles per gallon (m.p.g.), typically between 30 and 45, with a current (1995) UK average of about 35, considerably dependent on driving conditions. (For international comparisons we should really be using kilometres per litre, km/l; the above range corresponds to about 10–15 km/l.) German, Swedish and US manufacturers have demonstrated prototypes up to four times better than this over the last decade, but most of the lessons learned have been applied, in production cars, to higher performance. The number of new cars in the UK with top speeds over 120 m.p.h. trebled in the 80s. At the same time average fuel efficiency rose by only 5%; the best currently available is about 55 m.p.g. (nearly 20 km/l). It is clear that more efficient cars *can* be made: the world record achieved in the Shell Mileage Marathon in 1988 (admittedly for a very small low performance vehicle) was 6049 m.p.g.! This could surely be translated into a practical 100–200 m.p.g. if the manufacturers perceived a demand. Powerful tax incentives, such as progressive increases in fuel taxes (greater than those of current UK policy) and a steeply graded excise duty related to m.p.g. performance, were

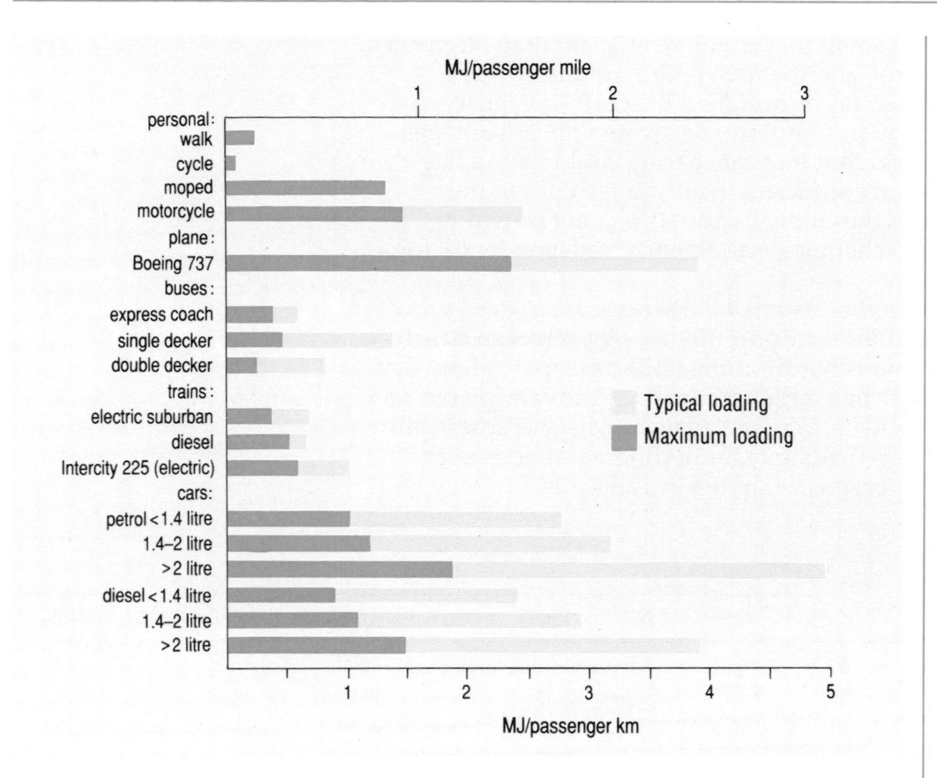

◀ *Figure 3.3 Energy efficiency of different transport modes.*

recommended by the RCEP's 1994 report to encourage changes in user outlook about environmental effects and manufacturers' perceptions about demand. There would also be significant reductions in energy use if speed limits were enforced, and tax advantages for company cars were removed – company cars are on average larger and less efficient than those bought privately.

A more general comparison of energy efficiency between different *modes* of transport is shown in Figure 3.3. In each case, the shorter bar represents fully loaded and the longer bar represents normal occupancy levels. Note that electric vehicles actually consume about three times as much primary energy if conventional power stations provide the electricity.

Q What are the most and least efficient forms of car transport?

A A fully loaded small diesel and a normally loaded >2 litre petrol car respectively.

Q How do planes, buses and trains compare with cars?

A Large jet aircraft are broadly comparable to large cars according to this measure of energy efficiency (when fully loaded, better than all normally loaded cars). Buses and trains are in general better than cars, much better than large cars, with a fully-loaded double-decker bus best of all. *Electric* trains of course have the *primary* energy disadvantage mentioned (though still better than large cars).

Q What is the most energy-efficient form of transport of all?

A Cycling.

It is clear, if unsurprising, that small cars are more efficient than large ones, diesel more efficient than petrol, and that perceptions of improved performance or prestige have so far outweighed the costs of vehicles, fuel and tax at present levels, although company car use and tax advantages confuse the picture. It is also evident that car sharing could have a large effect, but only if incentives were sufficient to encourage it, as in the Californian scheme of special lanes for full cars – this is not providing *exactly* the same service, but exchanging one amenity, independence, for another, shorter journey times.

This question of the *comparative* value of the service arises even more for the much more efficient public transport modes. (Air travel so clearly provides a different service that is hardly comparable, except to show that its fuel efficiency is no worse than a large car; the speed advantage over a fast train for internal trips has to be weighed against costs and accessibility.) The UK response to the perceived advantages of different modes over recent decades, whatever the reasons, is shown in Table 3.1.

Table 3.1 Travel trends by mode (billion passenger km)

Mode	1955	1960	1965	1970	1975	1980	1985	1990	1994
Bus, coach	91	79	67	60	60	53	49	46	43
Car, van, taxi, motorcycle	91	150	238	301	337	396	449	594	600
Cycle	18	12	7	4	4	5	6	5	5
Total road	200	241	312	365	401	453	504	645	648
Rail	38	40	35	36	35	35	37	40	35
Air	0.3	0.8	1.7	2.0	2.1	3.0	3.6	5.2	5.5
Total, all modes	239	282	349	403	438	491	544	690	689

Source: Department of Transport (1995), Transport Statistics Great Britain, HMSO, London.

Q How has the total travel, in 10^9 passenger-km/yr, changed from 1955 to 1994?

A It has increased by nearly 3 times, from 239 to 689.

Q How have rail and air travel varied over this period?

A Rail traffic has remained roughly constant, in the range 35 to 40 – so the percentage has fallen from 16% to 5%. Air travel has risen steadily from 0.3 to 5.5 (still $<1\%$).

Q How has the percentage of travel by road varied?

A From 84% to 94%.

Q How have the bus, cycle and car components of road travel varied?

A Bus and cycle have both declined in real terms (91 to 43; 18 to 5) and even more in percentage terms (46% to 7%; 9% to 0.89% of total road travel). Car travel has increased by over 7 times (91 to 600), from 46% to 92% of total road travel, 38% to 88% of all travel, and is essentially responsible for the increased total travel, and the growing share of total energy use.

This shows such an overwhelming preference for car travel in the prevailing UK conditions that it will take very determined measures to reverse the trend for resource or environmental reasons. Fuel shortages will sooner or later be reflected in price rises, but this may cause severe economic disruption if the commitment to road transport continues. If greenhouse gas emissions, or other environmental problems, are to be controlled, taxation, subsidies or regulations will need to be modified before the perceived balance of amenity between private and public transport shifts.

There are some differences in car ownership and use in different developed countries which reflect their different traditions. Car ownership varies from 0.58 per capita in the USA to 0.31 in Japan, with the UK, at 0.38, rather low among European countries. But 88% of travel in the UK is by car, almost the highest in Europe. Italy, for example, with ownership at 0.5 per capita, has only 79% car travel, and Japan, with its lower ownership, only 53% (1991 figures). This no doubt reflects varying measures in different countries to promote public transport and/or restrict car use in cities; various forms of road pricing are now being explored. However, these differences are variants on a general problem that can only increase if the developing countries follow a similar path. At the very least, energy, transport and environmental policy have to be considered together.

Satisfying transport needs is not an end in itself, but is about providing *access* to people and services. Land use planning can seek to group homes, work, shops, schools, hospitals, entertainment in such a way that more journeys can be made by low energy means – on foot, by bicycle or public transport. The UK Royal Commission on Environmental Pollution in its 1994 report on transport and the environment represents a significant attempt to change this response. It identifies six perspectives on transport policy: 'letting congestion find its own level' (which it dismisses), 'predict and provide' (the traditional UK policy, which it considers to have failed), 'greening the way we live', 'collective action', 'selling road space' and 'relying on technology'. The last four are discussed in considerable detail and lead to over a hundred policy recommendations involving various aspects of them. Similar studies elsewhere will perhaps eventually lead to profound changes in transport policies worldwide.

Activity 2

Consider your own use of transport. How far does it provide what you want, or is it constrained by work, housing patterns and availability of alternatives over which you have no control?

3 Energy efficient futures

3.1 Low energy strategies

After the oil crises of the 1970s, a number of workers in different countries made detailed studies of possible low energy strategies – not predictions or blueprints, but possible scenarios for energy demand reduction under sympathetic government policies.

To avoid confusion between energy policy and discussion of future lifestyles, official government projections for economic growth were used, and proposals for more efficient energy use and renewable supply were limited to the technology that was commercially available or at pilot/prototype stage. (This leaves aside the growing conviction of recent years that the standard economic measure, gross domestic product (GDP) is not a good measure of national prosperity, since it takes no account of environmental degradation.)

Three of the most detailed studies, in the USA, West Germany and the UK, are listed at the end of this chapter. Similar analyses in Denmark, Sweden, Canada and other countries covered 95% of OECD energy use.

The US study assumed an increase in national wealth, as measured by GDP, of 1.5 times by 2000. It was shown that energy demand could fall to 50% by 2010, 80% of it coming from renewable sources (SERI, 1981).

In West Germany, GDP was assumed to increase 2.3 times by 2025. Although it is already one of the world's more energy-efficient countries, its national consumption could fall by 45%, and renewable sources could supply 40% of the reduced demand (Krause *et al.*, 1980).

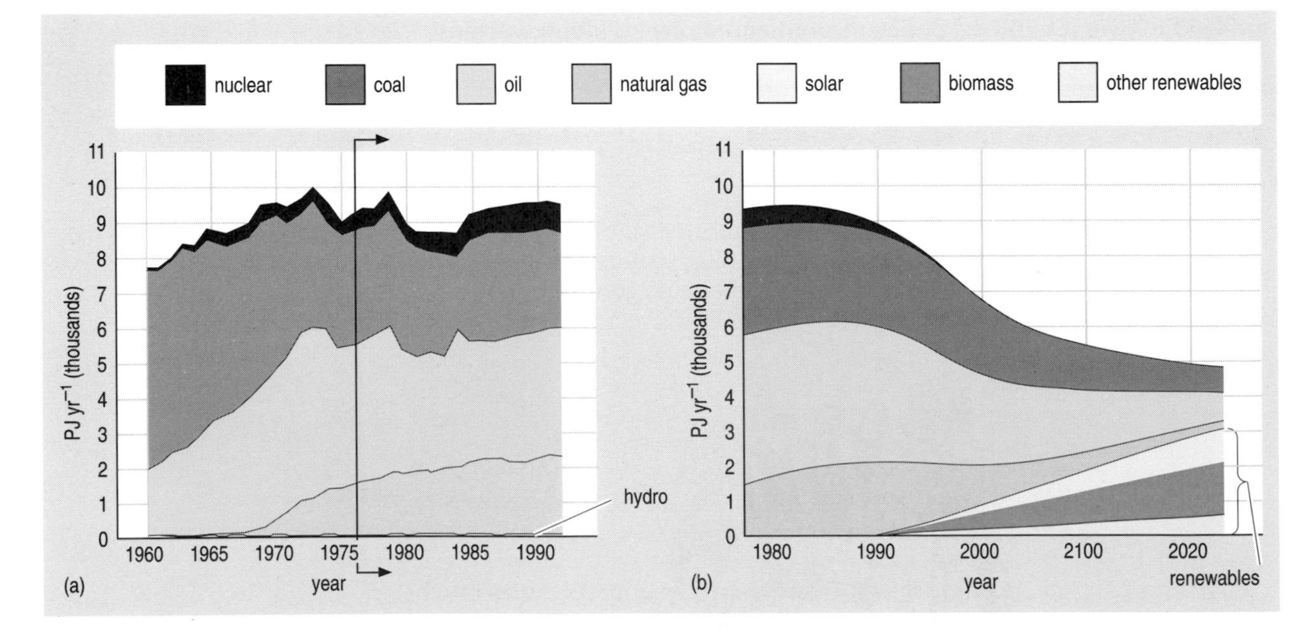

▲ *Figure 3.4*
UK total energy 1960–92 and projections of Earth Resources Research made in 1982.

In the UK, a 2.9-fold increase in national wealth was assumed, and a 40% reduction in demand by 2025 was found, with 50% from renewable sources. A comparison of these projections with actual energy use in the 80s is shown in Figure 3.4. These show that the formerly rising energy use has indeed been checked, as was projected. The *falling* energy use projected increasingly from about 1985 has not occurred so far, however. It has to be said that such similarity as can be seen is almost entirely a coincidence, as the main cause has been reduced economic activity, not the widespread adoption of the efficiency measures proposed in the ERR report (Olivier *et al.*, 1983). There are now, however, environmental pressures for reduced energy use, in addition to the resource shortages driving the 70s studies, so perhaps the reductions may be achieved, if for different reasons.

To achieve efficient energy use, a mainly renewable energy system was projected for all these countries, even the least favoured like Germany, the Netherlands, Denmark and the UK, which are cloudy, densely populated and highly industrialised. More recent studies, in the mid- to late-1980s, have been extended to the regional and global scale, and have included environmental aspects that were not considered in the earlier work. The Brundtland report of 1987, already mentioned, is an exhaustive account of the general possibility of 'sustainable development', with world energy policy based on a combination of demand reduction by efficiency measures, and supply essentially from renewable resources.

3.2 *Making it happen*

There can be no single worldwide prescription for the best way forward, but more efficient energy use seems an essential component of any strategy, as it automatically reduces environmental impact, whereas all supply options have some effect on the environment, considered below. The potential is unquestionably enormous for all countries. In effect, a vast 'source' of cheap energy is available, larger than the entire output of OPEC, and much more permanent. If energy use can indeed be halved or more in the developed countries, there could be a real objective of a more evenly balanced world energy use per head of 1–2 kW, which could accommodate a significant increase in the developing world. (Such a reduction in the US from 10 kW per head seems improbable, but in Europe a reduction from 4 kW is more conceivable, though massive attention to energy efficiency would be needed.)

Patterns of energy use, and energy waste, are deeply ingrained. Progress since 1973 has been patchy, and no country has tapped even a small fraction of the known potential. The most striking results have been confined to particular economic sectors in a restricted number of countries anxious to move away from dependence on oil. For example, in the Danish building stock, the primary energy used for space heating per unit floor area fell by 55% between 1973 and 1988, averaging 5% per year; in Sweden it fell by 50% between 1980 and 1988! Refrigerators sold in Denmark and the USA in 1988 consumed 50% less electricity per unit volume than those sold in 1973. The primary energy used to produce a ton of finished steel in Japan fell by 50% between 1973 and 1986.

The overall reduction in primary energy use per unit GDP was much less, however, as other sectors have had less vigorous or successful programmes. This is more true of other countries, like the UK, Norway and Italy, where there has been less, or less effective, government intervention.

The rate of progress also slackened after 1982, when energy prices fell. Energy use in Japan and the USA had been almost static since 1973, while economic activity rose by 40–50%, but in the mid-80s it began to rise significantly. In western Europe, primary energy consumption was well above its 1973 level by 1994 (though it wasn't in the UK, where GDP rise was low).

Countries each have their own social and cultural traditions, which have to be taken into account in developing energy policy, but there are some common strands. Some governments in Europe and North America own the energy suppliers – the gas and electric utilities, oil and coal industries. Even those that don't own them recognise the need to regulate the natural monopolies, like gas and electricity supply, on behalf of the consumer. It is important to ensure that funds for users to introduce energy efficiency measures are available on the same terms as those for suppliers, and that utilities carry out 'least-cost planning', supporting demand reduction by improved efficiency when this is cheaper than increasing supply.

Three conditions seem to be required for a programme of efficiency improvement to succeed: users need (1) up-to-date technical information tailored to their personal circumstances, (2) equitable access to funds and (3) equitable access to the necessary skilled work. These have not usually all been met at once, with disappointing results. However, those countries that *have* set up a combined advice, financing and installation service have had notable success.

Sweden is a shining example. In the late 1970s, the government set a target of insulating and weatherstripping old buildings to reduce their space heating requirements by 30% by 1990. The state provided initial start-up funds, but local government administered the grants and loans, and provided inspection, technical advisers who could be phoned or visited, and builder training programmes. At first, 100% grants were available, and later, as familiarity with efficiency techniques grew, these were replaced by loans. The results, monitored by a parliamentary committee, exceeded the 30% target.

Mandatory efficiency standards can be set for new vehicles, electrical appliances, industrial boilers, etc. The United States has for some years required appliances such as refrigerators to be labelled with their energy consumption, so that buyers can make an informed choice; surely this should extend to other countries. Minimum standards eliminate the most inefficient items from the market, but incentives are needed to encourage the use of the best available designs.

There will always be arguments about the extent to which a 'free market' will provide these incentives – if it *were* free. On the one hand all users do not have access to the necessary information, often quite technical; on the other hand prices of different forms of energy are often manipulated so that they do not reflect the relative costs of providing them, and there are long time delays, so that a decision to invest in a particular method forces the choice of what is available later. More fundamentally, costs have not usually reflected environmental effects, either of using up a limited capital resource or of pollution and waste disposal. These underlying large-scale, often political, influences in the energy supply market imply a need for balancing intervention to influence demand.

3.3 Summary

An outline of the innumerable detailed possibilities for energy efficiency improvement was given in Section 2, showing vast potential for reduced energy use in most sectors without loss of amenity. The way these can be reflected in perceived economic advantage is a more open question, seeming to require much improved information and/or government intervention by regulation, taxation or subsidy. Transport looks to be an increasingly critical issue. General studies of total national energy strategies, initiated in the first instance by the oil crises of the 70s, but now supported by environmental concerns, were described in Section 3. Some local success has been achieved, though not very much in world terms; the possibilities are clear, but the politics of their realisation remain problematic. Energy policy is, however, now firmly on national and world agendas, and as environmental concern as well as likely fuel shortages are increasing, it will remain of vital importance.

4 Energy supply

In this section, the options for primary energy supply, and the 'vectors' used to transfer energy from one place to another are briefly reviewed. The problems with fossil and nuclear fuels are outlined, and then the possible renewable (or continuous) resources are discussed in more detail.

4.1 Non-renewable and continuous resources; energy vectors

The ancient Greeks introduced the myth of Prometheus, who stole fire from the gods for human use, and was punished for his presumption. What was then meant by 'fire' we might now call 'energy', with wood, fossil fuels, fission and fusion getting progressively nearer 'divine' origins, and correspondingly more serious punishments. (You can speculate for yourself about the symbolism of what happened to Prometheus – chained to a rock and having his liver eaten out by an eagle each day, to grow again each night, until he was rescued by Hercules.)

More practically, we now make a basic distinction between **'renewable'** and **'non-renewable' resources** – really a distinction about time and space scales. The renewables depend on the large-scale position of the Earth in the solar system, which is bathed in radiant energy from the sun and experiences gravitational effects from sun and moon that have some regular fluctuations but only change significantly over millions of years. The non-renewables are the Earth-bound treasure chests of stored solar energy in fossil fuels, and stored nuclear energy of earlier origin in radioactive, fission and fusion materials; these stocks are necessarily limited, though currently very large.

Humans are readily seduced by offers of unlimited wealth – with a modest inheritance it is usually considered prudent to live off the interest, but if it is large enough it doesn't seem to matter if the capital is spent as well. We have taken this attitude with non-renewables, spending the capital of fossil fuels and first generation fission, but after only a few hundred years the end is in sight. Fast reactor fission, and even more so, fusion, offer a renewed dream of vastly more capital to use up; if they offer thousands of years they tempt humans again not to worry about resource conservation. In the case of fusion the energy source is believed to be just that which powers the sun, so the choice is between using the fusion that is already happening there, and struggling to tame it on Earth (making the Prometheus image particularly apt, with who knows what punishment to look out for).

More like a source of income are the renewable resources, though the term 'renew*able*' is not entirely appropriate, as the primary resource is not under our control. No doubt the term has its origins in the particular case of wood, and other biomass, where we can make a choice to produce more if we wish from the effectively continuous supply of solar energy (**continuous resources** is perhaps a better description). The same applies to tidal (gravitational) and geothermal energy. The latter does have a finite terrestrial origin, the limited stock of internal radioactive minerals continuously heating the Earth, but we cannot change the rate at which they do it, so if we take heat away we are only modifying the rate at which it reaches the surface. All the energy supplied from these three continuous sources is steadily re-radiated to space, with the Earth adopting the mean temperature which keeps this balance. If we 'make use' of some of the energy we modify the route by which it degrades to low temperature, but we don't change this overall balance, so the temperature changes we make will be local, not global. The release of energy from *non*-renewable sources *does* change the overall balance, though so far apparently not significantly. The indirect effects of greenhouse gas emissions more immediately make a difference to the way surface temperatures relate to those higher in the atmosphere. This indicates the general importance of understanding the delicate feedback processes near the Earth's surface which maintain temperature and other conditions favourable for life, rather than perturbing them further. Another Greek myth, that of the Earth-goddess Gaia, has been invoked in recent years to stand for the hypothesis that life on Earth influences the environment so as to ensure its survival – in an extreme form, a view of the whole Earth as, in some sense, resembling a living organism. While the supernatural overtones of this idea seem to antagonise as many people as are attracted by them, there is much more recognition of the existence of stabilising feedback processes in the relationship between biosphere and geosphere than there used to be, and a growing interest in trying to understand them better.

Another energy supply distinction is between these *primary* energy sources and the **energy vectors** used to transfer energy from the point of origin to the point of use. The original examples are hot water, as in a district heating scheme, and fuels: wood, coal, oil and gas storing chemical energy, and transported in bulk from forest or mine to the consumer – or, in the case of vehicles, *by* the consumer. There may be considerable processing in between, e.g. at oil refineries, and the transport may be by pipeline as well as road, rail or water; and there will be losses on the way which make the *delivered* energy less than the primary. Conversions from coal or biomass to gas or liquid fuels may become convenient for heating and transport uses, and will have their own plant and energy costs and losses to influence the stage at which they become economic.

Alternatively, hydrogen, already used as a rocket fuel, has been strongly advocated as a future chemical energy vector for fuel in the 'hydrogen economy'. Various 'safe' handling methods have been developed, which would remain to be proved in large-scale use. Hydrogen would be produced by the electrolysis of water using whatever primary energy source was available to generate electricity.

Undoubtedly hazardous would be the plutonium 'nuclear energy vector' of a fast-reactor based nuclear economy. Plutonium is a commodity analogous to a fuel, but highly toxic, and much sought after for weapons use – strong arguments, it might be thought, against its widespread adoption, though they have not deterred governments from extensive study of this possibility.

The energy vector we now all take for granted is *electricity*, unique in its ease and speed of transport, along the wires of the grid system. This lies alongside the 'mass' handling of the conventional transport system and the 'information' handling of the telecommunications system to form the three crucial networks of industrialised society. The grid has been associated with the development of large-scale electrical power *generation*, but this is not an essential link. If there are fair tariffs for purchase as well as supply of electricity, it can be seen as a 'common carrier' for diverse suppliers and users of all sizes, balancing local and temporal fluctuations. Improved electronic communications about 'spot-prices' to reflect rapid supply and demand changes may bring about a more responsive 'market' system. This should have a bearing on the development of CHP generation, since the supplier who also has a market for otherwise wasted heat should be able to supply electricity more cheaply. (The difference between primary and delivered energy is particularly large for non-CHP power plant, as some two-thirds of the primary energy is wasted.) Renewable energy of all forms is often readily converted into electricity, so its development would also be encouraged by a grid with a fair tariff system.

Electricity can of course also be transported by conventional 'mass' methods when stored as chemical energy in batteries. While this is obviously convenient for small portable equipment (lamps, electronics) it is limited for supplying energy to vehicles by the energy/weight ratio of present batteries compared with chemical fuels. This may change with progress in electrochemistry, along with complementary improvements in fuel cells. In this case the energy vector would still be a suitable chemical 'fuel', but motive power would be obtained by the consumption of this fuel in a fuel cell to generate electricity to drive an electric motor. This would replace the present combustion of fuel to produce heat, which is converted into mechanical energy in a heat engine (with its thermodynamic efficiency constraints). The 'fuel' supplied to the fuel cell might not be immediately recognisable, as the electrochemical reaction used might not be the reaction with oxygen we call combustion. It would again have to be produced specially as an energy vector from some primary energy source.

Large vehicles like trains do better not to carry their energy with them but to pick it up from fixed wires or track – or more ingeniously from distributed electromagnets providing both levitation and propulsion. Still more fancifully, we can speculate about chemical energy vectors chosen to energise 'chemical motors' giving direct mechanical output as muscles do, not via heat or electricity. Maybe there will be analogous nuclear batteries, fuel cells or motors; or even some completely novel energy vector derived from improved understanding of nuclear processes. It should be recalled that chemical combustion and heat engines were known about long before nineteenth-century exploration of more subtle processes like electrolysis led to our present understanding of electricity, with all its subsequent applications. Perhaps there will be some equivalent advance in the application of nuclear processes to replace their present relatively crude use to produce heat.

Activity 3

Look back over Section 4.1 to make sure you can distinguish between the various forms of primary energy (non-renewable and continuous) and energy vector (existing and potential future types).

4.2 Fossil fuel limitations

The fossil fuel industries (coal, oil and gas) are deeply interwoven into the economic and employment structures of world trade and industry, so they will inevitably continue to contribute significantly to energy supply over the next century (with controls on their polluting effects), however much we theorise about it. We have already seen that resource limitation is bound to create shortages, which will be reflected in prices for oil and gas, less so for coal. One way or another the developed world will adapt, though conflict with developing countries will be hard to avoid as their internal demands increase and willingness to export their own resources reduces. A new factor in the manoeuvring is the enhanced greenhouse effect. Current concerns about the link between carbon dioxide emissions and climate change are leading to strong institutional pressures within and between countries to reduce emissions, by some combination of regulation, taxation and subsidy. This is a further source of international tension, as the developed countries are primarily responsible for present greenhouse gas levels, but the likely increase from developing countries (unless different energy policies are adopted) will dominate in future – wise counsels will be needed. A short-term effect of a carbon tax, say, to penalise carbon dioxide emission, will be to shift fossil fuel use, where possible, from coal to oil to gas, as that reduces emission per unit energy release, because of the different carbon–hydrogen ratios in the three types of fuel. Because this is just the direction in which available resources reduce, on present evidence, such a shift will make shortages worse.

A separate issue about fossil fuels is their value, not simply for their total energy content, but for their particular chemical structure, which makes them useful as chemical feedstocks. A large part of the chemical industry (dyestuffs, pharmaceuticals, and particularly plastics) is dependent on coal and oil for its raw materials.

The data on UK use of energy resources in Chapter 1 indicate that only some 1.6% of total delivered energy, 7.4% of that to 'other industry', is for chemical feedstocks. Thus this 'premium' use for these materials is only a small, but significant, proportion of industrial energy use. If there are to be shortages it would seem preferable to reserve them for this purpose, when they would last much longer. However, price will be the determining factor: if it rises it will of course be reflected in the resulting product prices.

4.3 Nuclear energy limitations

Three general possibilities were outlined in Chapter 1. Two are based on the **fission**, or splitting, of the nuclei of heavy elements. In burner reactors, the only type in commercial use, the essential fuel is a particular isotope of uranium, only present as 0.7% of natural uranium. In breeder reactors, demonstrated in principle but nowhere commercial, the uranium is better used by converting its main isotope into plutonium, which then becomes the essential fuel. The third possibility is **fusion**, the combining of the nuclei of light elements, such as the deuterium and tritium isotopes of hydrogen. This has not yet been demonstrated in a controlled form.

Fission

In spite of vast research and development spending by developed countries worldwide, and the euphoria of the 1950s about 'electricity too cheap to meter', the 1980s saw a general disenchantment with nuclear power as a solution to energy supply. This in spite of the established position of the nuclear power industry, with strong links to government decision-making (not least because of the overwhelming political significance of nuclear weapons). If nuclear power is losing favour it will not be for lack of encouragement and funds.

Doubts about economics had already led US utilities to stop ordering reactors before the Three Mile Island accident (1979). The Chernobyl accident (1986) strengthened opposition in Europe. The UK withdrew its existing stations from the 1989 electricity privatisation on cost grounds; the renewed proposals of 1996 only involve the most recent stations, and new construction now seems unlikely. Only France among European countries, with some 75% of its electricity produced by 55 nuclear reactors (in 1990), has a continuing strong commitment (Figure 3.5) – it will be interesting to observe developing views of reliability and costs there too.

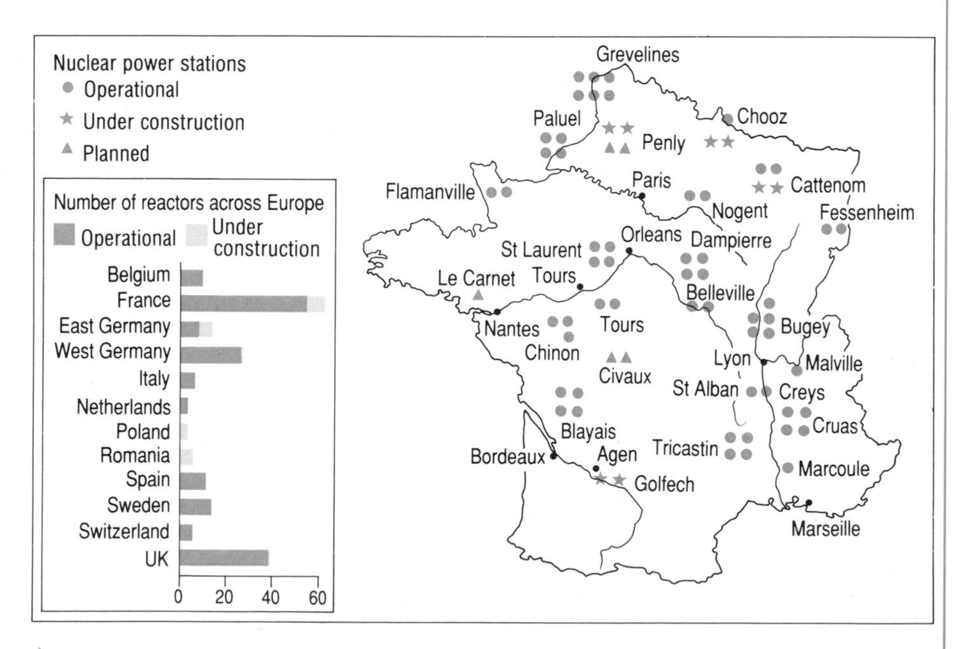

Figure 3.5 French and European nuclear power in 1990.

More generally, as already noted, the uranium resource for this 'burner' use is less than that of oil – Sir Alan Cottrell in 1988 put it at 'about half that of the world's oil'. Assuming continuing growth, 'it is quite possible by 2030 and virtually certain by 2050 that any remaining pressurised water reactor stations will then have to stand idle because of prohibitively expensive fuel costs.' Without the anticipated growth, of course, the availability of uranium is not a present limitation. At the other end of the process, the safe disposal of radioactive wastes continues to be a problem everywhere, as Chapter 7 will consider in more detail.

The limited uranium resource was long ago the reason for proposing the 'fast breeder', which certainly stretches the resource to be as good or better than coal, but introduces a further set of cost and environmental hazards. The UK research and development (R&D) spending on its fast reactor programme since 1955 is shown in Figure 3.6, where it is compared with total renewables R&D over the same period. In spite of this massive effort, funding was cut in 1988 and there are no plans for a practical reactor. Quite apart from these evident technical difficulties in producing a safe and economic design, there are the major implications already mentioned of reliance on a system involving the production and trading of large quantities of the highly dangerous plutonium.

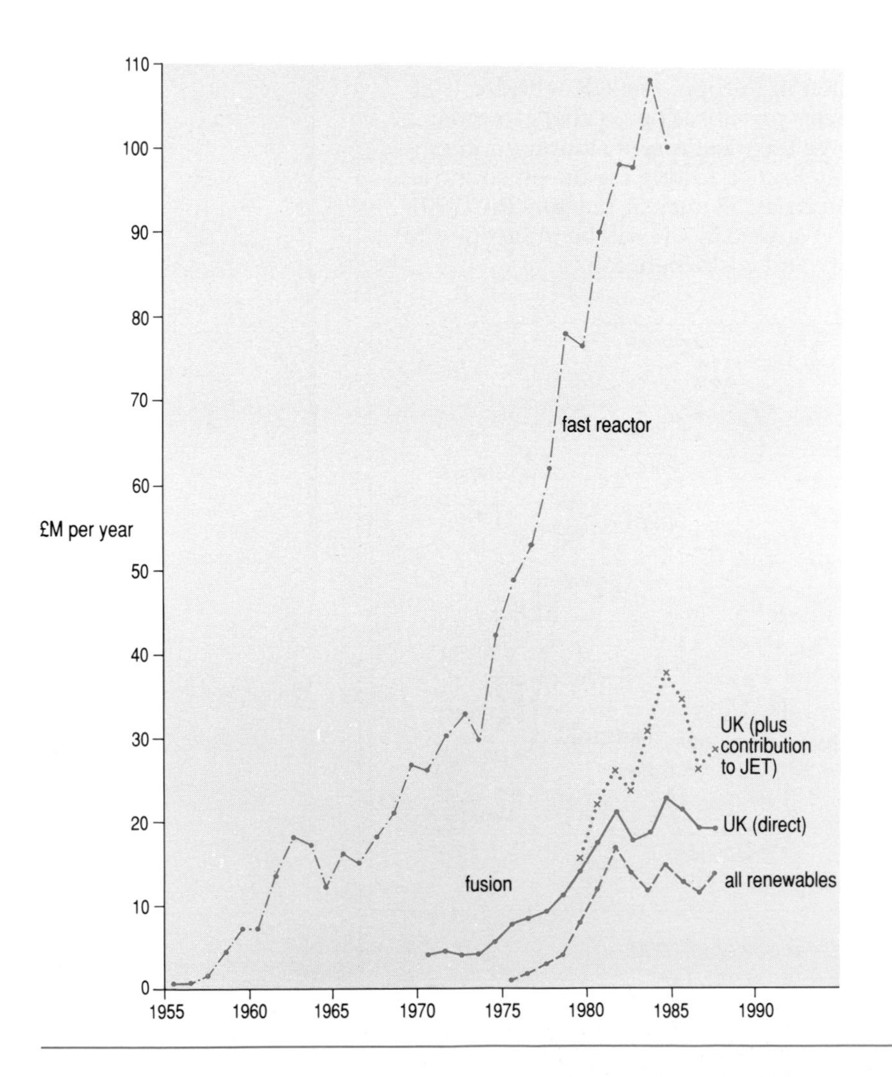

◄ *Figure 3.6*
Some UK Department of Energy research and development budgets.

Fusion

This is in a different category from all the other options, as there is as yet no evidence that a practical reactor can be made at all, let alone any details of its economics and unforeseen difficulties. The attraction is a theoretical one, which has again provided massive research funding worldwide. The UK figures are also shown in Figure 3.6. The appeal is so powerful to a society dependent on large-scale electricity generation from concentrated energy sources that it is unlikely that work on the problems will, or should, cease, but policy priorities should surely regard it as a long-term possibility meriting basic physics research rather than engineering commitment. The assumption that its application would inevitably be in thermal power stations has focused emphasis on large expensive installations as well as exploratory studies. A typical large research machine, the Joint European Torus, JET, is shown in Figure 3.7. (The surge of interest in 'cold fusion' in 1989 should perhaps also be mentioned. Although this particular idea seems to be a dead end, it emphasises that all is not understood, and that some completely novel application of nuclear energy may yet emerge – also that hope springs eternal for some magically simple energy source.) In contrast, renewables, requiring inventive engineering using *established* physics principles, have found the funding to move from exploratory studies to larger scale prototypes hard to obtain.

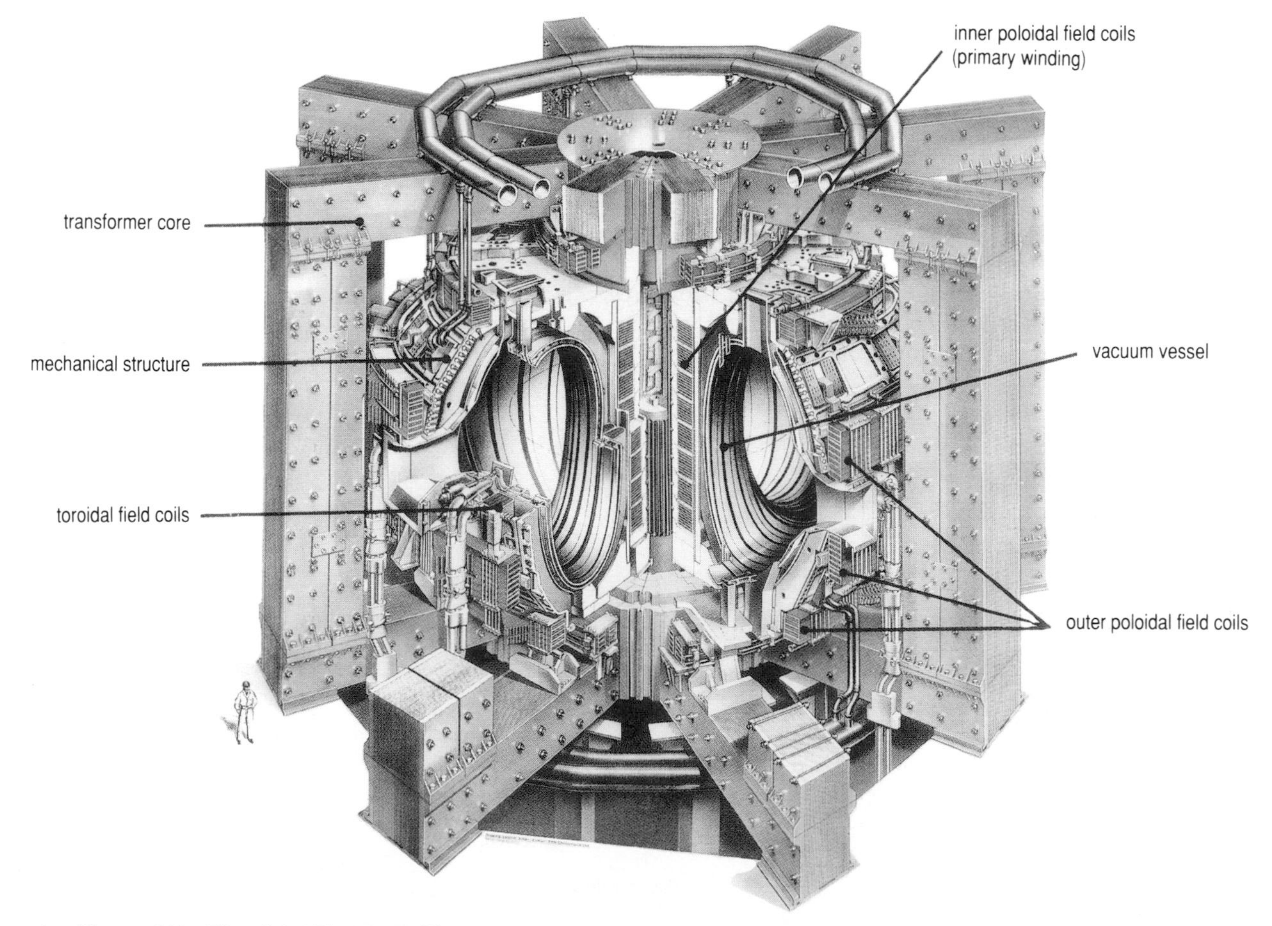

▲ *Figure 3.7 The Joint European Torus.*

4.4 Continuous (renewable) energy resources

We now come to a more detailed review of the continuous resources. They have always been available and used in traditional ways; they were displaced by the exploitation of fossil fuels in the industrial revolution and of nuclear energy in the twentieth century for what seemed like good reasons of cost and convenience. Can they be reintroduced in a twenty-first century industrial, or post industrial, society?

Solar energy

Direct

Radiation from the sun (infra-red, visible light, ultra-violet) can be used directly to produce heat, as has always happened, or less obviously to produce electricity. The most basic form of heat production is achieved by re-examining traditional practices of building design with more scientific analysis and modern materials – the **passive solar** design, already mentioned above under efficiency. Orientation to the sun, window design and heat flow through the structure are arranged to make the best use of the winter sun without summer overheating. It is closely linked to insulation and ventilation control in an integrated low-energy design, the importance of the 'solar gain' part obviously varying with climate.

When solar heating systems also include moving parts such as pumps, they are called **active solar**, the simplest example being the 'flat plate' collectors for water heating often seen on roofs in sunny countries (Figure 3.8). More advanced (and expensive) versions have focusing mirrors or lenses and computer controlled tracking to make better use of the sun. They are proving economic for heat production even in relatively cloudy parts of north-western USA, and in the sunnier conditions of south-western USA they can raise steam for electricity generation (Figure 3.9) competitively with fossil fuels.

A much simpler idea, with its own design subtleties, is a **solar pond**: essentially a pit with a dark coloured base filled with brine of carefully controlled concentration. Because the density of brine increases with its

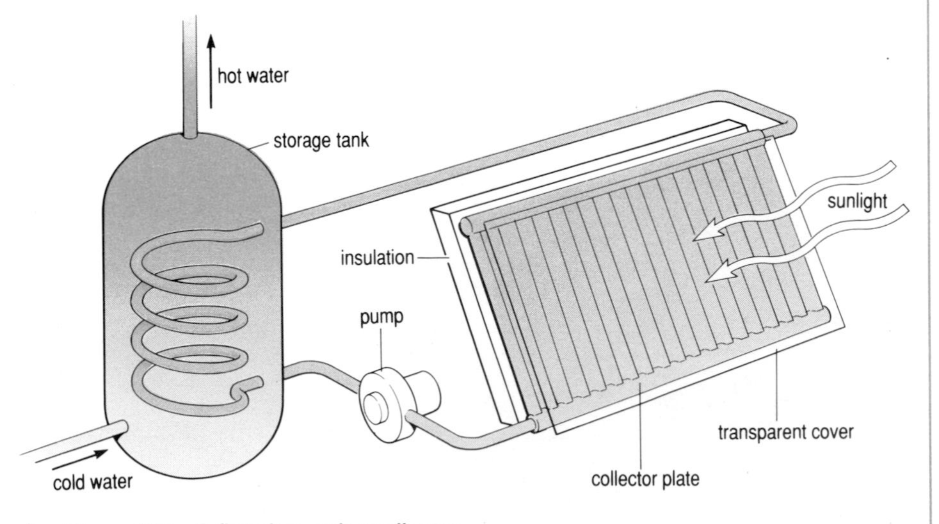

▲ *Figure 3.8 A flat plate solar collector.*

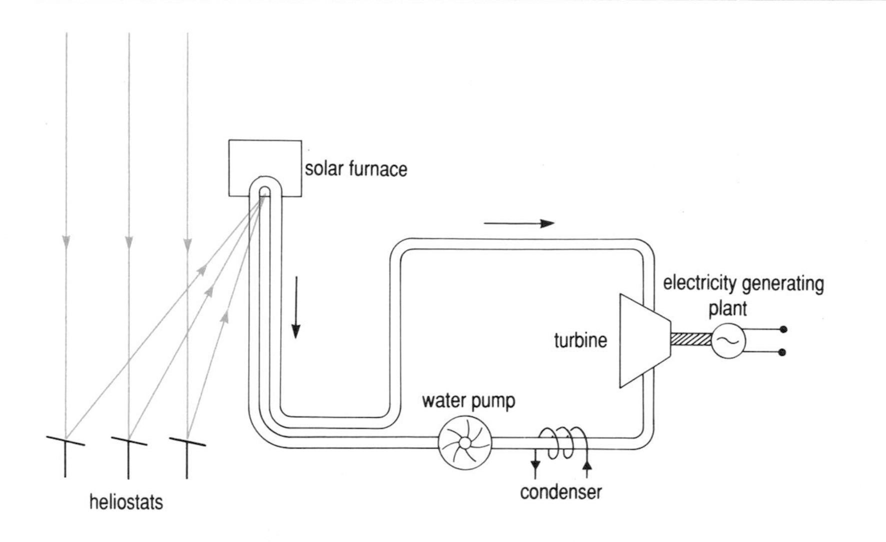

◄ *Figure 3.9*
A solar power plant in California.

concentration, it is possible to maintain a variation in concentration, from very high at the bottom to pure water at the top. Incoming solar radiation is absorbed at the dark bottom surface, heating the lower brine layers to 100 °C or more. In a suitable design this heated brine remains at the bottom, without convection; the only heat loss is by conduction through the purer water above. This is low enough for the whole structure to act as an integrated heat collection and storage system from which heat may be extracted as required. In sunny regions such as Israel and southern USA, well-designed solar ponds appear to be competitive heat sources, and it is claimed that electricity generation from them will also be competitive if fossil fuel prices rise again to the levels of the early 1980s.

A more distinctively high-technology development is the direct generation of electricity from sunlight using **photovoltaic devices**. These use semiconductor materials resembling those of electronics, and are used extensively for small-scale applications like exposure meters and pocket calculators, and in space, where cost is not a limitation. A great deal of development, particularly in Japan and the USA, has been directed to reducing the cost of manufacturing large-area panels of high enough efficiency to meet more general energy needs – the economics are beginning to be favourable in special circumstances, e.g. remote sites with modest energy demands in sunny climates. It seems likely that further cost reductions will gradually extend these circumstances. Hybrid devices in which *both* heat and electricity are produced in a domestic scale 'solar CHP' system are also possible; again costs will determine whether, or when, these become available.

Indirect

The longest tradition in the indirect use of solar energy is in the burning of fuelwood, still believed to account for some 15% of world energy use. More generally, photosynthesis produces plant material, and hence animal material, of all kinds, collectively known as **biomass**.

Q How much of the incoming solar energy is converted into biomass?

A Solar radiation = 5.4×10^{24} J/yr = 1.7×10^{17} W (from Chapter 1).
Photosynthetic energy stored in biomass = 1×10^{22} J/yr
= 3×10^{14} W = about 0.2% of solar energy.

Q How does this compare with total world energy demand?

A About 30 times greater – present demand is about 10 TW = 10^{13} W.

So this is a large resource, though, like the primary solar energy, it is dispersed over the Earth's surface, and requires concentration to meet our needs – at its least satisfactory, by individuals collecting wood over a large area. More systematically, 'energy crops' (fast-growing trees or other plants) can be grown in forests or plantations, harvested mechanically, and marketed directly, compressed or converted into gaseous, liquid or solid fuels of higher energy density by controlled heating, anaerobic digestion by bacteria, etc. A wide range of possibilities is being explored in different countries, the growth of sugarcane in Brazil to produce alcohol as a vehicle fuel being a well-publicised example. As always, economic factors will determine their success, in ways dependent on the system chosen, alternative fuel prices and competing land uses. Energy crops are usually seen as competing with food crops, and may also be visually unattractive, as with timber forestry and intensive agriculture. Agroforestry systems seek to integrate timber, food and energy production in ways that take advantage of complementary species as in natural ecosystems, and aim to be at once economic and attractive – an appealing objective that may have increasing impact.

A large amount of biomass suitable for conversion into heat and fuels is already available as wastes from agriculture, industry and the home: 'dry biomass' as straw, wood and refuse, 'wet' as animal slurries, sewage, and food industry residues. For the UK this is estimated as 10% of fossil fuel demand. Some countries already make substantial use of this resource: Denmark, for example, uses about 75% of available wastes. The UK has a large potential for avoiding waste and environmental damage which is only just beginning to be realised. The burning of straw or the dumping of sewage are direct nuisances. Burying refuse leads to the hazard of producing methane, which is potentially explosive and is a 'greenhouse gas' but can be recovered and used as 'landfill gas'. Refuse incinerators can also be polluting, but well-designed plant of various types can produce heat, methane and/or refuse derived fuel (RDF) without adverse effects (Figure 3.10).

The second group of indirect solar methods derive from the hydrological and weather systems: the evaporation and precipitation of water, the vertical and horizontal movements of air masses, driven by solar heating and modified by the Earth's gravity and rotation, i.e. water and wind power. The first of these has a long history, as the water power that

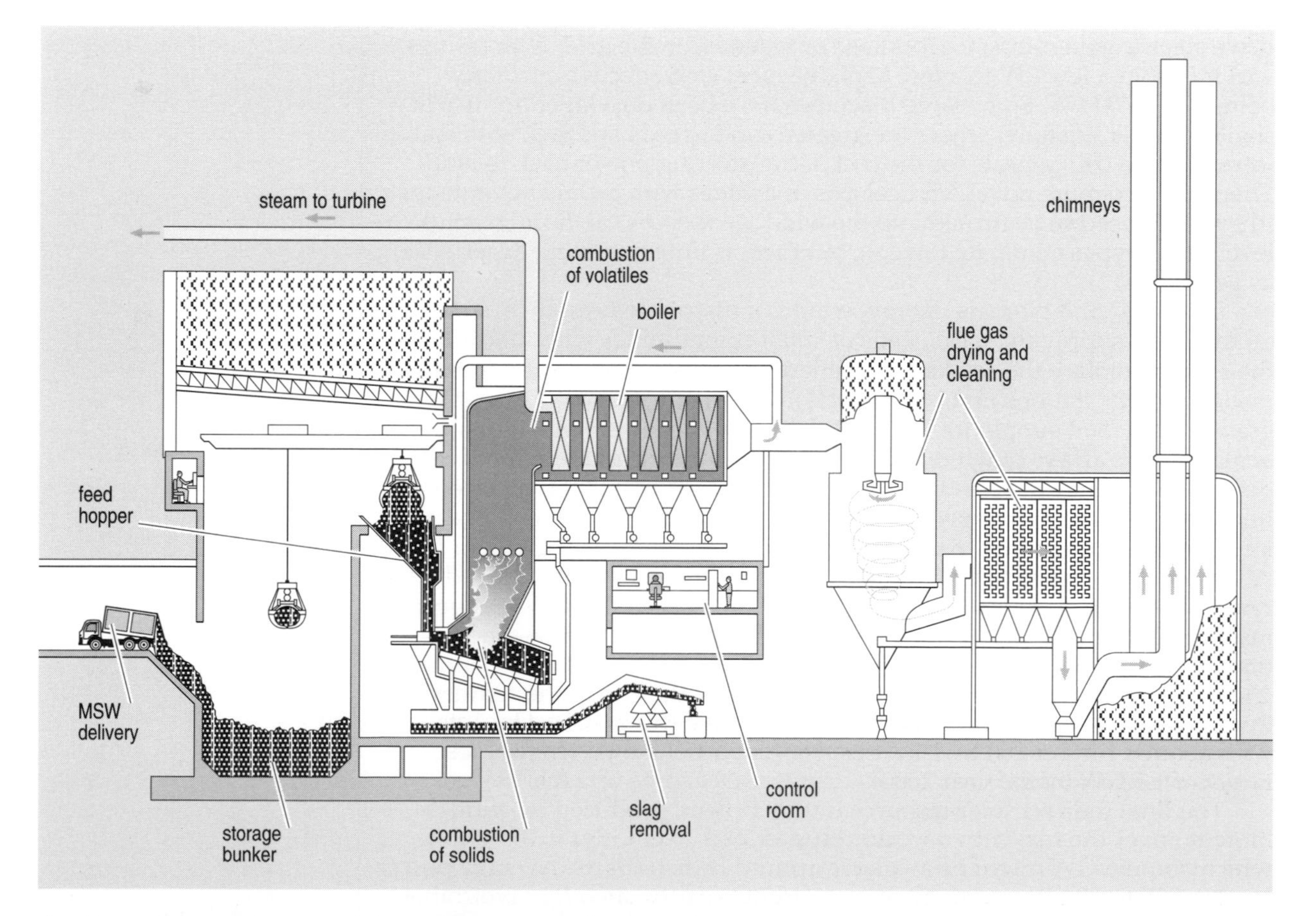

Figure 3.10 A large municipal solid waste combustion plant.

preceded fossil fuels in the industrial revolution, and more recently as the **hydroelectricity** that currently supplies some 4% of world energy needs. These depend on the kinetic energy of falling water, in turn derived from its potential energy in a mountain lake or reservoir. The power available depends on the 'head', the height through which the water can fall, as well as the volume of water available, so mountainous countries like Norway and Switzerland make more use of hydroelectricity. The high costs and engineering skills involved in large schemes have concentrated them in Europe and North America, which have already exploited most of the available sites. Many more remain elsewhere: a world resource of some 2 TW, 20% of energy demand, has been estimated. Large dams have figured significantly in aid schemes to the developing world, though they have been much criticised for their environmental effects and inappropriateness to local needs. In developed and developing world alike there is also considerable scope for the use of smaller turbines in fast flowing rivers, which may satisfy local needs better.

The second weather-derived energy source again has a long tradition, as windpumps for water and windmills for corn, though the use of modern forms of **wind turbine** for electricity generation is comparatively recent. The total energy involved in the world wind pattern is about 3×10^{23} J/yr $= 10^{16}$ W, but of course much is higher in the atmosphere than can be tapped by Earth-bound turbines: a maximum UK resource of 2×10^{18} J/yr = 60 GW has been estimated, for example. Most current machines are 'horizontal axis', with blades influenced by aircraft propeller design; they drive electric generators, for local use or to feed into the grid, with ratings varying from a few kW to a few MW, the most cost-effective currently being 200–1000 kW. Some large machines have been developed for use in shallow water offshore, where the greater wind speeds and lack of visual intrusion can compensate for the cost of transferring power back to land. There are also more novel 'vertical axis' machines with certain advantages (they don't need to be turned into the wind, generators can be at ground level, some types eliminate towers), but there is little operating experience as yet (Plate 12).

Like solar and biomass energy, wind is a distributed resource, so while all three can supply the local needs of small communities without difficulty, their use to replace the concentrated energy supplies provided by fossil and nuclear power stations involves collection over large areas. We take it for granted that food supply for urban communities requires large farms; by analogy large arrays of wind turbines are called **wind farms**. A modern power station may have a rating of 1.3 GW, which can only be replaced by hundreds of large, or thousands of medium-sized wind turbines. Several large arrays of medium-sized turbines have been built, particularly in the Altamont Pass region of California, where tax concessions encouraged such renewable developments (Figure 3.11); by 1991 a total of 10 GW had been installed in the USA. It was the operating experience with these schemes that led to the confidence now placed in medium power designs. Denmark and Holland plan to have over 1 GW installed by the end of the century. Recent UK developments mean that some 60 MW was installed by 1994, with about a further 200 MW in prospect (as part of a total renewables target of 1.5 GW by the year 2000).

The final indirect solar resource is *wave* power, produced by wind blowing over the sea, with a world resource of 3×10^{22} J/yr $= 10^{15}$ W, of which about 5 GW might be available around British coasts. Average power levels of 15 to 70 kW per metre of wave front are estimated for favourable regions around the UK, so tens of kilometres of wave energy conversion

Figure 3.11
A wind farm in California.

machines are needed to match the output of large power stations. The machines must be able to withstand adverse and very variable conditions (peak power levels hundreds of times higher than average). A number of promising designs have been explored ('nodding duck', hinged raft, oscillating water column), generating electricity which can be transmitted to shore by cable (Figure 3.12). Various prototypes on scales up to a megawatt or so have been tested in the UK, Norway, Japan and the USA on or near the shore, but the feasibility of a full-scale deep-water system remains to be proved. Models of deep sea systems on a 1/10 scale were tested in the UK in the early 80s, but work stopped in 1982. A 1992 reappraisal by the Department of Trade and Industry concluded that current designs were not likely to become economically competitive in the short term; continuing R&D is believed to be justified in the longer term.

◄ *Figure 3.12*
Wave machines:
(a) raft,
(b) duck.

Tidal (gravitational) energy

The exploitation of tidal energy resembles that of hydropower – a high tide is trapped behind a dam-like structure or 'barrage' across a suitable estuary, the water then being passed through turbines to generate electricity. The effective 'head' of water is much smaller than in hydroelectric schemes, but the volume of water is much greater. The water flow can be arranged to generate on the ebb (falling) tide only, or on the flood (rising) as well. Successful pilot schemes have been demonstrated in the Russian Federation and France, where a 240 MW plant has operated for

Tidal scheme performance (estimates, 1986)

	Mean tidal range/m	Barrage length/m	Installed capacity/MW	Annual energy output/GWh	Cost of energy/(p/kWh)
Severn – Inner line	7.0	17 000	7 200	12 900	3.7
Severn – Outer line	6.0	20 000	12 000	19 700	4.3
Morecambe Bay	6.3	16 600	3 040	5 400	4.6
Solway Firth	5.5	30 000	5 580	10 050	4.9
Dee	5.95	9 500	800	1 250	6.4
Humber	4.1	8 300	1 200	2 010	7.0
Wash	4.45	19 600	2 760	4 690	7.2
Thames	4.2	9 000	1 120	1 370	8.3
Langstone Harbour	3.13	550	24	53	5.3
Padstow	4.75	550	28	55	4.2
Hamford Water	3.0	3 200	20	38	8.5
Loch Etive	1.95	350	28	55	11.7
Cromarty Firth	2.75	1 350	47	100	11.8
Dovey	2.90	1 300	20	45	7.2
Loch Broom	3.15	500	29	42	13.9
Milford Haven	4.5	1 150	96	180	10.0
Mersey	6.45	1 750	620	1 320	3.6

Source: A. C. Baker, ICE Symposium Paper, Oct. 86.

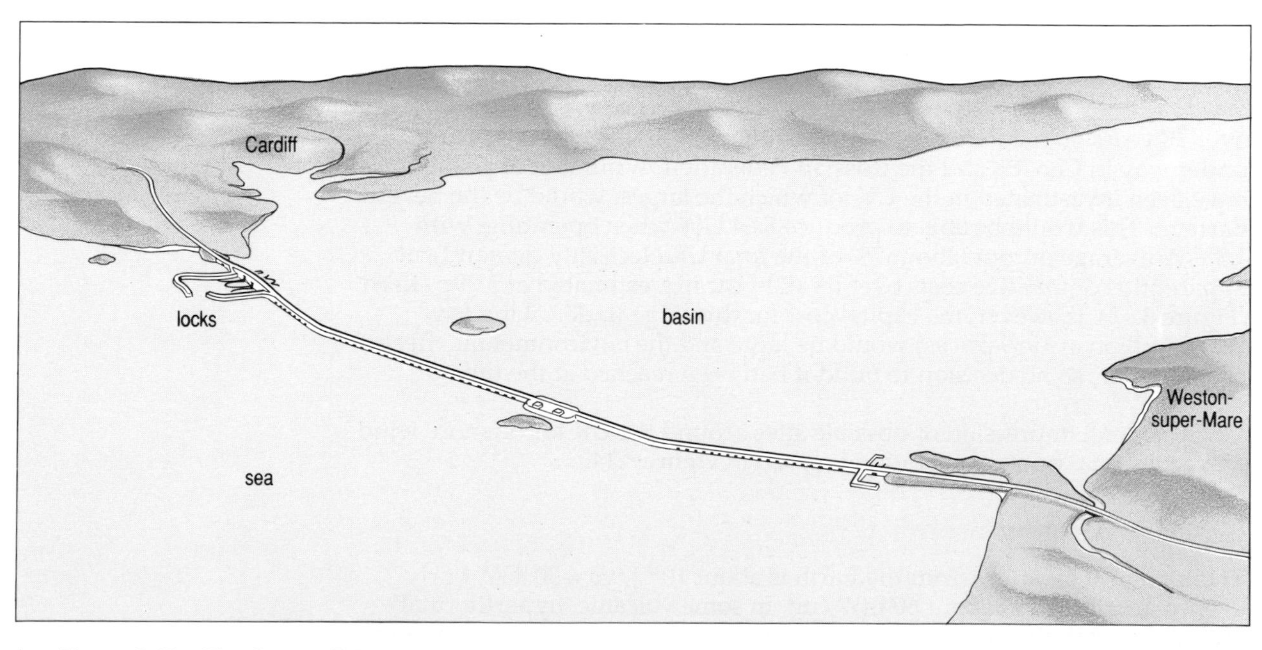

▲ *Figure 3.13 The Severn Barrage.*

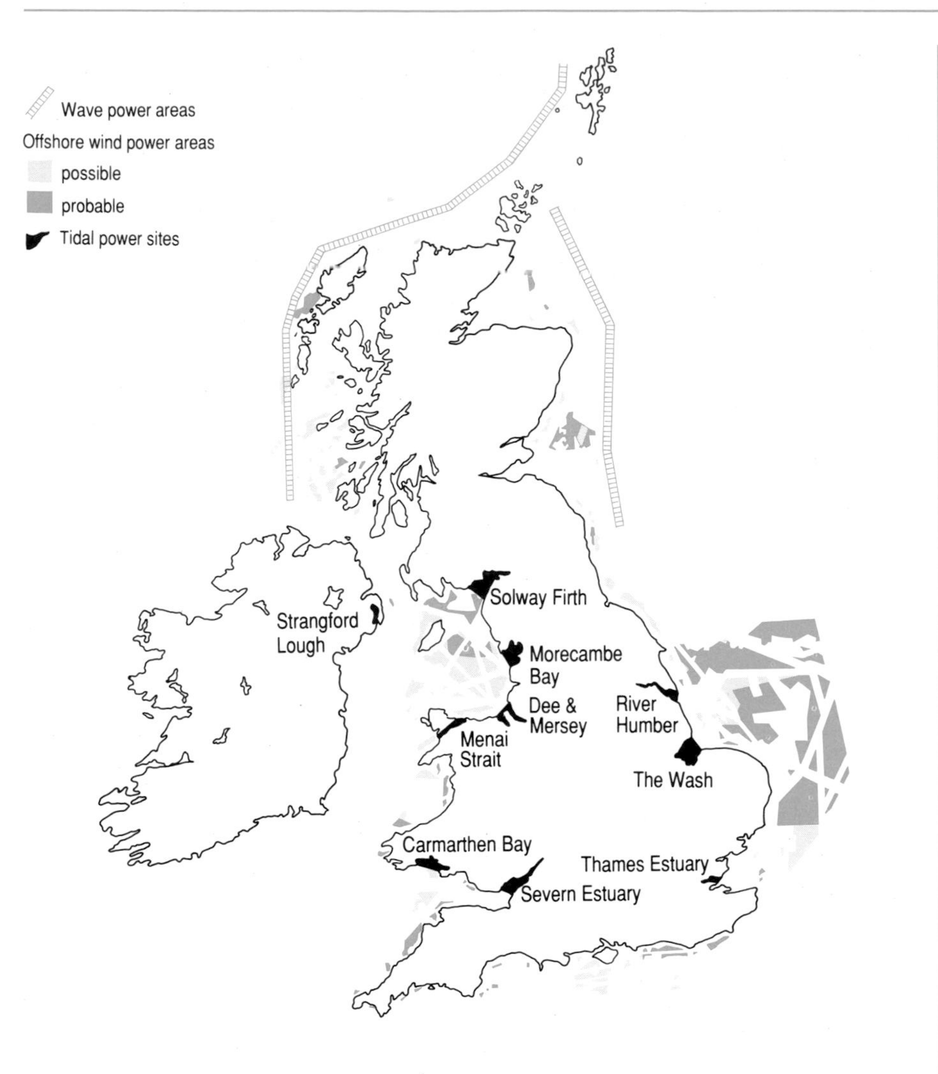

Figure 3.14 *Possible sites for offshore wind, wave and tidal sites in the UK.*

over 20 years on the Rance estuary in Brittany. Major new projects are under way in Canada and the Russian Federation. A number of possibilities have been investigated in the UK, of which the largest would be the Severn Barrage. This would be able to produce 8.64 GW when operating, with 1.3 GW average output (about 7% of the total UK electricity demand), at apparently competitive costs over its 120-year life, estimated at 3.79p/kWh (Figure 3.13). However, the capital cost for this large undertaking (say £8500 million at 1989 prices) would be large and the environmental effects considerable, so no decision to build it had been reached at the time of writing.

A general impression of possible sites around the UK for offshore wind, wave and tidal power generation is given in Figure 3.14.

Geothermal energy

The total heat escaping from the Earth is about 10^{21} J/yr = 30 TW, fairly evenly distributed at about 60 mW/m^2. In some volcanic 'hyperthermal' areas, however, this rate can be much greater, so much so that steam can be tapped from boreholes and used for space heating, industrial processes and

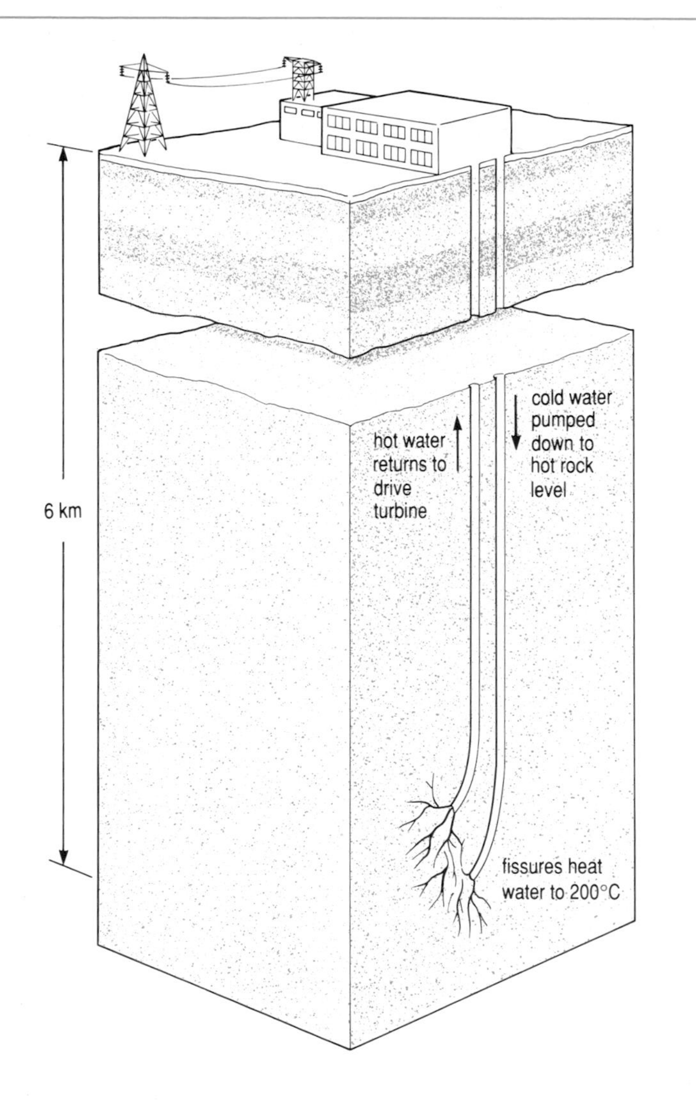

Figure 3.15
A hot dry rock geothermal station.

in power stations. The total world electrical capacity from such systems in 1990 was almost 6 GW, and non-electrical about 4 GW; the ultimate potential is perhaps 360 GW. In other 'non-thermal' regions, favourable geological conditions can provide useful heat from below the surface. If the surface rocks have better than average thermal insulation, underlying porous rock may contain hot water – such aquifers 1–2 km below the surface have been exploited for district heating in Paris. Alternatively, underlying rocks may have a higher than average amount of radioactivity and therefore an enhanced temperature; these 'hot dry rocks' lie at about 5 km depth below the surface in south-west England, for example. Techniques have been explored for fracturing such rocks at depth and pumping water through them to make this a usable resource (Figure 3.15). The total energy available in the UK from this source is said to be comparable to the coal reserves.

Potential and costs

A much more detailed account of all these renewable techniques and their economic potential is to be found in the Open University course

T521 *Renewable Energy* (Open University, 1994), but is outlined here. Their relative merits are hotly argued by their protagonists. The economics vary in different countries and with time, depending on the scale of R&D funding. The earlier comparative UK spending on fast reactors, fusion and renewables, for example, has already been noted (Figure 3.7); a comparison between 1986 government R&D spending in various developed countries on renewables and other forms of energy is given in Table 3.2, showing the UK level then to be exceptionally low. So it is unlikely that there will be any final merit order; it can change with new developments worldwide. The *total* resource is in many cases a significant proportion of total demand; where opinions and estimates differ is in the rate at which renewables can be introduced in competition with existing methods, and the reduction of demand that might be achieved by better efficiency, so that the renewables proportion increases.

An estimate of the potential of renewables for electricity generation (not other energy requirements) in the UK was provided in 1992 by the government's Renewable Energy Advisory Group (REAG, 1992). The 'accessible potential' of various different techniques is shown in Figure 3.16, giving a total three times larger than present conventional supply. However, the amount that will actually be introduced depends on cost factors – one estimate of the potential contribution by 2025 is shown in Figure 3.17 – evidently a significant proportion of total supply will only be reached if prices of 6–8p/kWh are acceptable, which are considerably above conventional prices. It was felt that a 20% contribution might be more realistic, and the allowance for the environmental disadvantages of conventional methods might be made by some system of subsidy for renewables (see the discussion of the NFFO below).

The European Union ALTENER (Alternative Energy) programme produced the targets for Europe, for both electricity and heat production, shown in Table 3.3, with a 7.8% share for renewables by 2005. On the world scale, a UN report to the Rio UNCED conference in 1992 (Johansson *et al.*, 1993) for a 'Renewables-Intensive Global Energy Scenario' (RIGES)

Table 3.2 Government R&D spending on renewables in selected countries, 1986

Country	Renewables R&D spending/$\$10^6$	Share of energy R&D budget/%	Spending per capita/$
Sweden	17.3	21.8	2.06
Switzerland	10.2	14.7	1.57
Netherlands	17.0	10.6	1.17
West Germany	65.9	11.6	1.09
Greece	9.7	63.2	0.97
Japan	99.2	4.3	0.82
United States	177.2	7.8	0.73
Italy	29.5	3.9	0.52
Denmark	2.6	17.8	0.51
Spain	19.4	27.6	0.50
United Kingdom	16.6	4.4	0.29

Sources: International Energy Agency (1987) *Energy Policies and Programmes of IEA Countries: 1986 Review*, Organisation for Economic Co-operation and Development, Paris; Population Reference Bureau (1986) *1986 World Population Data Sheet*, Washington, D.C.

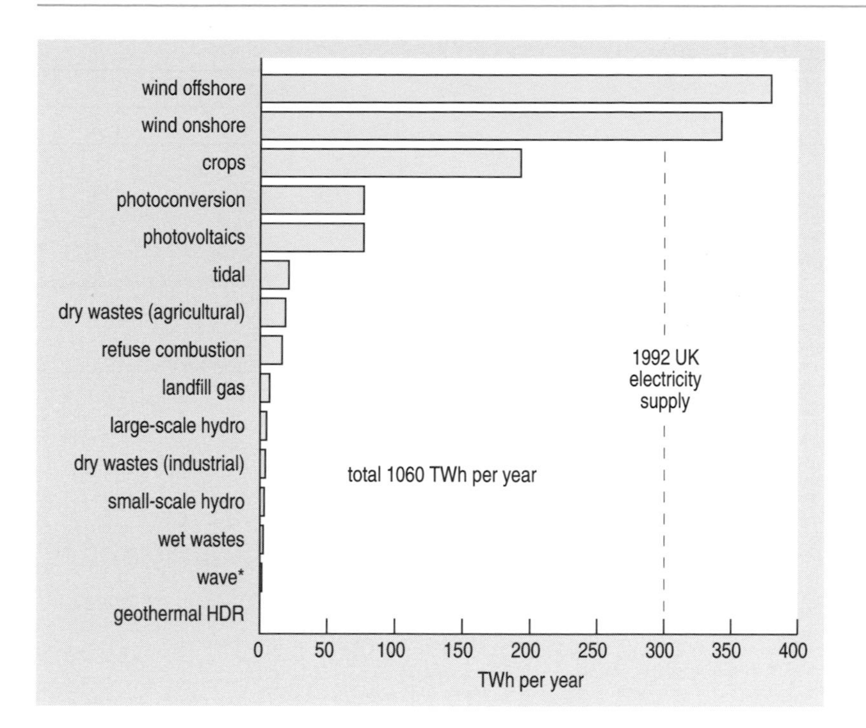

Figure 3.16 Accessible potential of electricity-producing renewable energy technologies in the UK.

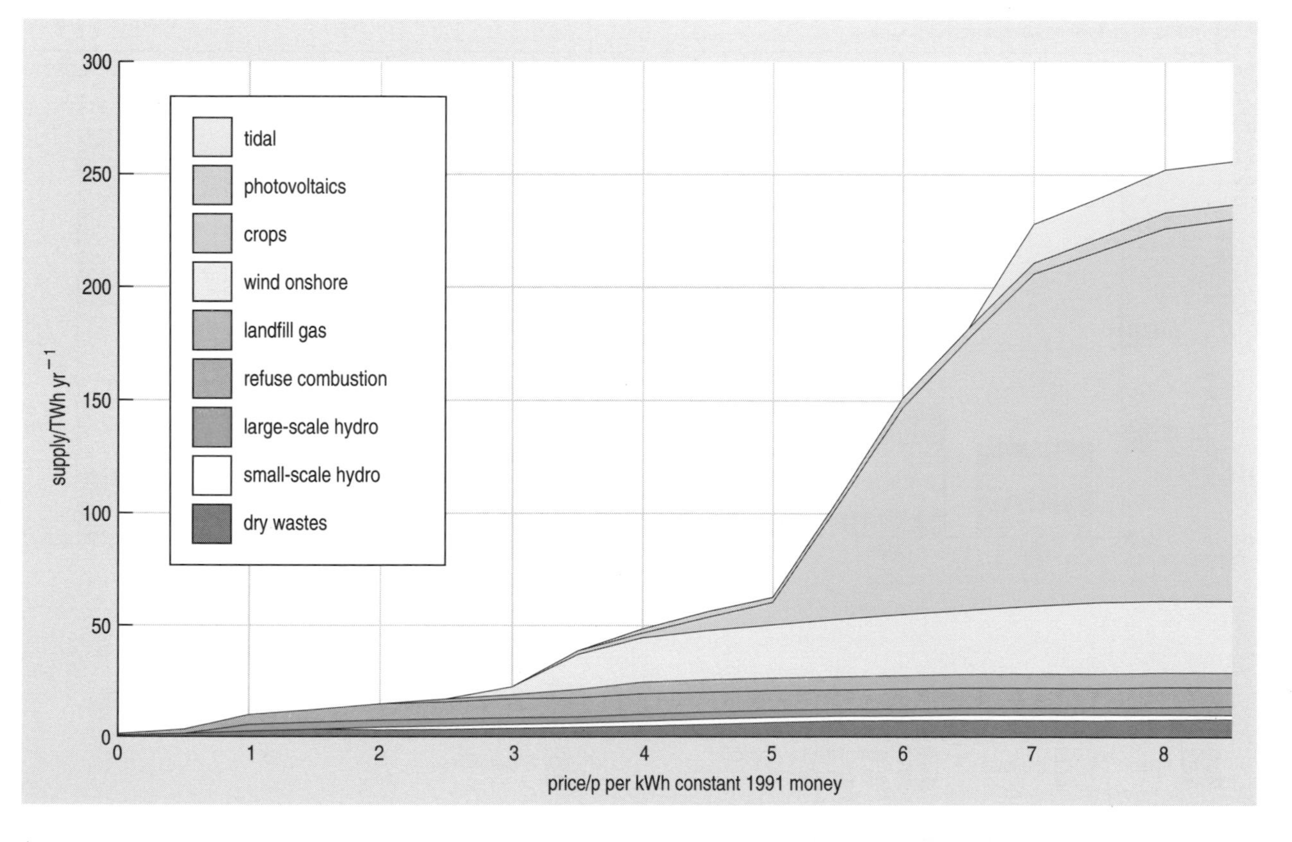

Figure 3.17 Estimated contribution of renewables to electricity generation in the UK in 2025.

Table 3.3 ALTENER new and renewable energy targets for 2005 in the European Union

	Production 1991			Objective 2005		
	GW	TWh	mtoe	GW	TWh	mtoe
Electricity						
Small hydro (under 10 MW)	5.0	15.0	1.3	10.0	30.0	2.6
Geothermal	0.5	3.0	1.9	1.5	9.0	5.4
Biomass and waste	2.0	6.3	2.7	7.0	20.0	8.6
Wind	0.5	0.9	0.1	8.0	20.0	1.7
Photovoltaic	0.0	0.0	0.0	0.5	1.0	0.1
Total electricity, excluding large hydro	8.0	25.2	5.6	27.0	80.0	17.6
Large hydro	74.8	154.5	13.3	88.6	198.5	17.1
Thermal						
Fuelwood			20.0			50.0
Other biomass (biogas, waste, etc.)			2.7			2.7
Geothermal			0.4			0.4
Solar			0.2			0.2
Total thermal			23.3			62.2
Biofuels			0.0			
Total contribution, excluding large hydro			29.3			91.6
Total contribution of renewables (including large hydro)			*42.6*			*108.7*
Total energy consumption			1160.0			1400.0
Percentage share of renewables			*3.7*			*7.8*

Source: CEC, 1992

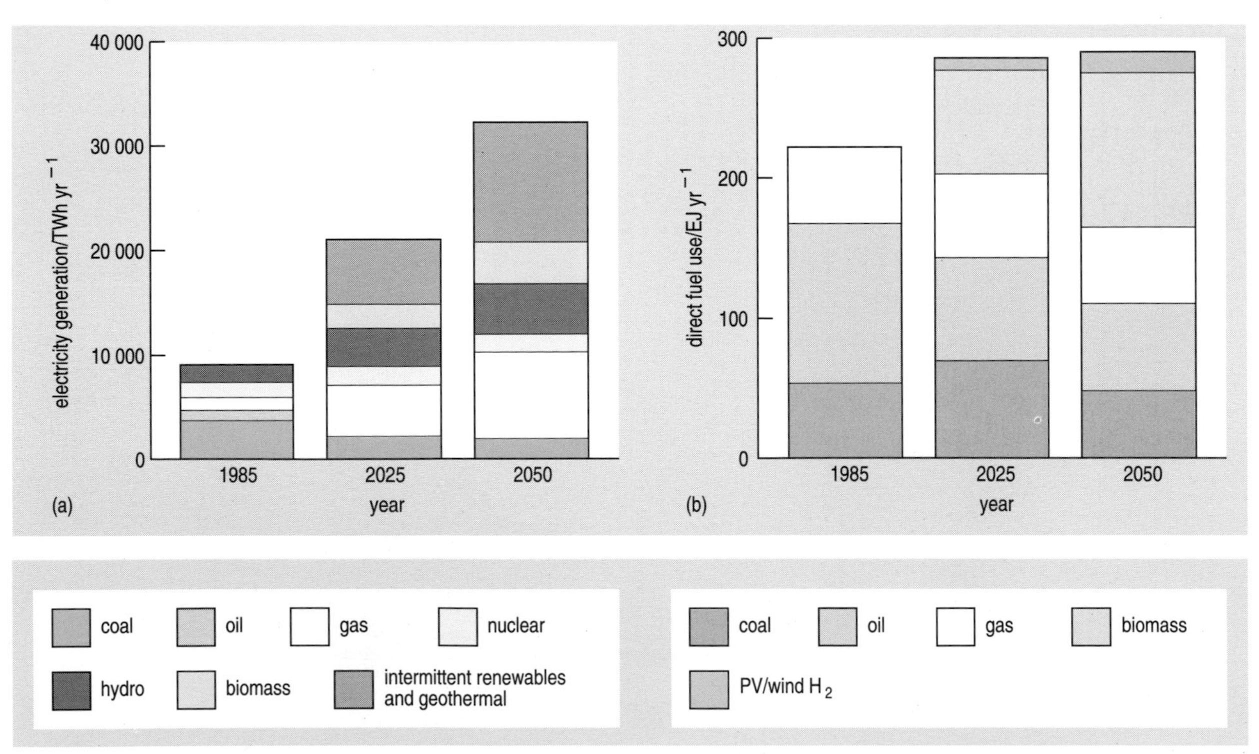

▲ *Figure 3.18*
The Renewables-Intensive Global Energy Scenario (RIGES): (a) electricity generation; (b) direct fuel use.

provided the estimates shown in Figure 3.18, again with significant renewables contributions in the next century. Thus renewables are no longer marginal, but are part of the accepted agenda of energy planning. Of course the rate of introduction will continue to be controversial and the possibilities for demand reduction by efficiency improvements need to be considered at the same time. Much will depend on developing perceptions of environmental effects and the need for government intervention in energy markets to allow for them.

The Non-Fossil Fuel Obligation (NFFO)

One form of such government intervention is provided by the NFFO scheme. This was introduced as part of the 1989 privatisation of the UK electricity industry, initially primarily to subsidise nuclear power, but increasingly used to support renewables projects as well. A levy on electricity consumers (set at 11% in 1992) compensates the industry for the extra costs involved in purchasing electricity from non-fossil sources. For projects supported before 1994, a premium price of 6–11p/kWh can be paid until 1998, compared with the 2.5–3p/kWh for large-scale fossil fuel generation. (This time limit of 1998 arose from EU objection to a longer period of subsidy for *nuclear* power and did constrain the renewables projects that could benefit to those with short payback times.)

More recent renewables support extends for 15 years and the system is continuing to evolve. As a result of it, by the end of 1994, 141 projects producing some 320 MW had been completed in England and Wales (of the 197 projects, 624 MW contracted in the first two NFFO 'rounds' of 1990 and 1991). The third NFFO order of 1994 involves a further 141 projects contracted to produce some 627 MW in England and Wales. Further orders for Scotland and Northern Ireland involve an additional 92 MW, and the scheme will continue with more orders at least up to 1998. All of these are contributing to the current (1996) renewables target of 1.5 GW by 2000. The financial conditions have meant that most of the projects have been for energy from wastes, landfill, sewage gas and hydro, but with significant amounts from wind (some 60 MW commissioned, a further 200 MW or so contracted). This is providing direct experience of reactions to these forms of generation to replace earlier speculation.

5 *Environmental impact of renewables*

> It is a little ironic that environmental groups here and abroad are the most vociferous in questioning these (renewable) technologies. The Severn and Mersey barrage proposals are said to have potentially serious impacts on bird and marine life, and some people are suddenly seeing windmills as being noisy and obtrusive, at least on the scale that would be required to provide significant amounts of electricity.
>
> On a recent visit to Denmark, which has a large number of windmills producing electricity, I found tremendous controversy among environmental groups, some of which were vociferously for and some of which were equally vociferously against wind power as an energy source. That environmental uncertainty is one of the strongest reasons for the Government's continued involvement in research into this area.
>
> (Michael Spicer, UK Parliamentary Under-Secretary of State for Energy, 30 October 1987)

Activity 4

Consider the environmental effects that have been identified for traditional fossil fuel and nuclear power generation, and compare the effects that can be foreseen for the renewable methods described in Section 4.

5.1 *A general comparison*

The massive environmental effects on all scales of existing power sources, from the local visual effects of open-cast coal mining to the regional scale of pollution and the world scale of possible climate change are obviously not shared by the renewables. But in tackling one set of problems we introduce new ones, mostly local in scale. They must be considered sympathetically, weighing the environmental disturbance against the social value and cost of the energy produced, and the interests of local communities against more remote beneficiaries. Most of the solar sources (whether direct, biomass, wind or wave) are essentially diffuse, of low energy density, so any attempt to use them to produce the large concentrations of power associated with existing plant involves collection over large areas. Thus, a 2-MW wind turbine has *larger* blades than those of the steam turbine shown in Figure 3.19, but the vastly greater power of the jets of steam striking these blades than that of the wind gives a steam turbine a rating of some 500 MW, emphasising the concentration achieved with fossil or nuclear fuels. (This disparity is less pronounced for hydro, tidal or geothermal schemes, which take advantage of geographical factors to give energy concentration, and correspondingly greater power production.) All of the diffuse sources can more readily meet distributed local needs than those of large urban populations. In the long run this might influence

◀ *Figure 3.19 Steam turbine blades.*

settlement patterns, but at present we will have to accept, along with large farms for food production, similar large land areas for energy production by these means – though the uses are not incompatible: energy and food crops can be mixed, and farming can continue between wind turbines. An advantage of such distributed schemes is that they can be expanded or contracted without massive investment or decommissioning costs.

As the renewables involve the 'harvesting' of energy from natural flows, rather than the exploitation of concentrated energy stores, a possible primary measure of their environmental effect is the extent of disturbance of these flows; different methods also have individual 'secondary' effects of various kinds. (This approach is adopted in a study by Alexi Clarke listed under 'Further reading'.) Thus

- *solar energy* is normally absorbed and re-radiated at lower temperature; collection only modifies this process to redistribute the heat produced, so the primary effects are small; secondary effects might be visual intrusion or reflected glare;
- *biomass production* is the normal response of the ecosystem to solar input, though the natural system shows much species diversity; any agricultural system (for food, timber or energy) disturbs this process, with perhaps more risk to its sustainability the nearer it is to a monoculture, and the more agrochemicals are used; secondary effects might be visual, conflicts over land use and amenity, and the noise and pollution from mechanical harvesting and processing;
- the use of biomass *wastes* is in a sense a complementary process, as they only exist as a result of our earlier disturbance of the natural flow, and we *reduce* the environmental effects of burning or burying them by controlled energy production; secondary effects might be visual and other pollution from collection and processing;

- *wind energy* is extracted by wind turbines, so the natural air flow is reduced downstream, but only to a very small extent as so little of the total energy is accessible; secondary effects are visual, noise, etc., considered in more detail below;
- *wave energy* is likewise extracted by conversion machines, though soon replaced by the wind; coastal erosion might be *reduced*; the movements of shipping and marine organisms would be affected;
- *tidal energy* is extracted by significantly modifying natural water flows so the local effect is considerable, as it is with large *hydropower* schemes; secondary ecosystem and visual effects, and disturbance during construction, may be large, and are site dependent;
- *geothermal energy* is extracted from the natural flow of heat within the Earth's crust, originating from the fluid interior, or more local radioactivity; the relation between the rate of extraction and the rate of replenishment determines the temperature reduction, and hence the continuing use of the site, or conceivably, larger geological changes, as with other mining operations; secondary effects are like those of mines, but with no spoil.

The worst case for its immediate environmental effects is evidently tidal, with specific dependence on the particular site. Thus, the Severn Barrage scheme has raised concern about the effects of the modified water flow on mud flats which are feeding areas for birds, on the deposition of silt and on effluents which will no longer be swept out to sea (but should be cleaned up anyway?) – to be measured against 120 years of pollution-free electricity generation on a very large scale. The balance of local opinion seems to be in favour, but a decision to proceed depends on issues of private and public financing which have not yet been resolved. Similar debates will take place over each other new scheme (Mersey and Humber, for example) as it arises.

The environmental concerns now being expressed as wind farms are being introduced in the UK make this subject appropriate for a more detailed case study, which forms the rest of this chapter.

5.2 *The wind resource*

We have already seen that substantial wind farms have been installed in the USA, Denmark and Holland and are now increasingly being introduced in the UK. What conditions are required for wind power to be useful, and what difficulties are foreseen?

The fact that wind is intermittent means that operation can only be expected for 30–40% of the time whereas conventional power stations operate for 55–70% of the time, allowing for maintenance, mechanical problems and fluctuating demand. So up to twice the installed capacity would be required from wind to replace more conventional sources. But unless the proportion of wind energy were to exceed about 20%, the use of the grid to smooth out contributions from different parts of the country, and from existing plant, seems able to cater for local fluctuations. The total number of turbines of existing designs (around 1 MW) needed to replace a large (e.g. 1.3 GW) power station is very large. Where will these thousands of machines be sited, and how will they be regarded – as ugly intrusions, or as environmentally appropriate, since they use a continuous resource, and are non-polluting?

Without going far into the complexities of technical wind turbine design, it is worth trying to grasp two basic principles which influence their size and location. The first is that the power depends on the 'swept area' covered by the blades. You may remember that the area of a circle is πr^2,

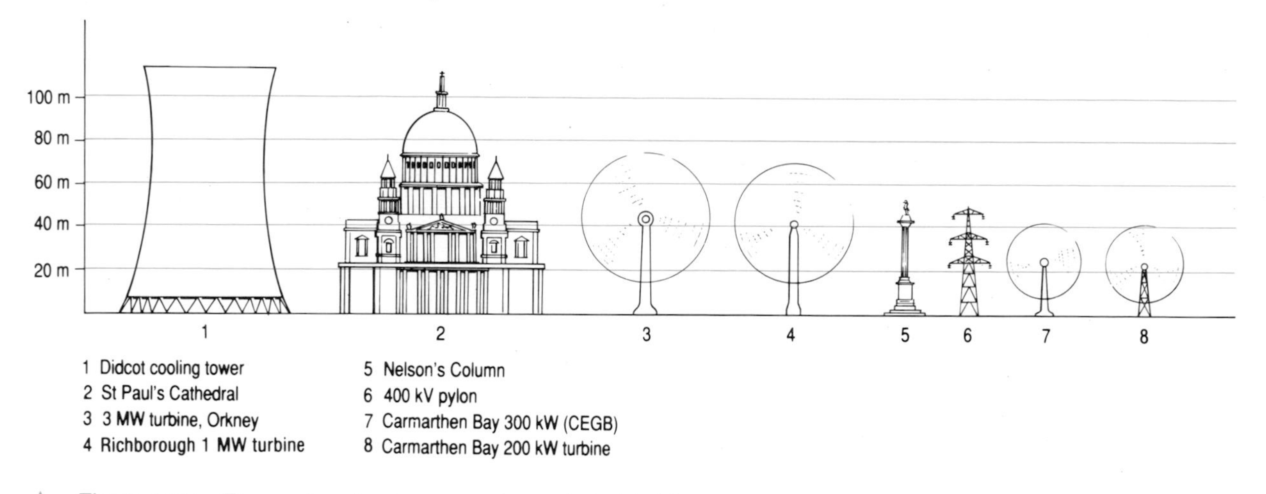

Figure 3.20 Figures in a landscape – sizes of wind turbines.

that is, a number, 'pi', times the radius 'squared' (i.e. multiplied by itself). What this means is that doubling the blade length multiplies the power by four, so there is a considerable advantage in increasing their size, with a corresponding increase in the height of the tower – which also enables them to reach levels of higher wind speeds. There are then increasing structural difficulties, which at present limit designs to a few MW, with debate about the most reliable and cost-effective size, perhaps 'medium-power' of about 300 kW. What this means in terms of relative size is shown in Figure 3.20, with heights of 40–70 m, spanning the height of grid pylons or Nelson's Column in Trafalgar Square.

The second point is that power is even more dependent on wind speed – in fact on speed 'cubed', v^3, that is, speed multiplied by itself twice over so that doubling the speed increases the power by eight times. So it is highly rewarding to seek out areas of high wind speed. High, exposed areas may readily have average wind speeds 1.6 times higher than surrounding lowlands (higher peak speeds and also fewer calm periods) so the power available is about four times greater.

The basic pattern of wind speeds over EU countries is shown by maps like Figure 3.21, taken from the *European Wind Atlas* prepared in Denmark in 1989. From this map it is clear that much of the UK is exceptionally suitable for wind power development in relation to the rest of Europe, but also from the table that in all areas the wind speed at 50 m above ground level is strongly dependent on the local topography. For this reason, and the 'speed-cubed' rule, more detailed local measurements and analysis are needed to determine the best sites. A 1993 study by the UK Energy Technology Support Unit provides estimates for each 1 km square of the UK Ordnance Survey. Taking an average wind speed of about 6 m s^{-1} (metres per second) as a typical figure above which wind turbines become practical, the study shows that many coastal regions, particularly in the north and west, and large highland areas in Scotland, Wales and the Pennines, are well suited for wind generation, and this is reflected in the sites so far being developed. The table also shows, alongside the average wind speed, the average power available per square metre of area swept by the blades (measured in watts per square metre, W m^{-2}).

Wind resources at 50 m above ground level for five different topographic conditions										
	Sheltered terrain		Open plain		At a sea coast		Open sea		Hills and ridges	
	$m\ s^{-1}$	$W\ m^{-2}$	$m\ s^{-1}$	$W\ m^{-2}$	$m\ s^{-1}$	$W\ m^{-2}$	$m\ s^{-1}$	$W\ m^{-2}$	$m\ s^{-1}$	$W\ m^{-2}$
	>6.0	>250	>7.5	>500	>8.5	>700	>9.0	>800	>11.5	>1800
	5.0–6.0	150–250	6.5–7.5	300–500	7.0–8.5	400–700	8.0–9.0	600–800	10.0–11.5	1200–1800
	4.5–5.0	100–150	5.5–6.5	200–300	6.0–7.0	250–400	7.0–8.0	400–600	8.5–10.0	700–1200
	3.5–4.5	50–100	4.5–5.5	100–200	5.0–6.0	150–250	5.5–7.0	200–400	7.0–8.5	400–700
	<3.5	<50	<4.5	<100	<5.0	<150	<5.5	<200	<7.0	<400

▲ *Figure 3.21 Wind energy resources over Europe (EU countries).*

Such studies imply that 10–20%, maybe even 100%, of UK electricity demand could be met by wind power, particularly if the huge offshore potential is included. More detailed environmental impacts, however, remain to be considered.

5.3 *Environmental constraints*

As usual with discussions of changes which impinge on the environment, the factors involved are many and interrelated. Consider first some technical matters which set the scale of the problem, before going on to more subjective questions about how we perceive their effects.

First, how big are the new turbines? Setting aside 'small' machines of less than 100 kW power, which will no doubt be scattered about without too much impact, the choice for substantial generation is between 'medium' (100 kW to 1 MW) with blade diameters below about 50 m, like pylons or Nelson's Column, and 'large' (>1 MW), maybe up to 100 m, like large cooling towers or St Paul's Cathedral. The emphasis in the practical wind farm schemes in the USA, Denmark and Holland has been on medium size. Research in the USA, and by the former CEGB in the UK, has looked at large machines – obviously fewer will be needed, but there is less operating experience, and the relative impact remains uncertain, depending on siting. At present the economics seem to favour medium size, and they have been adopted in most of the current schemes.

Second, how much *land* do they occupy? With one more bit of technical information you can estimate this for yourself: adjacent wind turbines do not interfere with each other if they are separated by from 5 to 15 times the blade diameter – say 10 times as a standard figure.

To get a general idea, let us keep to simple numbers – 1 MW turbines of 50 m diameter replacing a power station of 1 GW output.

Q How many turbines are needed to provide the same power (when operating) as the power station?

A 1 GW/1 MW = $10^9/10^6 = 10^3$ = 1000.

Q If the wind turbines only operate for 35% of the time, while the power station operates for 70% ('load factor'), how many turbines are needed?

A Twice as many, 2000.

Q If these were divided into 20 'farms' of 100 turbines each, arranged in a 10 × 10 square array, what would be the area of each farm, and the total area?

A Turbine spacing = 10 × blade diameter = 10 × 50 m = 500 m;
array length = 10 × 500 m = 5 km
(strictly, 9 × 500 m = 4.5 km between end turbines, but this allows an extra 250 m at each end as a 'buffer zone' discussed below).
Farm area = 5 × 5 = 25 km^2.
Total area = 20 × 25 = 500 km^2.

Obviously the exact figure depends on the assumptions about power/size relationships, load factors, array shapes and spacings, and can be a source of contention – the CEGB said that a wind replacement for Sizewell would occupy $27 \times 27 = 729$ km^2 whereas a wind enthusiast can get it down to $17 \times 17 = 289$ km^2 by making different assumptions. The *general* idea of the area involved is not in doubt, but a factor of two or more in detailed planning is quite important!

The two basic design points made above in Section 5.2 are relevant. What is the effect of blade length and wind speed on these estimates?

Doubling the blade length increases the power by four, but also doubles the required spacing, so increases the area by four, leaving the power per unit area about the same. So the choice between medium and large turbines doesn't have much effect on this question if the spacing *pattern* is kept the same; it might be possible to find a better detailed arrangement for fewer large machines.

On the other hand, the 'wind-speed cubed' rule means that using the same size of turbine at different sites can make a big difference to its power: an increase in wind speed from 5.5 m/s to 7 m/s more than doubles the power ($7/5.5 = 1.27$; $1.27^3 = 2.06$), turning our 1 MW model into a 2 MW, and halving the number of turbines and area of land required. This emphasises the importance of seeking out the windy sites, and the sensitivity of the calculations to site data which are not always available.

Even more important is the fact that these comparatively large land areas are not in fact used up by building wind farms on them. The proportion of the area occupied by the towers and necessary services has been variously estimated at 1%–3%, the rest being still available for agriculture. The land actually occupied is reduced to a few tens of square kilometres, not much more than a large power station site. It would not be expected to be all in one place, but distributed over several farms. This permits a site-dependent compromise about land use – lower wind speed areas where more turbines are needed may share use with arable farming; windy hill-tops with fewer, well-spaced large machines may be used for livestock, or not farmed at all, with amenity and visual impact becoming more important.

The next consideration is that of 'buffer zones' between wind farms and the public, both for general access and habitation. Two issues, noise and safety, fortunately point to similar requirements, and also help with the most controversial, visual impact, which can be assessed against the whole of this more technical background.

1 *Noise* from wind turbines comes from three sources: the regular rhythmic swish of the rotating blades, mechanical noise from the gearbox and an electrical hum from the generator. The latter two are internal, and in good designs are reduced by conventional sound-proofing. The blade noise is more fundamental, and improved aerodynamic design to reduce it is being eagerly sought. It is most noticeable in light winds; as the wind gets stronger its own noise in trees, buildings, etc. dominates. Turbine noise carries downwind, but reduces with distance. How much reduction is required depends on the background noise level, and the activities, around the particular site. Urban background levels are higher than rural; permanent homes obviously require quieter conditions than working areas, industrial or agricultural, or chance passers-by. The requirements for remote 'wilderness' sites with very low background levels may be particularly exacting. While a buffer zone of 1 km or so is usually completely satisfactory, this may drastically reduce the wind capacity available in some areas, so careful assessment is needed. Concern to avoid noise disturbance is essential, but it would be a pity to limit the use of wind power by restrictions which are unnecessarily stringent.

2 A similar buffer zone is also needed for *safety* reasons. Like all large engineering structures, wind turbines can fail if the design, construction, inspection or maintenance are faulty. The development period of modern machines, from the 1940s to the 1980s, did see some spectacular tower collapses and runaway in high winds causing blade break-up in the USA, Denmark, Holland and France, though without casualties or adjacent property damage. These have led to improvements in design and quality control, better understanding of blade fatigue, better control and braking mechanisms in high winds and tighter regulation. The risk from current established designs is considered to be negligible – very much lower than from road transport, for example.

Denmark, with much experience, has a proposed buffer zone 'formula' of >90 m + 2.7 × blade diameter from open public areas, roads and buildings.

Q What does this imply for 50-m and 100-m diameter blades?

A 50 m: 90 + 2.7 × 50 = 90 + 135 = 225 m
100 m: 90 + 2.7 × 100 = 90 + 270 = 360 m

There is thus little difficulty in providing adequate noise and safety buffer zones in remote locations, but in areas of scattered settlement they can restrict the available wind resource considerably. In a Cornwall study using 70-m diameter 2-MW turbines for reference, 350-m (5 diameters) spacings and buffer zones enabled sites of 1.4 GW capacity to be identified, whereas 700-m (10 diameters) zones reduced this to 0.132 GW. A Danish study of the national resource used 7 diameters, giving 200–300 m distances for different sizes of turbine, and based on *experience* of 100, 200 and 500 m distances rather than calculation or regulation. Distances greater than 300 m from buildings were considered to make siting in general impossible in the normal Danish landscape of scattered agricultural buildings.

Evidently, whatever formula is chosen for the size of buffer zones constrains the size of turbine that can be used on each site and the total resource available. A 1991 proposal from the European Wind Energy Association was that 100 GW of wind power, 10% of European electricity demand, could be installed by 2030, taking up a land area equivalent to that of Crete, but with 99% of the land still available for agriculture.

Further technical factors in choosing sites are ease of *access* to the electricity grid and the road system, and *electromagnetic interference* disturbing radio and television reception, microwave communication links and radar – this may require the introduction of additional 'booster' transmitters, which are, however, comparatively cheap in relation to turbine costs.

With all these practical matters dealt with, we are led to the most controversial and subjective question, of the *visual impact* of wind turbines. They are large structures on unobstructed hill and coastal sites, moving so that they attract attention, and possibly spread out in large numbers as wind farms. Many of the most suitable sites are in the highlands of the north and west of the UK, much valued as National Parks and unspoilt landscape. Individual reactions can obviously vary from outrage at ugly intrusions to enthusiastic support for glorious symbols of environmentally appropriate engineering; which reactions dominate will be dependent on the site and the imagination and sensitivity shown in its detailed use.

It seems to be accepted that we have an exaggerated perception of vertical objects compared with horizontal, regarding anything more than 50 m high as 'large', which is consistent with the distinction between medium and large turbines. 'Extreme dominance' is said to be felt within distances of three times the height, dominance within ten times. (So the buffer zones already required go most of the way to removing 'dominance'). Features are dissociated if separated by more than three times their height, which again will occur with the spacings required to avoid interference, though not if you can see through several staggered rows, with their blades forming a 'thicket'. Most experience is with medium size turbines. The comparative effect of fewer, more widely spaced large machines with fewer access roads may be less; at sufficiently large distance they will be less likely to coalesce. Conditions in the UK rarely permit visibility of slender structures beyond 5–8 miles. Rotation speeds of less than 45 r.p.m. are thought to be restful. Solid towers are preferred to open gridwork. (Why are traditional windmills picturesque? Are slender modern aerodynamic designs equally attractive?) There are 22 000 pylons and many aerial masts that are already tolerated up to a point, though whether similar structures should be added will again depend on the site. Substations and access roads have to be included in the landscaping.

This all adds up to a complicated set of judgements to be made about each site in turn; most important is that the visual effects of proposed schemes should be clearly presented in local consultation.

A separate concern that has been expressed about wind turbines and the environment is their effect on wildlife – specifically, *birds*, the only form that seems to be involved. Turbines are large moving structures with high blade tip speeds, so are potential obstacles and hazards. It must be said immediately that this has not deterred us from putting up power cables, pylons, masts and lighthouses, which are similar stationary objects with which birds frequently collide in bad weather, or from building motorways and aircraft, which kill large numbers of mammals and birds. Bird strikes by aircraft seem to be treated more as hazards to the machines than to the birds. Nevertheless, it is appropriate and natural that a development proposed as environmentally benign should be scrutinised carefully for adverse effects.

A number of studies have been done on turbine sites in Holland, Denmark, the USA and with the Royal Society for the Protection of Birds (RSPB) in the UK. The dependence of bird behaviour on species, site, number of turbines and weather conditions continue to be examined, but the major concern that has emerged so far is about the disturbance caused by increased human activity around wind farms, not the machines. While any doubts remain it may be better to avoid known breeding grounds and migration routes for wind farms, and use underground cables for grid connection where there is much birdlife. At coastal sites a recommendation has been made to site turbines as far inland as possible.

The exploitation of the substantial resource still available after all these factors have been considered, together with the even greater offshore possibilities, can meet perhaps 20% of our electricity needs. Conflict will however only be avoided with sensitive consideration of each site, and full consultation with the people who will be affected. Apart from the areas specifically designated as National Parks or Areas of Outstanding Natural Beauty (AONB) all areas will raise some questions of visual intrusion. The reactions to recent UK wind power developments will now be considered.

5.4 *UK wind power development in the early 1990s*

In 1988 the then CEGB concluded that the economics of wind power had become sufficiently attractive to justify a £28m programme to develop three

8-MW 'wind parks' of some 25 medium power turbines each in various sites in England and Wales over the period 1990–93. This would provide a test not only of technical performance, but also of public reaction; it was CEGB and government policy to monitor this carefully.

In the event, the scheme was overtaken by the privatisation of the electricity supply industry in 1989, and only one of these schemes, in Cornwall, went ahead, being completed in 1993. Instead, the non-fossil fuel obligation (NFFO) scheme (see Section 4.4) provided a mechanism for subsidising renewable energy (which was eagerly seized on by private developers), of wind power alongside projects for energy from waste, landfill gas and so on.

In the first two NFFO 'rounds' for England and Wales, 9 projects to generate 12 MW were contracted in 1990, and 49 to generate 84 MW in 1991 (evidently these were on average smaller than the CEGB schemes); 31 (some 60 MW) had been completed by the end of 1994, about 10% of the total power from all renewables. Acceptance of projects meant that the electricity supply companies would be required to pay an agreed price (some 6–11p per kWh) for electricity from these sources (rather than 2.5–3p per kWh from fossil fuel stations), but only until 1998. This constrained the projects to those with rapid recovery of capital costs, and therefore to sites that were very favourable in wind regime and accessibility, and so particularly likely to be environmentally sensitive. The third NFFO round for England and Wales in 1994 extended the period of supported prices to 15 years, at levels of 4–6p per kWh, nearer conventional prices – on balance easing the capital recovery problem. In this round, 31 large schemes (over 1.6 MW each) and 24 smaller schemes were contracted, in all generating some 165 MW. These wind schemes accounted for 26% of the contracted capacity in NFFO-3.

The first schemes for Scotland and Northern Ireland were also announced in 1994. In Scotland wind accounted for 45.6 MW, 60% of the total, and in Northern Ireland 6 wind schemes with a capacity of 12.6 MW provided over 80% of the total.

It is a further principle that an NFFO contract does not override the requirement for schemes to obtain local planning permission in the normal way. The development of 'planning guidelines' for wind projects has been a continuing issue alongside the technical and financial ones.

5.5 *Public reactions to wind farms*

Before any schemes were in place, there was a fairly strong polarisation of views. Environmentalists had an almost utopian commitment to renewable energy in all its forms. Traditional energy suppliers, both the CEGB and the nuclear industry, were inclined to dismiss renewables in general, wind in particular, as romantic and irrelevant. As real wind farms have emerged, there has been a shift to a more complex position, with some national conservation groups opposing wind farms, at least in some cases, on environmental grounds, and overtures developing from the nuclear industry for an alliance against fossil fuel generation.

The first wind farm to be completed in the UK was at Delabole in Cornwall, involving ten 400 kW turbines that started up in December 1991. A systematic study of 'before' and 'after' responses carried out in 1990 and mid-1992 showed a considerable shift to favourable opinions with experience: for example 86% had expected noise to be a problem in 1990,

80% felt that it was not in 1992; 40% were concerned about visual intrusion 'before', only 29% thought it was 'after'. Early negative warnings from government energy ministers about 'noise equivalent to a helicopter' [Peter Walker 1986] or 'very ugly turbines' [Cecil Parkinson 1989] were replaced by a response from Prime Minister John Major after a visit to Delabole in 1993 that he was 'most fascinated to see such an effective and practical example of renewable energy supply' and that the turbines were 'more attractive than I thought and less of an intrusion on the landscape'. Nevertheless, with the dramatic rise in the number of projects after 1993 – some of which were perhaps pushed ahead because of the financial pressures and lacked sufficient environmental sensitivity – planning difficulties and adverse reactions increased.

Draft planning guidelines were issued in 1991, and full ones emerged in 1993, which left detailed assessment to the local planners, though recommended a basic policy to support wind power for global strategic reasons. This left local planners unsure of their stance – one response to the draft guidelines was that 'Local planning authorities are not responsible for energy generation'. Some conservation groups, like the Countryside Commission and the Council for the Protection of Rural England, felt that the Government was being 'soft' on developers; others, like the Campaign for the Protection of Rural Wales, thought that the inadequate financial incentives were leading to invasive siting. The Government insisted that the issues were of planning, not finance, and 'require planners to balance the Government's policies for renewables with those of the countryside.'

The first public inquiry, in 1991 on a 24-turbine project on the edge of Snowdonia, led to positive recommendations, which were accepted by the Secretary of State for Wales, but the next, for a 15-turbine project in Cumbria, was turned down by the Inspector on the grounds of visual intrusion, to be overruled by the Secretary of State for the Environment. Most proposals do not lead to inquiries, of course, and the majority have obtained planning permission, sometimes after Environment Secretary intervention. Conversely, the Welsh Office turned down an appeal against rejection of a scheme in Anglesey, and local planners have rejected others. However, on balance, environmental groups began to feel that too many schemes were being rushed ahead without enough time to consider all the implications. A factor in the mounting opposition was the setting up in 1993 of an anti-wind lobbying group called Country Guardians, which received considerable media support and a reaction from the British Wind Energy Association that 'a small but vociferous number of people have generated a disproportionate amount of press coverage'.

The debate will obviously continue, with a growing recognition of the need for careful siting and consultation with local opinion – in favourable cases this can undoubtedly be very supportive. Perhaps the biggest issue is about scale and ownership. Small locally owned schemes of obvious benefit to the local community are much more readily supported than large intrusive developments in valued environments imposed to provide large amounts of power to distant urban populations. This is part of the general issue of matching diffuse renewable energy sources to the expectations of vast power concentrations associated with fossil fuel and nuclear stations. Our landscape once comfortably embraced 10 000 traditional windmills; can it not do so again with the newer turbine designs?

5.6 Conclusion – renewable energy futures

The many renewable forms of energy described in Section 4 are together able to provide a supply system that can gradually replace existing methods. Whether and how quickly it will do so will (apparently) depend on costs, but these will in turn depend on underlying factors of availability, technical ingenuity, political intervention and environmental effects – expressed both by better accounting for existing methods and careful scrutiny of alternatives. We have already seen that all energy supply involves *some* environmental effect; how this should be assessed raises difficult questions of values. Current ideas of 'environmental accounting' propose that money values should be attached to the environmental costs of policies, including energy policies. Certainly this seems better than neglecting them, but you will have your own view of whether money is the right measure in every case. How do you weigh a view of Great Gable in the Lake District against more convenient commuting to work – or Third World starvation?

It is important that *some* judgement be made about new policies, and we have seen the need for careful consideration of their effects, and consultation of local opinion, before introducing them. In the case of renewable energy, sensitive use of natural energy flows may even be seen as *enhancing* the environment, unlike the affronts produced by thoughtless exploitation of natural energy stores.

6 Sustainable energy futures

The methods of energy demand reduction described in Sections 2 and 3, and of alternative supply described in Sections 4 and 5, offer a general way forward to developing sustainable energy policies. The balance between them will vary country by country, and in response to changing perceptions of environmental issues and political responses to them.

We have emphasised immediate UK policy, but in the longer term and world wide the same issues arise, with the additional pressing inequity of large differences in energy use between different countries, more urgently calling for sustainable policies. A general point about industrial society is that it places in individual hands a large number (millions) of technically complex energy *consuming* devices (motor cars, refrigerators, television sets) and these are no doubt sought everywhere. Although some of the large energy demand comes from the industrial and service infrastructure that provides these goods, as much of it comes from their widespread use, and is sensitive to efficiency improvements, as we have seen.

Energy *production* on the other hand is traditionally concentrated in large units (mines, oil refineries, power stations) and the centralised financial and governmental institutions which control them. A shift of emphasis to renewable energy production, particularly that based on the distributed solar energy resource, lends itself to the distribution of many small units, down to the local or even domestic scale (small hybrid solar, or biomass fuelled, CHP; small wind, wave and hydro machines). The experience with energy-consuming devices suggests that there is no reason why these should not be technically complex if they are well designed for ease of use. As well as meeting local needs, they could all be coupled to the grid to provide an energy trading and support system. This would limit the required centralised institutions to control of the transport, grid and communications networks, and of specialised manufacture, distribution and maintenance, rather than the present monolithic energy suppliers. Along with an increasing use of information (rather than goods and energy) transfer, this raises implications for social and political organisation in all parts of the world which may be very different from those of existing industrial societies.

References and further reading

BLACKMORE, R. & REDDISH, A. (eds) (1996) *Global Environmental Issues*, Hodder & Stoughton/The Open University, London, second edition (Book Four of this series).

CLARKE, A. (1993) *Comparing the Impacts of Renewables*, Technology Policy Group Occasional Paper 23, Open University Technology Policy Group, Milton Keynes.

COMMISSION OF THE EUROPEAN COMMUNITIES (CEC) (1992) Briefing Paper on ALTENER Programme.

EUROPEAN WIND ENERGY ASSOCIATION (1991) *Time for Action: wind energy in Europe*.

JOHANSSON, T. R., KELLY, H., REDDY, A. K. N. & WILLIAMS, R. H. (eds) (1993) *Renewable Energy: sources for fuels and electricity*, Island Press, Washington DC.

KRAUSE, F. *et al.* (1980) *Energie Wende: Wachstum und Wohlst ohne Erdol und Uran*, S. Fischer Verlag, Frankfurt (M).

LOVINS, A. B. (1988) *The State-of-the-art: Lighting*, Rocky Mountain Institute, Colorado.

OLIVIER, D. *et al.* (1983) *Energy-efficient Futures: opening the solar option*, Earth Resources Research, 258 Pentonville Rd, London N1 9JY.

OPEN UNIVERSITY (1994) T521 *Renewable Energy: a resource pack for tertiary education*, The Open University, Milton Keynes.

RENEWABLE ENERGY ADVISORY GROUP (REAG) (1992) *Renewable Energy Sources: report to the President of the Board of Trade*, Energy Paper No. 60, HMSO.

ROYAL COMMISSION ON ENVIRONMENTAL POLLUTION (1994). *Transport and the Environment*, 18th Report, Cm 2674, HMSO, London.

SERI (SOLAR ENERGY RESEARCH INSTITUTE FOR US DEPT OF ENERGY) (1981)
A New Prosperity: building a sustainable energy future, Brick House Publishing, Andover, Massachusetts.

UK ISIS (1987) *Proceedings of the 1st UK ISIS Conference on Superinsulation*, UK ISIS, London.

WORLD COMMISSION ON THE ENVIRONMENT AND DEVELOPMENT (1987) *Our Common Future*, The Brundtland Report, Oxford University Press, Oxford.

Notes on Activities

Activity 1

World energy consumption has increased from a rate of about 1 to about 10 TW (i.e. from 10^{12} to 10^{13} W) during the twentieth century. With a current world population of about 5×10^9 this makes the *average* rate of consumption now about 2 kW per person, but this ranges from <1 kW in less developed countries through about 4 kW in Europe to over 10 kW in the USA. With a doubled population in the next century, and some increase of average consumption, say to 3 kW each, the total would be 30 TW, with extreme possibilities such as: worldwide attention to energy efficiency bringing the average *down* to 1 kW each, so a total of 10 TW, as now, or world profligacy matching the US at 10 kW each, so a total of 100 TW. These *rates* of energy consumption imply an *annual* consumption of around 10^{21} J (there are 3.1×10^7 seconds in a year, so 30 TW = 3×10^{13} J/s = $3 \times 10^{13} \times 3.1 \times 10^7$ J/year = 9.3×10^{20} J/year). A resource of 10^{22} J would last about 10 years, 10^{23} J about 100 years, at this rate. Some impression of the pressure on the various fuels is gained by comparing their 'reserves' and 'resources' with these figures; exceeding 'proved reserves' produces price increases as less accessible resources are brought into use. Of course this is a *total* consumption estimate, so shouldn't strictly be compared with resources of any one fuel, as they are to some extent interchangeable, though conversions to less favourable fuels again lead to increased costs. However, a comparison with each fuel in turn gives a broad picture of availability. In these terms, it is clear that the oil and gas reserves, of 5 and 4×10^{21} J, are under severe stress, and their total resources, of a few times 10^{22} J, are not much better. Uranium, with reserves of about 10^{21} J for use in burner reactors, will be even less readily available. Coal, with quoted reserves of 2.6×10^{22} J and resources more than 10 times higher, might be available for a century or so; uranium used in breeder reactors (providing some 10^{23} J) is rather better – but in both cases the environmental price is high. The only terrestrial resource apparently almost unlimited is the deuterium in sea water, estimated to provide over 10^{30} J if it could be successfully used in a fusion reactor – this is yet to be proved possible. Solar energy on the other hand arrives continuously at the Earth's surface at a rate of some 2.4×10^{24} J/year, some 2000 times larger than this projected demand – though much of it arrives where it is not wanted, making its systematic use depend on changed priorities. Even the small percentage of solar energy converted into biomass energy, some 10^{22} J/year, is well above the total demand, though the same problem of its wide dispersal arises.

The environmental problems identified for fossil fuels include the gross local pollution of uncontrolled coal burning; the SO_2, NO_x, CO and hydrocarbon emissions, leading to acid rain and photochemical smog, from coal and oil combustion, now under increasingly tight control; and the inevitable CO_2 production from all fossil fuels, with its projected effects on the climate from the enhanced greenhouse effect (gas is somewhat preferable to oil, in turn preferable to coal).

Nuclear power does not produce CO_2 directly, though it has been argued that increased nuclear generation would imply increased energy use, in fuel production and construction, that would at present be derived from fossil fuels, with their direct CO_2 output. Nuclear has its own environmental problems – radioactive hazards, of catastrophic accidents and routine waste disposal; 'proliferation' of nuclear weapons from stolen materials, particularly plutonium in a breeder-based system.

The economic effects of direct supply and demand fluctuations for fossil and nuclear fuels are already considerable, but environmental effects are not included. GDP is often used as a measure of economic prosperity, and energy intensity (energy use per unit GDP) as a measure of energy efficiency, but it is increasingly recognised that high GDP can be achieved by destruction of environmental 'capital'. New 'environmental economics' measures are under discussion, but are not yet in general use.

Activity 2

There can be no general answer, as it will depend on your personal circumstances. No doubt you will have noted something about the relative locations of your home and work, and perhaps domestic, financial or amenity difficulties in bringing them closer together. You very likely find a car convenient for you and your family because of the ease of door-to-door transport and carrying shopping it provides; perhaps public transport is inadequate in accessibility, availability or cost. Maybe you are a model user of cycle or moped – or perhaps you would be if you didn't feel so vulnerable, to other vehicles or the weather. What changes would you like to be made?

Activity 3

The primary energy sources are the non-renewable fossil fuels (coal, oil and gas) and nuclear fuels (uranium and potentially deuterium), and the continuous (or renewable) solar, gravitational or geothermal energy. Energy vectors are traditionally hot water, refined fuels, wood and electricity. A potential chemical energy vector is hydrogen, and a nuclear one plutonium. More speculative chemical fuels might be transported for use in fuel cells, or more hypothetically in chemical motors; there might be nuclear equivalents.

Activity 4

The environmental effects of fossil fuels and nuclear power, on all scales from the local to the global, have already been discussed under Activity 1. Renewables do have environmental effects, discussed at more length in what follows. They are essentially *local*, the visual effects of machinery in the landscape or changed land use. The most severe are those from large tidal and hydroelectric schemes, which might influence not only the local landscape and ecosystem but also water flows at greater distances.

Chapter 4 Mineral resources

1 Introduction

As you read this chapter, look out for answers to the following key questions.

- What kinds of minerals are used in industrialised societies?
- Where are they extracted and by what methods?
- What kinds of environmental impacts result from mineral extraction, processing, use and disposal?

In the developed world people's lives are highly dependent on mineral resources that are extracted from the earth. They provide the substance of the places where people live and work: homes, hospitals, factories. They contribute to the infrastructure of the means of getting around – roads, railways and airfields – as well as to the vehicles, trains or aeroplanes themselves. They provide the fuel, because oil, gas and coal are minerals too, as Chapter 1 made clear. Indeed, minerals appear in almost everything people use, from cups and saucers to computers. Their contribution to high output agricultural production systems via machinery and fertilisers has been stressed in the previous book in this series (*Sarre & Blunden, 1996*). Developed countries have come to depend more and more on a whole range of minerals, which are finite in quantity, and in the time since the industrial and agricultural revolutions, the needs they serve have increased at an exponential rate.

Sometimes a mineral requires little processing beyond that concerned with its extraction from the ground in order to meet a particular need: for example, the gravel to refurbish a drive-way, or the raw salt applied to a road to stop it from freezing in cold weather, or the ground limestone scattered over a field by a farmer to improve the pasture. In other cases minerals may need to be refined through sophisticated processes and then manufactured into items of equipment which may be further assembled into a complex machine before people's needs can be met; the car is an example of this.

Q Can you list some of the many items the car contains which are made from minerals?

A To begin with, it contains many different metals which are fabricated into items such as the body shell (steel – which is made from iron plus smaller amounts of other minerals such as manganese, phosphorus and silicon, and also carbon); the wheels (steel alloys); all the parts of electric motors used in the starter mechanism, the windscreen wipers and automatic window winders (iron, steel, copper, aluminium, etc.); the engine (iron, steel alloys, aluminium, zinc, etc.); the exhaust system (iron and steel, but if it is of modern design, also the precious metal, platinum, used as an agent of pollution control in catalytic converters); the battery (lead); and the electric wiring (copper and aluminium). Then there is the glass, made from sands, for the windows, and a mass of plastic materials derived from the chemicals industry that go into the production of headlining, seats, tyres, light covers, wiring insulation, etc. Finally, there is the paint on the exterior of the car which may include several chemicals, a plastic base, china clay and even powdered slate! Of course, you may add other items. You might, for example, include the petrol in the tank and the oil in the sump, along with the instruction manual which will contain a considerable amount of china clay, which produces the surface of the glossy paper on which it is printed.

Don't worry if you were not able to match the list provided. You will become more familiar with many minerals and their uses as this book progresses. The important point here is that all these manufacturing processes bring with them the possibility of environmental impact. The fabrication of steel and the manufacture of glass inevitably involve the venting of gases, and the processes of the chemicals industry lead to a wide range of wastes for disposal, some of which are highly toxic. The federal government of the USA records 12 000 of these toxic wastes, of which 15% are known to cause cancer. Although the release of all these effluents is now subject to some legislative constraint, there are still accidents and mistakes, and even sometimes a wilful disregard for the regulations, when personal gain is put above public health interests. Moreover, attempts at regulation are sometimes seen as inadequate. For instance, paint gases vented to the atmosphere from a car body shop was, until recently, an environmental *cause célèbre* in parts of Oxford, where those affected alleged damage to property and personal discomfort.

Apart from the impact on the environment caused by any one of these manufacturing and assembly processes, whether they involve the construction of a car or other manufactured goods, such items eventually reach the end of their useful lives and will need to be disposed of. In the case of cars, this is usually about ten years, and this, too, will have its environmental dimension, for they are either dumped in some refuse tip, or taken apart and the metals and plastics obtained fed back into the production process – in other words, the raw materials of the component parts will be recycled. Unfortunately, from an environmental point of view, such recycling is still not a matter of normal practice in what remains essentially a 'throw away' society.

In the chapters that follow, the use of minerals in manufacturing, the methods of waste disposal and the possibilities for recycling will be examined in relation to their environmental impact. But this is only one part of the story. The initial provision of minerals for use in industry also has a strong environmental aspect, so the rest of this chapter considers the range of minerals, how they are removed from the Earth's crust and the initial processes of refining. It also examines what determines *where* such activities occur and the extent to which they are stimulated or constrained, because the interplay of the supply of minerals and the demand for them does not operate in a free market. Governments can and do give incentives to mining companies engaged in minerals extraction in order to maximise production within their own state borders, usually by way of modest tax concessions. But in a climate of increasing environmental awareness, they may also wish to restrict or even prevent development at sites of great ecological or landscape value, especially if alternatives are available. At the same time it has to be understood that if there is a choice of sites and it is left for companies to decide, those sites of least cost are usually worked, or those that are unlikely to be subject to political interference of one kind or another. The need for heavy capital investment in a mine over several years makes a secure political environment for the company a factor of major significance. Thus we will see that the interplay between sets of factors influences the location of mines, the nature of activities at such locations and therefore their likely impact on the environment.

But let us first consider the ways in which minerals are extracted and undergo the initial process of refinement. To assist in this task a case study is used which involves one of the United Kingdom's most valuable mineral resources.

2 *Identifying key issues*

2.1 *The landscape of china clay*

For those who live in Britain, the landscape of Cornwall must be among the most familiar and appealing. But anyone who has travelled west by rail or road eagerly anticipating the beaches of Newquay or St Ives, cannot fail to have been struck by the awesome nature of the 'Cornish Alps'. These are not the aftermath of some gigantic earth movement of distant geological time, but the results of human activity – the extraction from the Earth's crust of china clay, i.e. the mineral called kaolin.

To appreciate just what has occurred as a result of human intervention in this, one of Britain's most extensive mining landscapes, it is necessary to do no more than turn a few kilometres north from St Austell on the coast or south from the village of Roche along the A30 road. Here, in a vast area of over 8000 hectares, the entire landscape, apart from the villages, has been transformed by mining. A few hours spent examining this area cannot explain everything about the impact of the extraction and processing of minerals on the environment, but it is a spectacular place to start.

As an introduction, it is not necessary to turn to a technical description of the area: that of the Cornish writer Daphne du Maurier (1907–89) will serve as well. Look at Box 4.1 and the extract from her book *Vanishing Cornwall*. When you read it, also consider carefully the photograph in Plate 1, which rounds out the picture she conjures up in words.

Box 4.1

In the china clay country is the strange, almost fantastic beauty of the landscape, where spoil heaps of waste matter shaped like pyramids point to the sky, great quarries formed about their base descending into pits filled with water, icy green like arctic pools. The pyramids are generally highest and the pools deepest, on land which is no longer used.

These clay heaps with their attendant lakes and disused quarries have the same grandeur as tin mines in decay but in a wilder and more magical sense, for they are not sentinels of stone or brick constructed to house engines, but mountains formed out of the rocky soil itself, and the pools, man-made, are augmented by water seeping from underground sources and by the winter rains. Sites in full production may work close at hand, cranes swing wide, trolley-buckets climb to the summit of a waste-heap . . . before unloading and returning to base, lorries pass in and out of entrances to the road, the precincts barred by wire and DANGER notices; but the discarded pyramids and pools seem as remote from industry near by as any lonely tor upon the moors.

Wild flowers struggle across the waste, seeds flourish into nameless plants, wandering birds from the moorland skim the lakes or dabble at the water's edge. Seagulls flying inland hover over the surface. There is nothing ugly here. Cornishmen are wresting a living from the granite as they have done through countless generations, leaving nature to deal in her own fashion with forgotten ground, which being prodigal of hand, she has done with a lavish and careless grace. (du Maurier, 1972)

These images in words, although colourful, underline several important issues concerning the winning of china clay; these issues are also appropriate to many other minerals that are important to the lifestyles of those living in the developed countries. At the same time they identify some key environmental impacts that are the consequence of minerals extraction.

2.2 *A valuable resource*

In the extract in Box 4.1, the writer suggests that kaolin is not the only mineral substance to be extracted from the rocks of Cornwall. Tin mining might have been the most important Cornish industry in the eighteenth and nineteenth centuries, the relics of which are still to be seen in the derelict engine houses, but china clay has now come into its own. Even so, this, too, has been exploited for two centuries. It took the discoveries of William Cookworthy (made some time between 1745 and 1748) to turn what was a useless material that had lain in the earth for 290 million years into a valuable material which is ideal for making porcelain. And it took the discovery in this century of its value as a surface coating in the manufacture of high-quality paper to turn the extraction of Cornish china clay into a multi-million pound industry of major international significance, second only to North Sea oil in the value of exports.

The realisation that a mineral can perform a valuable function makes it worthy of exploitation and therefore it becomes a resource. But it need not remain of intrinsic value for ever. Grimes Graves near Thetford in Norfolk was prized for a thousand years by Neolithic peoples as a site for flint axe heads which were essential to their lives. However, the discovery of more efficient bronze substitutes meant that the value of flint as a resource fell away to insignificance by 1400 BC. From being a centre of industry, pocked with shafts from which flint was hewn, Grimes Graves reverted to an uninhabited area of heather and birch forest.

Q What is a mineral resource?

A Defining the word 'mineral' presents no problem. A good dictionary will tell you that it is any substance which is not animal or plant, although it could have originated from once living material, as will be seen below. But for it to become a resource it must be subject to human appraisal and to be judged as having a value. Remember that resources are not fixed over time or space. The Bushmen of the Kalahari desert survive in a world in which minerals that are valued by industrial societies play no part. Their resources, which people in developed countries would hardly consider as such, are the succulent plants, grubs and insects which make their lives possible in an arid environment. Thus resources are as dynamic as society itself.

2.3 *Methods of extraction*

Of more significance in environmental terms are the images purveyed in the extract in Box 4.1 and in the photograph regarding the magnitude of china clay operations. In one year in this area alone (1993), the operations yielded 2.6 million tonnes of china clay and 20 million tonnes of waste rock – all from pits which can be as deep as 85 m and cover 40 hectares. The method of excavation normally used here, **open-pit extraction**, is that most commonly employed in the extraction of minerals. It is a method well suited to any shallow deposit, whether it is china clay or sand and gravel. A metal ore that (for reasons which will become apparent later) has been disseminated through some other rock material will often be extracted in this way, as will hard rocks such as granite. But this mode of extraction is suitable only so long as the depths involved are not too great.

Open-pit extraction of the kind in Box 4.1 offers the advantage of operating in a flexible way. Because the working surfaces of the mineral are exposed over a wide area, it is easier to expand or contract production. Open pits are also economically more attractive. This is because they allow the application of large earth-moving machines on a scale which makes for maximum cost-effectiveness. Not surprisingly, it is now the most commonly used method of winning minerals, even if it is the most environmentally disturbing.

Before a new pit is begun in Cornwall, as elsewhere, all the **overburden** (overlying unwanted rock) must be cleared to expose the valuable material beneath. When the overburden is added to other forms of waste which come from the processing of china clay (quartz, and a fine mud-like substance known as micaceous residues), the ratio of waste extracted to valuable material can rise to between 7:1 and 8:1. For open pits where metal ore is extracted, ratios can vary widely; indeed for the rare metals they can be between 100:1 and an extreme of 500 000:1. For open-pit operations extracting coal, ratios can be 15:1. These coal workings, referred to as open-cast pits, are unusual in that the overburden is not removed before the extraction of the mineral. Instead, **open-cast extraction** involves the *progressive* removal of the overburden, prior to the extraction of the mineral and the return of the overburden to the site. Coal workings of this kind became increasingly familiar in the East Midlands, north-east Yorkshire and in central Scotland during the expansion of this form of production from the 1980s.

Strip mining, for example for coal in the USA, is similar, but the overburden and minerals associated with the coal are left exposed to the weather.

Underground mining, the other major form of minerals extraction, entails removing minerals from the end of often narrow tunnels where there are obvious limits to the application of more human effort and machines (Figure 4.1). Underground mining could not be used to extract china clay because of the nature of the material and its location in deep pockets of surrounding granite which are close to the surface of the land. Where underground mining *can* be used, one of two methods is followed. In the open-stoping approach, the rock body from which the valuable material is extracted must be sufficiently strong to allow underground cavities to remain after the ore has been removed through drilling and blasting. The other method, that of filled-stoping, requires that waste materials are used to fill those portions of the mine that have been worked out. Where metallic ores have been extracted, the waste will include the **tailings** (that is, the fine waste rock left after the valuable ore has been processed, using chemical or mechanical methods). Sufficient ground support is thus provided to allow further extraction of the ore, but it also has the added environmental advantage of reducing surface waste tipping.

Another means of extracting certain minerals involves drilling **boreholes** from the surface to permit the removal of the valuable material. In the case of oil and natural gas, where these are trapped in sedimentary rocks deep below the land surface, they usually emerge under pressure and without pumping. Many of you will have seen oil gushers, even if only on

▲ *Figure 4.1 Mechanical mining underground. Even though the use of a coal cutting 'shearer' can greatly increase productivity, such operations cannot approach the cost effectiveness of open-pit extraction. They also result in more waste material that will need to be disposed of.*

film! For the winning of minerals such as salt and potash, however, the injection of water via the borehole creates a solution which can then be pumped to the surface. Sulphur can be extracted by the injection of super-heated water under pressure. This melts the sulphur which escapes to the surface via another borehole.

The environmental disturbances created by these methods of extraction in relation to the winning of the various minerals are discussed later in Chapter 5. For the moment it is worth emphasising that because the amount of wastes produced from underground workings is small compared with that from open pits, the former produce less surface disturbance as a result of their disposal. Coal mining is perhaps an important exception insofar as the surface tipping of wastes can be a major environmental disturbance.

2.4 Minerals processing and wastes

The method of winning china clay from massive open pits is among the simplest of all those used in extraction. Unrefined kaolin is removed from the pit walls by a powerful water jet. The resulting clay slurry, together with its attendant quartz fragments and micaceous residues, is pumped away to tanks where the valuable mineral settles out from the waste. Only the method for sand and gravel, which involves washing and perhaps some crushing after extraction, is more straightforward. This simple procedure contrasts with the methods used for the extraction of metallic ores. For these, the means by which the unwanted constituents of the ore body are removed (**minerals processing**) is carried out on site because of the low value of untreated ore compared with ore in its refined state. Only in the case of the most valuable minerals (or those mined in high concentrations such as iron-ore) would it be economic to transport untreated ores over long distances for processing at locations which would be convenient for their further fabrication.

The extraction of metallic ores from the parent rock is carried out in two stages in the **milling** process. First, the mined rock containing the metal ore is reduced in size by crushing or grinding. Secondly, it is concentrated, a means by which the desired metallic ore in the rock body is separated from other materials which are discarded as waste. Separation may be carried out by a number of different methods depending on the properties of the mineral. Gravity separation, the means used for china clay, can also be used to concentrate some metals (such as gold and silver), but for other minerals it has been largely replaced by **flotation**. This method of concentrating minerals is based on the principles of surface chemistry and is most widely used with metal sulphide ores. The chemicals which are used to 'float out' the ore from the fine rock particles are selected not only for their effectiveness, but also for those qualities that may do least harm to the environment should they be discharged into local water courses. Although most flotation plants are able to recycle the liquids used for processing, they all contain a **settling pond** stage which allows solid particles, including metals, to settle at the bottom by gravity, as well as permitting the treatment of any undesirable acid formed from sulphides (salts created by sulphur combining with metals) which may have been present in the parent rock. These are important issues which are discussed further in Chapter 5. A third approach to the extraction of metallic ore from the fine rock material uses a magnetic separation process, especially for iron-ore. Finally, where the ore is dry, electrostatic techniques (which involve the attraction or repulsion of electrically charged bodies) can be used, particularly on iron and titanium ores.

In the quotation in Box 4.1, no mention is made of the tailings. Later on in her book, though, du Maurier refers to the disposal of the fine materials left after refining the kaolin which caused 'the streams to run from the watershed milky white'. Because the quantities of these tailings long remained relatively small, the practice of discharging them into the local watercourses continued until the 1970s. Only then did increased environmental awareness bring about the ending of the practice. In the processing of metallic ores, whether they have been extracted by open-pit or underground techniques, the majority of waste rock finishes up in the pit known as a **tailings pond** as the residues of the milling process.

As the metallic ores extracted have become less rich in metal content over recent years, the amount of tailings to be disposed of has grown – indeed, it is evident that the lower the metal content of ore that can be worked economically, the higher the waste content. Thus, for example, uranium winning operations produce about 99% waste, most of which is in the form of tailings, compared with a chemical mineral such as potash where the percentage of waste is about the same as that of the valuable mineral. The fine materials resulting from processing minerals are now less densely packed than they were as part of the parent rock and are between 20% and 40% greater in volume. This exacerbates the disposal problem and frequently adds to the environmental problems they may cause (Figure 4.2).

▲ *Figure 4.2 Tailings at the Inco nickel mines, Sudbury, Ontario. Rock waste from the milling of the nickel bearing ore deposited in a tailings pond at Inco's Copper Cliff operation. The total area of these ponds in 1989 stood at 1100 ha but this has now been expanded to double the size.*

China clay tailings are inert and their discharge to the wider environment, in whatever way, is largely a problem of visual disturbance. However, because tailings from the milling of metal ores are impregnated with residual heavy metals and may contain sulphides or even radioactive materials, a number of problems can be created. In particular, when the tailings site reaches the end of its useful life and is abandoned, the pollution of both air and water in the vicinity of the mine may occur. Also, the further processing of some metallic ores beyond the milling stage to the point at which they become metals can create a problem of environmental significance.

2.5 *Metals production and wastes*

Smelting is the process by which most concentrated metal ores emerging from the mill are transformed to a pure metal. The range of environmental impacts that can arise varies greatly depending on the nature of the ore itself.

Box 4.2 lists the most important metals and their uses. Although a few of these, such as platinum, gold, silver and occasionally copper and mercury, are found in the metallic state, they mostly occur in nature in combination with oxygen as oxides, with sulphur as sulphides, with oxygen and carbon dioxide as carbonates and with oxygen and silica as silicates.

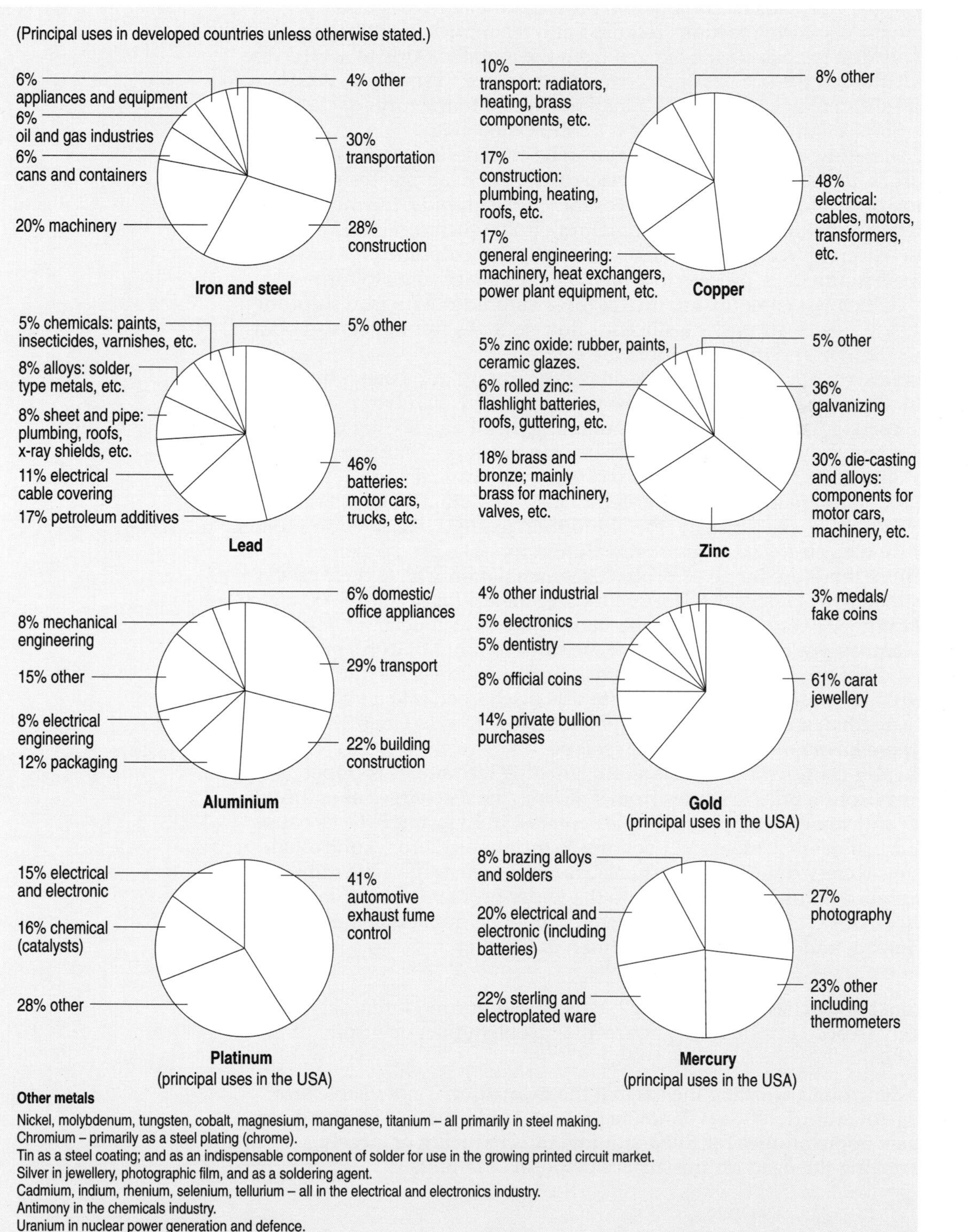

Other metals

Nickel, molybdenum, tungsten, cobalt, magnesium, manganese, titanium – all primarily in steel making.
Chromium – primarily as a steel plating (chrome).
Tin as a steel coating; and as an indispensable component of solder for use in the growing printed circuit market.
Silver in jewellery, photographic film, and as a soldering agent.
Cadmium, indium, rhenium, selenium, tellurium – all in the electrical and electronics industry.
Antimony in the chemicals industry.
Uranium in nuclear power generation and defence.

The metallic oxides, such as those of iron, zinc and tin, can then be reduced to a metal in a smelting furnace. The charge of the smelter consists of the metallic oxide, fuel and a flux. The fuel, which is usually coal or coke, serves a dual purpose. Its combustion provides the necessary heat to promote the chemical reactions required and it supplies the reducing agents needed to abstract the oxygen from the metallic oxide to set free the metal. Although the ore has been concentrated prior to smelting, waste material still remains. The **flux** is a material added to the furnace charge that combines with the waste to form a solid called **slag**.

The smelting processes, can of course, be considerably more complex and vary with the kind of ore under treatment. For example, from some sulphide ores, such as lead sulphide or mercuric sulphide, the metal may be liberated by smelting with a flux containing iron which combines with the sulphur, forming iron sulphide. Copper and nickel sulphide ores can undergo a multi-stage process in which the ores are smelted to form 'matte', which is a complex mixture of these sulphides with iron sulphide. This is blown with air under a siliceous flux, so that sulphur is given off as sulphur dioxide and the iron enters the slag.

There are a number of other means of separating out a metal from an ore, but electrolysis is by far the most important. This involves passing an electric current through a solution containing metal ions so that metal is deposited on the cathode. It is mainly used to produce aluminium. Pure aluminium oxide is separated from bauxite with a caustic soda solution. At the reduction plant, usually located to take advantage of cheap and plentiful supplies of electricity, the aluminium oxide is dissolved in molten cryolite and electrolysed to give pure aluminium. The production of aluminium therefore involves processes which not only have their power requirements far greater than those for other metals (between 15 000 and 25 000 kWh for every tonne of ingot aluminium, or as much as 20 times the energy equivalent needed to produce the same amount of iron), but in the production of the initial oxide leaves residues of caustic soda and fluorides in a red mud. Until recently, this waste was discharged into a tailings pond and created major environmental problems. However, as Chapter 5 makes clear, these have now been largely resolved.

The slag produced by the standard smelting techniques is tipped, but of all the environmental outcomes from smelting, the discharges of sulphur dioxide and small metal particles (particulates) from smelter stacks cause the most problems. Unless their venting to the atmosphere is controlled, they can severely damage ecosystems in a wide area around the mine, and the sulphur dioxide can contribute to the wider problem of 'acid rain', already discussed in Chapter 2. The environmental impacts of smelting are given further and more detailed attention in Chapter 5.

Q From the discussion in Section 2, why is underground minerals mining likely to pose less of an environmental problem than open-pit extraction?

A Underground working means that the excavation is not visible and large overburden tips will not be present. Underground working also offers opportunities for filled-stoping, thus reducing or largely removing the need for the surface disposal of tailings.

Q What are the chief environmental impacts of the processing of metallic ores?

A The output of the mill by way of tailings has a major impact if these are surface-dumped in a tailings pond. The smelting of metal ores, which employs the heating of oxides with the fuel acting as a reducing agent, will give off carbon dioxide, now known as an atmospheric pollutant. Roasting sulphide ores and processing matte produces sulphur dioxide which, when discharged to the atmosphere, may contribute to the problem of acid rain. The venting of metal particulates also has adverse effects on ecosystems. The production of aluminium oxide from bauxite uses process chemicals which may remain in the residues of red mud, which were until recently a major environmental hazard. The reduction process makes considerable demands on electric power, which in itself may have environmental impacts since the fossil fuels used in power generation may result in the discharge to the atmosphere of both carbon and sulphur dioxides.

2.6 Healing the scars

What else does Daphne du Maurier state that is relevant? A striking feature of the author's vivid narrative not yet addressed is the idea that nature will take over to heal the scars of a defunct extractive enterprise, and to create a landscape of value, if only in its new-found beauty. Apart from the fact that her romantic view of this vast extractive area is not shared by all, the notion that such mineral workings can necessarily be abandoned to nature when their extractive life is over is hardly tenable; indeed, as has already been hinted, the aftermath of mining may not always be so benign.

In the late twentieth century many developed countries accept the idea that mining is a temporary affair and that land is a scarce resource. This is particularly true of the United Kingdom which has a large population compared with its size. This means that if minerals extraction is allowed to proceed, damage to the physical environment caused by extraction and minerals processing must not only be minimised during the working life of a mining operation, but totally rectified, insofar as this is possible, once it is all over. Indeed such a situation is one which companies winning the china clay now have to accept. Under the Town and Country Planning Act of 1971, they cannot legally walk away and leave a worked-out landscape to nature, even if relics remain of a time when this was possible, of which both our quotation and its accompanying photograph are a reminder.

2.7 Summary

In this section the extraction of kaolin has been used as a way into building up a broad picture of what is likely to be involved in the winning of a whole range of minerals, and some preliminary thought is given to what the environmental consequences of these activities might be. These likely effects, as they have been described, are largely of a local impact, but the secondary processing of minerals (for example, from ores into metals) indicates that these may have considerably wider effects, a matter to be considered in detail in the next chapter.

In Sections 3 and 4, consideration is given to those factors which determine the location and the circumstances under which minerals are exploited. Since the ultimate constraint upon where minerals are worked is that of geology, this aspect of the wider context for minerals extraction and their impact on the environment is examined first.

3 Patterns of location and extraction

3.1 A geochemical lottery?

Mineral resource analysts often use the idea of the average distribution or abundance (resources) of one particular group of minerals, the metallic ores, in the Earth's crust as a standard against which to evaluate the percentage of metal content that would constitute a resource of sufficient worth to win and process (reserves). Look at Table 4.1 which gives the average abundance of 20 metals, divided into percentages for the ocean and the continental crust. Any attempt to work a precious metal such as gold at 0.000 000 35% would plainly be absurd; indeed, economically workable deposits averaged 0.000 13% in the 1980s.

The reality of mineralisation, and the many different processes by which this has come about, is that many minerals have become concentrated in some places and not others, almost as if, at least to the non-geologist, it was the result of some great geochemical lottery. For a brief background insight into the processes involved, you should consult Box 4.3. What you need to understand as a result of your reading of this is that deposits of minerals (and this applies as much to metallic ores as any other kind of mineral) are only worked where geological circumstances have concentrated them in sufficient quality and quantity to make their

Table 4.1 The average percentage abundance of metals in the Earth's crust

Ocean crust		Continental crust	
aluminium	8.4	aluminium	8.3
iron	7.5	iron	4.8
titanium	0.81	titanium	0.53
manganese	0.18	manganese	0.1
vanadium	0.017	vanadium	0.012
chromium	0.016	zinc	0.008 1
nickel	0.014	chromium	0.007 7
zinc	0.012	nickel	0.006 1
copper	0.008 5	copper	0.005
cobalt	0.003 7	cobalt	0.001 8
lead	0.001	lead	0.001 3
tin	0.001 9	uranium	0.000 22
molybdenum	0.001 5	tin	0.000 16
uranium	0.000 1	tungsten	0.000 12
tungsten	0.000 094	molybdenum	0.000 11
antimony	0.000 091	antimony	0.000 045
mercury	0.000 011	mercury	0.000 008
silver	0.000 009 1	silver	0.000 006 5
platinum	0.000 007 5	platinum	0.000 002 8
gold	0.000 000 35	gold	0.000 000 35

Source: *Atlas of Earth Resources* (1979), Mitchell Beazley, London, p. 149.

extraction viable. The corollary of this is that the environmental impacts caused by the extractive methods and processing techniques that we have already looked at are also spread *unevenly* across the surface of the Earth. Indeed, this in itself helps to account for the dramatic appearance of the Cornish china clay workings in an otherwise intimate landscape of green rolling hills.

That minerals are found and exploited with great spatial variation is something that now must be considered in a little more detail. In order to do this more effectively, it is useful to work within a classification which groups minerals together and takes account of their degree of availability and their physical attributes.

3.2 Non-metallic minerals

It is sensible to start with a group with which most people are familiar – the **non-metallic minerals**. As this name implies, they are not exploited as an economic source of metals, although they may contain metals in small amounts, and are sometimes referred to as industrial minerals. This group includes a collection of widely found materials, mainly used in the construction industry for building and roads. These are called **aggregates**, and include sand and gravel, igneous and metamorphic rocks, sandstone and limestone. All tend to be used according to local availability since they can often be substituted with one another. Aggregates thus serve local markets.

Their abundance, plus the relatively modest amount of processing they require (usually no more than crushing and grading) means that they are generally of low value. Sand and gravel, for example (Figure 4.3), may cost as little as £3.50 per tonne (ex-works 1995 UK prices). The only exceptions

◄ *Figure 4.3*
A typical sand and gravel working seen through a gap in an amenity bank designed to screen extraction from public view. Compared with hard rock quarries, such operations are shallower, more extensive and yield much less material. However, they are more easily worked using earth-moving machinery.

in the widespread non-metallics group are brick clays where they have been transformed into bricks, and limestone used for the production of cement. In both cases the application of much more sophisticated processing has led to the addition of substantial value to the basic raw material. It thus follows that for most construction materials transport costs are of particular significance, with the delivered price doubling at distances of 50 km from their points of production. The spatial distribution of such workings has therefore tended to reflect road transport costs. They remain operations of medium size and show a dispersed pattern related to major

Box 4.3 Formation of mineral deposits

Human needs depend on relatively rare geological processes where the low amounts of metals in ordinary rock are concentrated to create a richer deposit which is economical to mine. There are two main types of processes, external and internal, that create a mineral deposit. Some of the external – those taking place near or at the Earth's surface – are shown in (a) to (c), together with examples of the minerals involved; those in (d) to (g) are internal, taking place within the Earth, and have been responsible for the concentration of a wide range of metals in the form of oxides or sulphides. In practice many mineral deposits have been produced by more than one process, each taking place at a different stage in the deposit's formation. An ore deposit, for instance, may have been formed by hydrothermal processes but on reaching the surface it may have undergone further concentration by secondary enrichment. Internal processes are associated with earth movements and with molten rock (magma) welling up from the interior into the crust.

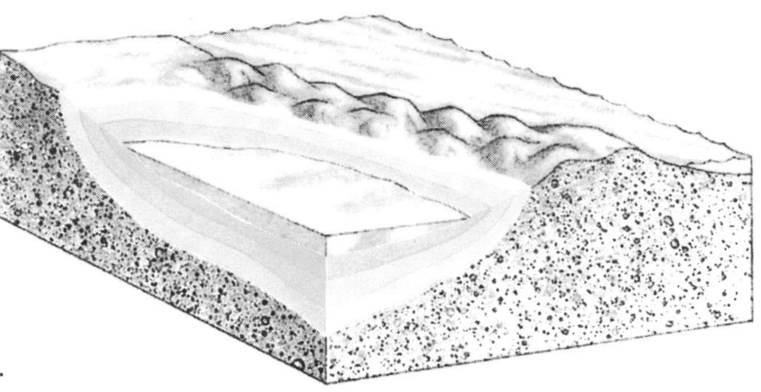

(a) Mineral salts form when the sun evaporates sea waters trapped in a barred basin. Deposits tend to settle in layers; the more insoluble are deposited first, the most soluble last. Sodium chloride (common salt) has been formed in this way. However, many other mineral resources have been formed by deposition beneath continental shelf seas, including iron-ores, clays and limestones.

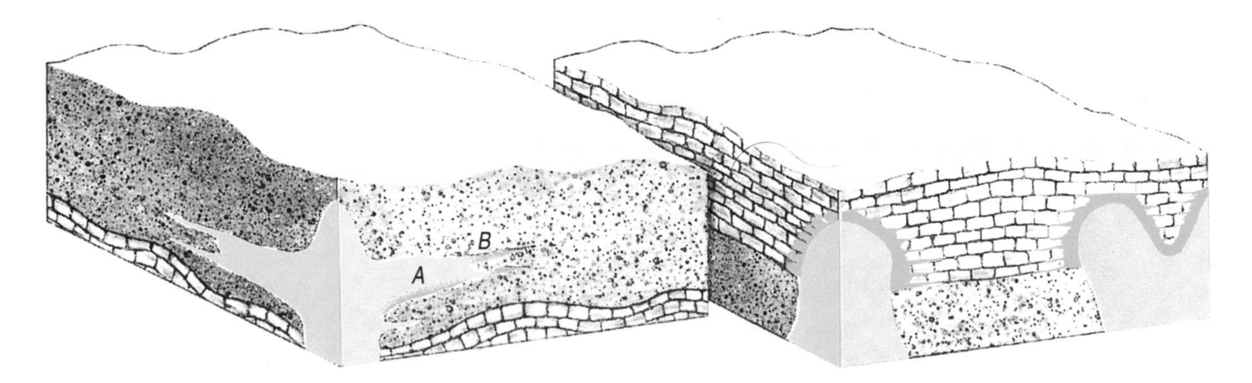

(d) Main methods of magma separation: A, minerals crystallise and settle to the bottom of the magma chamber; B, the magma solidifies, forcing the molten minerals into fissures.

(e) An ore deposit may result when gases and fluids escaping from the magma react with the rock face and later the composition of the rocks by adding valuable elements.

centres of population. This is in spite of a marked increase in quarry size over the past 25 years, with the costs of serving a wider market offset by economies of scale, improved transport and greater on-site processing to increase the value of the product. However, the movement of aggregates by ship from quarries located on a coast to areas of high demand can be economical, as research carried out for the Department of the Environment has indicated. The economies of scale achieved in superquarries producing up to 20 million tonnes a year, allied to low-cost shipping, enables materials to be moved many hundreds of miles, as exemplified by the mammoth

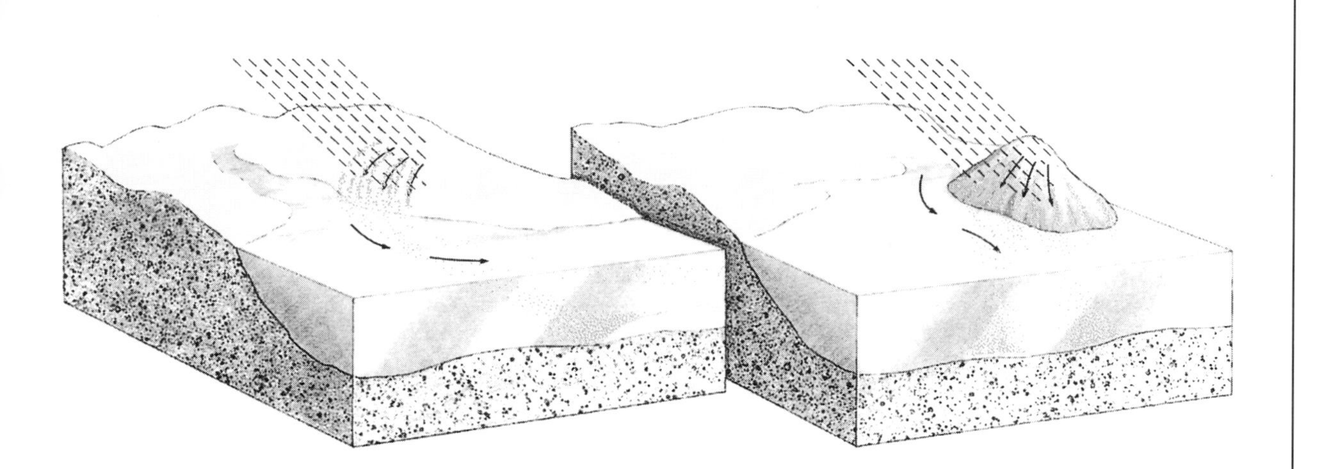

(b) Placer deposits occur where eroded particles of rock material pass into a water body. There, the heavier metal-bearing particles settle out early, leaving the lighter rock particles to be transported and deposited elsewhere. Gold placers were the objects of the Klondike gold rush in the nineteenth century.

(c) Residual deposits occur where minerals concentrated in a rock are left behind when other material within the rock is removed by mechanical or chemical weathering. Weathering under humid tropical conditions causes the residual concentration of aluminium hydroxide (bauxite), the raw material for aluminium manufacture.

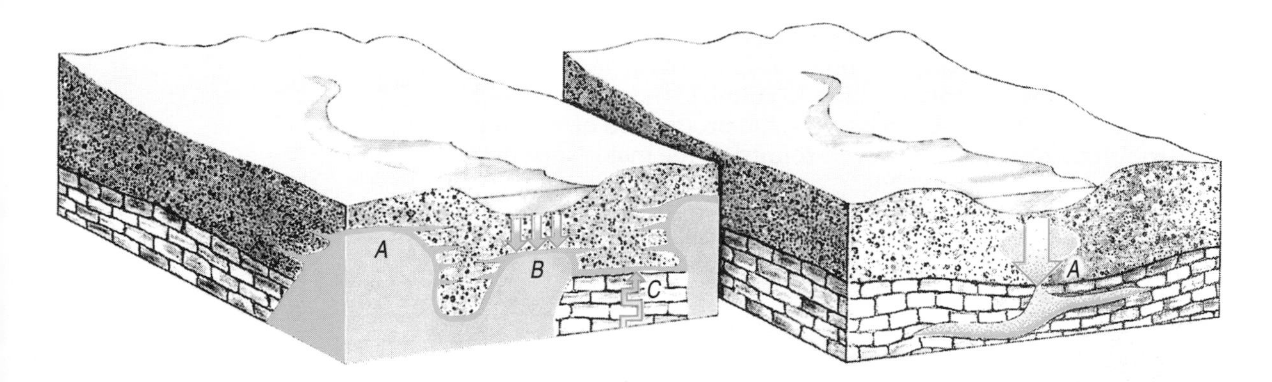

(f) Dissolved minerals are carried into fissures by: A, hot fluids left when magma crystallises; B, surface water heated by magma; C, water from the hotter mantle.

(g) In some cases, mineral deposits are formed when an acidic water containing a metal reaches an alkaline environment, or where there are changes in oxygen content (A).

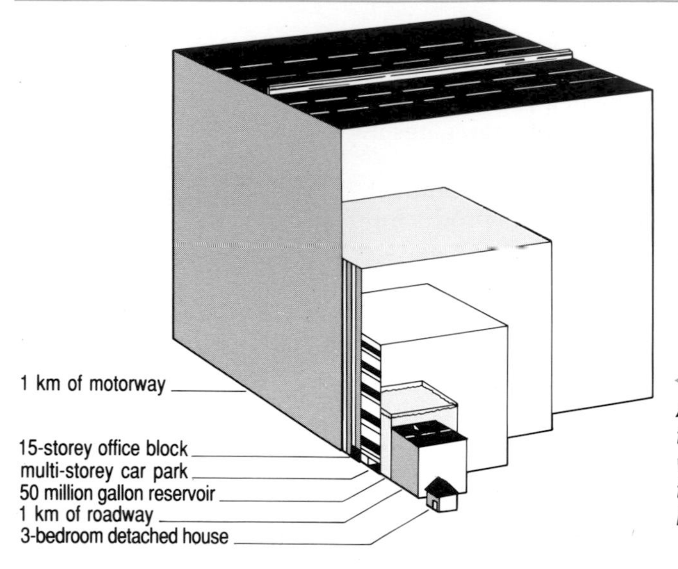

Figure 4.4 Aggregate and construction. Aggregates are the cheapest of all minerals to exploit, and vast quantities are used in construction work: 50 tonnes on average are needed for a three-bedroom house and 62 500 tonnes for every kilometre of a six-lane motorway.

quarry near Oban on the west coast of Scotland which supplied granite aggregate to the Channel Tunnel operation. But the demand for hard rock aggregates has in some areas been such that much more modest operations in terms of size can export their output over long distances, provided transport is by sea. One coastal quarry near Penzance in Cornwall has developed substantial trade links with Belgium and Holland. Even so, such materials are generally not traded internationally. Construction materials also share one other characteristic – whether they are used in crushed form or as stone blocks, they are required in vast quantities. Figure 4.4 illustrates this point.

In contrast with the widespread non-metallic materials, *local* non-metallic materials are much more restricted in terms of the quantity available, and, as their name suggests, are less widely found. These factors, of course, lead to enhanced values and thus an ability to stand high transport charges to more distant markets. The minerals in this group are mostly raw materials for the chemical and fertiliser industries. Examples are shown in Box 4.4. Although not as widespread as the aggregates, production of these minerals is in many cases distributed across the continents and among the industrial powers. Thus, of the leading chemical materials such as sulphur and salt, some 30 countries have access to their own internal sources of each. Even barytes, the production of which is least widely distributed, is available in seven countries, the major suppliers being the USA and the former USSR.

3.3 The metallic ores

Activity 1

A consideration of metallic ores can usefully be started by comparing the prices per tonne for zinc, lead and copper with those quoted earlier for sand and gravel. You will be able to find these in the business section of many newspapers under 'commodities'.

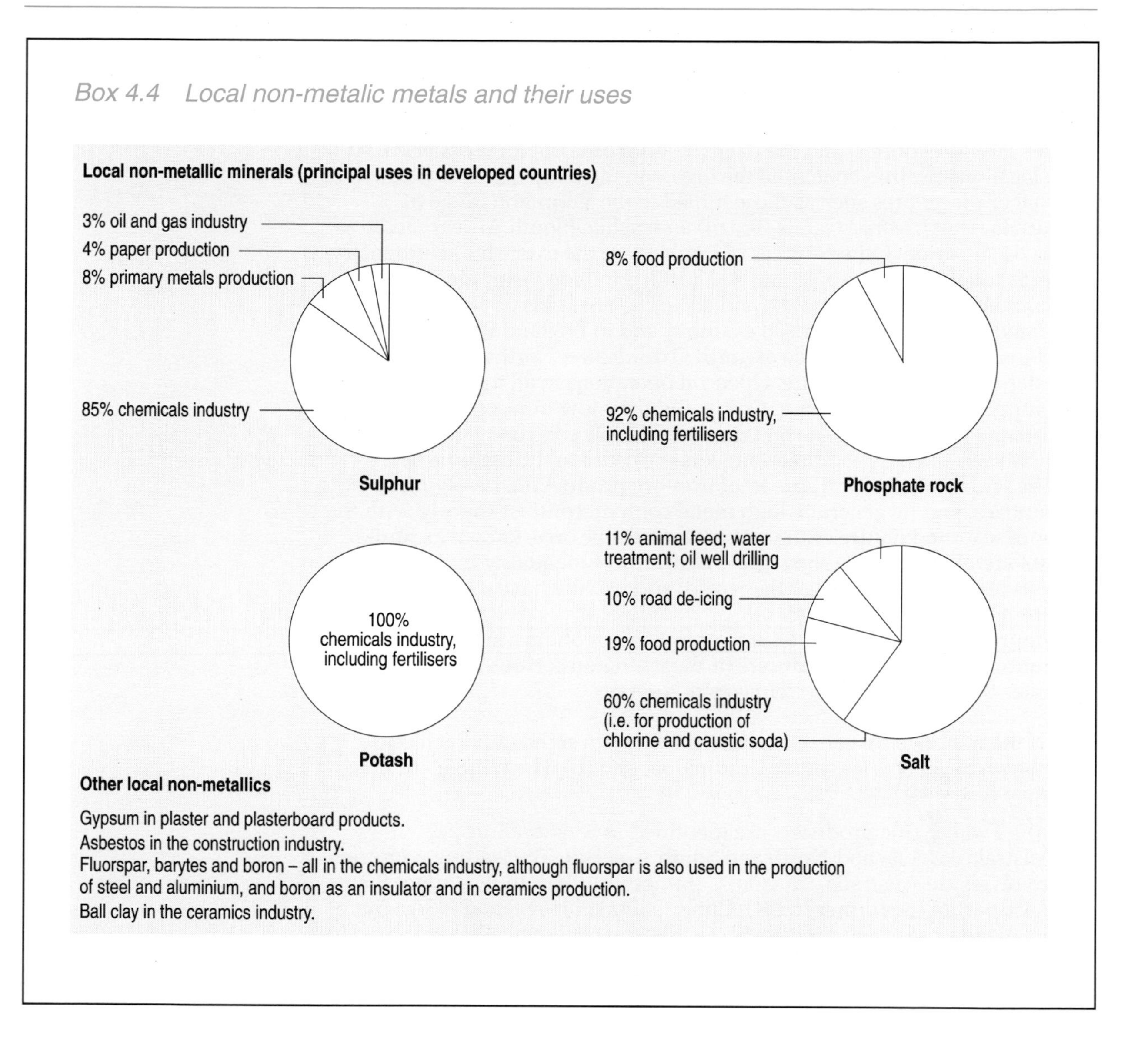

Q Why do you think metals are so much more costly per unit than aggregates?

A Metallic materials are much less widely available than aggregate materials. This, together with the sophisticated processes involved in extracting the ore from the host rock, as mentioned in the previous section, makes them of relatively high value.

Although values are generally high for **metallic ores**, there is just one exception, that of **ferrous metal ores**, i.e. the iron-rich rocks known collectively as iron-ore. Once processed, iron-ore is without doubt the most indispensable of all metals and is the basis of modern industry. It has many uses, as you have already seen in Box 4.2. The much greater accessibility of iron-ore compared with other metal ores stems from the fact that it may be formed during a number of different geological circumstances.

Two of the chief iron-ores, haematite and magnetite, have been extracted on a large scale from massive deposits in the oldest of rock formations. Examples are Kiruna in Sweden, Krivoi Rog in Ukraine (now the most important production areas in the world), Pilbara in Australia (now the world's largest exporter) and the Lake Superior area of North America. At these locations the iron content of the ores is in the range 50–60%. Elsewhere, residual or placer ores such as those mined in the Vermilion range of Minnesota (USA), Minas Gerais (Brazil) and Sishen (South Africa) can be as rich as 64–68% iron. Other sources of iron-ore are the extensive sedimentary deposits laid down between about 300 and 100 million years ago, but their iron content is only between 20% and 40%. The ore fields of western Europe associated with limestones are an example, and in England they are to be found in the extensive deposits of north Oxfordshire, Northamptonshire, Leicestershire and Lincolnshire. Open-pit operations in all these counties once supplied the steel-works at Corby. But their low iron content compared with other sources of iron-ore and the unavoidable environmental disturbance created by their working led to closure in the early 1970s.

The wide geographical spread of iron-ore production, involving some 40 countries, and its generally high metal content, contrast sharply with the far more scattered nature of the rest of the metallic ores, known as **non-ferrous metal ores**. These show great distributive inequality in their deposits among the nations of the world, and usually have a low metal content, frequently below 1%.

Figure 4.5 further illustrates this point since it offers a summary of the distribution of the major producers of the main non-ferrous metal ores and identifies those with high percentages of reserves.

Q Of the minerals that are extracted in their own right, (a) which one enjoys the most widespread production; and (b) which three are the most restricted?

A (a) Twelve zinc producers are identified in western Europe, Australasia, Asia and North and South America. The leading countries involved are Australia, Canada, Commonwealth of Independent States (CIS, part of the former USSR), Chile, China, Portugal and USA. There are also 11 cadmium producers. As it is often geologically associated with zinc it is not surprising to find it produced in Australia, Canada, CIS and USA.

(b) The production of tungsten is heavily dominated by China, followed by the CIS but with less than one-third of the output of China. Platinum is the commonest of a group of six precious metal ores, all of which are among the rarest metals in the world. Of the two producers of any significance, only one is of major importance and that is South Africa. Mercury, a metal which in its refined state is the only one that is liquid at ordinary temperatures, has the extraction evenly spread across the CIS, China and Spain, but production totals are small and its value high – thus a very large percentage is recovered and reused.

One of the minerals with a restricted production range is manganese, which in 1993 was dominated by China and the CIS, However, it is estimated that there are around 1.5 trillion tonnes in the form of nodules lying on the floors of the major ocean basins. These nodules, which are to be found at depths of between 30 m and 3000 m, contain not only 24% manganese, but also 1.6% nickel, 1.2% copper, 0.21% cobalt and 0.05% molybdenum. The success of their recovery has depended on two factors:

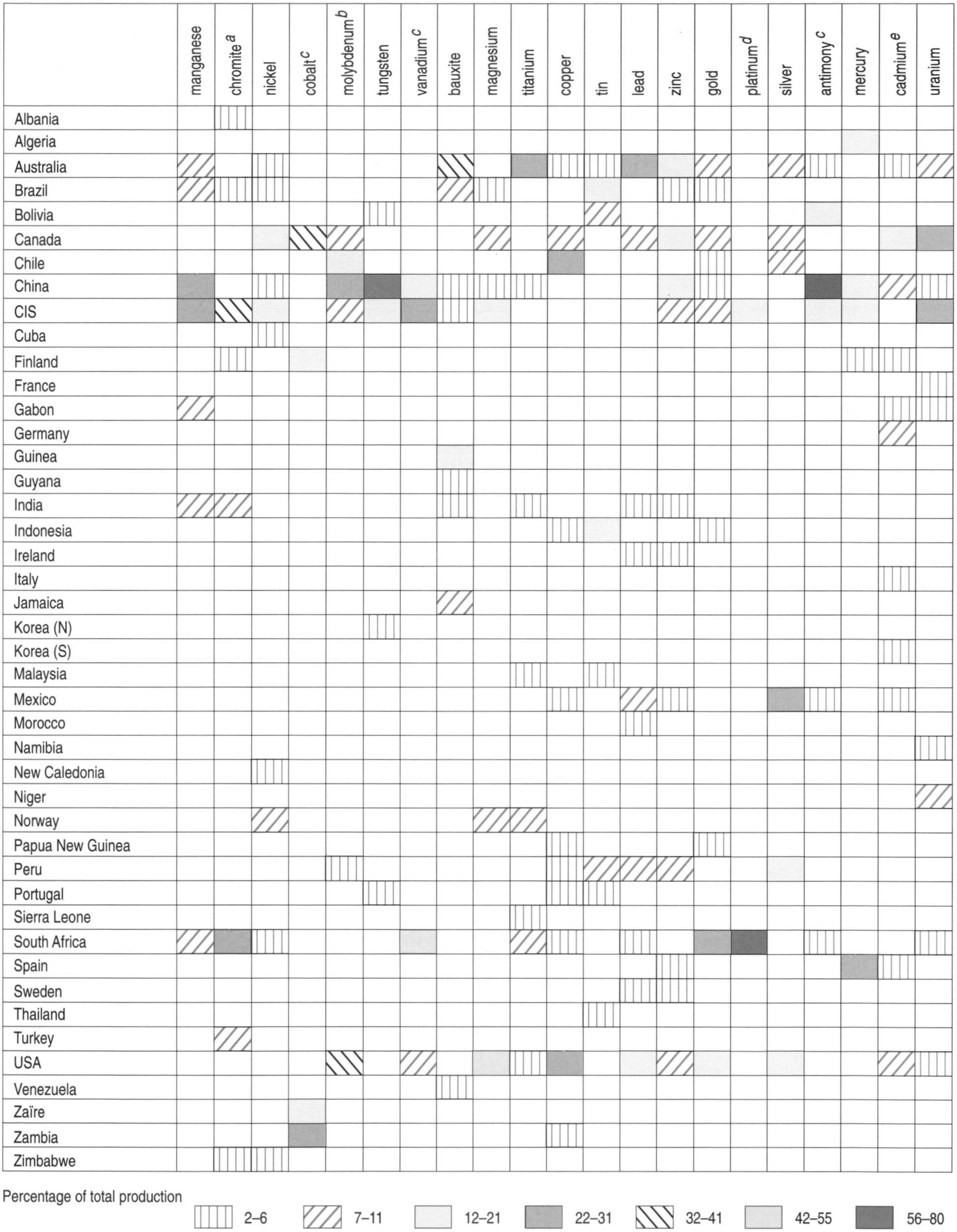

	manganese	chromite[a]	nickel	cobalt[c]	molybdenum[b]	tungsten	vanadium[c]	bauxite	magnesium	titanium	copper	tin	lead	zinc	gold	platinum[d]	silver	antimony[c]	mercury	cadmium[e]	uranium
Albania		2–6																			
Algeria																			12–21		
Australia	7–11		2–6					32–41		22–31	2–6	2–6	22–31	12–21	7–11		7–11	2–6		2–6	7–11
Brazil	7–11	2–6	2–6					7–11	2–6			12–21		2–6	2–6						
Bolivia						2–6						7–11						12–21			
Canada			12–21	32–41	7–11				7–11		7–11		7–11	12–21	7–11		7–11			12–21	22–31
Chile					12–21						22–31				2–6		7–11				
China	22–31		2–6		22–31	56–80	12–21	2–6	2–6	2–6				12–21	2–6			56–80	12–21	7–11	2–6
CIS	22–31	32–41	12–21		7–11	12–21	22–31	2–6	12–21					7–11	7–11	12–21		12–21	12–21		22–31
Cuba			2–6																		
Finland		2–6		12–21															2–6	2–6	
France																					2–6
Gabon	7–11																			2–6	2–6
Germany																				7–11	
Guinea								12–21													
Guyana								2–6													
India	7–11	7–11						2–6		2–6			2–6	2–6							
Indonesia											2–6	12–21			2–6						
Ireland													2–6	2–6							
Italy																				2–6	
Jamaica								7–11													
Korea (N)						2–6															
Korea (S)																				2–6	
Malaysia										2–6		2–6									
Mexico											2–6		7–11	2–6			22–31	2–6		2–6	
Morocco													2–6								
Namibia																					2–6
New Caledonia			2–6																		
Niger																					7–11
Norway			7–11						7–11	7–11											
Papua New Guinea											2–6				2–6						
Peru					2–6						2–6	7–11	7–11	7–11			12–21				
Portugal						2–6					2–6	2–6									
Sierra Leone										2–6											
South Africa	7–11	22–31	2–6				12–21			7–11	2–6		2–6		22–31	56–80		2–6			2–6
Spain														2–6					22–31	2–6	
Sweden													2–6	2–6							
Thailand												2–6									
Turkey		7–11																			
USA					32–41		7–11		12–21	2–6	22–31		12–21	7–11	12–21		12–21			7–11	2–6
Venezuela								2–6													
Zaïre				12–21																	
Zambia				22–31							2–6										
Zimbabwe		2–6	2–6																		

Percentage of total production: 2–6, 7–11, 12–21, 22–31, 32–41, 42–55, 56–80

[a] Chromite is always found with other metals (usually iron) as an oxide.
[b] Molybdenum is sometimes produced as a by-product of copper.
[c] Cobalt, vanadium and antimony are produced only as by-products of other metals.
[d] Only South Africa and the CIS mine platinum in its own right.
[e] Cadmium is found with lead–zinc ores.

▲ *Figure 4.5 Non-ferrous metals – production and reserve potential by country.*

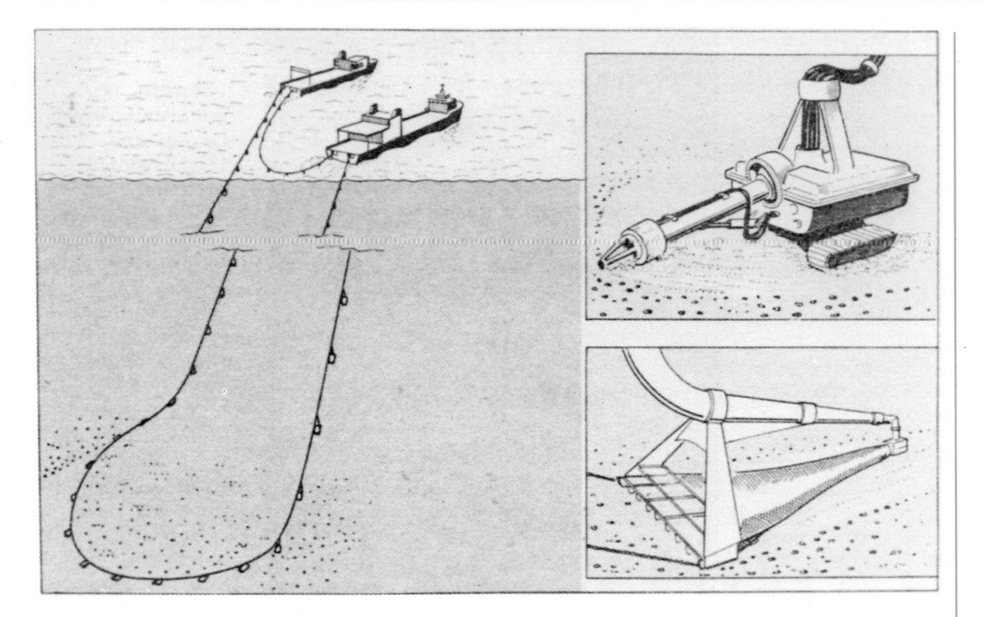

▲ *Figure 4.6 Harvesting the sea-bed. One method may be to trawl the ocean floor with the aid of continuous buckets (above left). An alternative method may use a suction device similar to a vacuum cleaner, either a sled which is towed through heaps of nodules (above right) or an active head which can be swept from side to side (top right).*

1 International agreement was needed to determine the division of such a rich harvest between countries outside internationally accepted limits to territorial waters. After 12 years of continuous negotiation this has now been achieved with the signing of the Law of the Sea Convention, which will be implemented through a new UN agency, the International Seabed Authority, inaugurated in November 1994.

2 Technological considerations required resolution. These involved the development of a reliable low-cost system capable of hoisting between 1 and 3 million tonnes of nodules from the sea-bed each year of the kind illustrated in Figure 4.6, and the development of efficient metallurgical processes for extracting the minerals from the nodules. At the time of writing (1995) the stage was therefore set to add substantially to the reserves of manganese and other minerals and exploit them in what will probably be a more environmentally friendly way compared with conventional mining techniques.

3.4 *Carbon and hydrocarbons*

These are the fossil fuels of Chapter 1 where the uneven distribution of present resources and the future reserve prospects were discussed. Two points need to be added here.

The first is that these minerals (especially oil and natural gas) are not merely used as a source of energy, but are also an important feedstock for the chemicals industry. A wide variety of industrial products depend on this source: for example, plastics such as polyethylene and polypropylene, together with artificial fibres, pharmaceuticals, resins, solvents, and a range of agricultural chemicals including ammonia and the fertilisers derived from it.

Secondly, the recent history of oil production deserves comment because of its strong political dimension and the considerable effect this has on the patterns of production. In the early 1970s, the Gulf States, Nigeria, Libya, Algeria and Venezuela had the lion's share of oil production (apart from the eastern European countries, the USSR and the USA). These countries founded the group known as the Oil Producing and Exporting Countries (OPEC). This was in effect an oil cartel, i.e. a producers' union aimed at managing the market for the product. Acting in concert through OPEC they proceeded to control output and to raise the price of oil in two major steps in 1973 and 1979. These policies, however, had other impacts, for the major consumers sought sources of supply outside OPEC control. Once other sources were discovered and their long-term capacity calculated, their productive capacity was developed as quickly as possible. Located in the USA, Canada, and in the waters adjacent to the United Kingdom and Norway, these are largely sites that prior to the rise in prices would have been uneconomic to exploit. This clearly illustrates the significance of a situation where major sources of the supply of minerals are not in the hands of the chief consumers, and not least the use of this situation as an economic or even political weapon against the short-term interests of consumers.

3.5 Summary

This section has briefly offered a geological rationale for the occurrence of minerals, and has provided further evidence of their global distribution by an examination of the major sources of supply by country in the mid-1990s, together with some indication of the amount and location of reserves. You should begin to understand the environmental disruption created by the demand for new mineral supplies, especially after reading the more detailed account in the next chapter of the environmental impact that such minerals may have as a result of their extraction and processing. However, the patterns of production which might be expected to emerge as a result of any analysis of reserves are considerably modified by other factors. These are discussed in the next section, starting with a consideration of the attitudes of the companies whose business it is to extract and process minerals.

4 Constraints on points of supply

4.1 Transnational corporations and investment

Until the early 1970s any general comment regarding the structure of the minerals industry could not have failed to underline the significant control of minerals production by large international companies, mostly American based but operating in more than one country, often in the developing world. Indeed, despite the rise of OPEC, seven major transnational oil companies dominated the world petroleum industry. Six large producers of aluminium set the terms under which most of the world's bauxite was extracted. Three companies controlled the uranium industry. Two companies overshadowed the production of much of the world's nickel. Many of them had extended their control through their ownership not only of the mineral rights and extraction and processing facilities, but also smelting and primary fabrication. Such a command of output exercised over mines in more than one country, allied with their market intelligence, ensured for them a strong voice in dealing with consumers.

However, the rapid extension of territorial independence to former colonial countries, and the desire of these new states for national management of their resources, led to a considerable reduction in the foreign control of mining enterprises. Sometimes this occurred through sudden expropriation by newly independent governments, as happened in Latin America and Africa during the late 1960s. In the case of the central African countries alone, 1969 saw the Zambian government take a 50% controlling interest in the two giant transnational copper mining groups, Anglo-American Corporation and Roan Selection Trust, later rising to more than 60% of the share capital. In Zaïre the government took over the interests of Union Minière. In Botswana the state acquired an interest in De Beers Consolidated Mines' diamond venture and in the copper-nickel mines run by a consortium of foreign companies led by the transnational company AMAX.

Such actions by host governments certainly caused the transnationals, partly or wholly dispossessed of their mineral holdings, to question their future patterns of mining investment. There was considerable reluctance on their part to invest further in territories where their capital would so readily be lost, especially when development costs for new mining ventures were escalating at a rate well in advance of general inflation. This was largely due to the greater sophistication of the minerals processing plant which was necessary to cope with the lower grades of ore generally being worked worldwide. At that time, an integrated copper mine, smelter and refinery could involve an outlay of around $US1 billion. But even where attempts were made to enter into profit-sharing schemes with less developed countries, the going was also not easy in what might be called the post-colonial environment. A case study illustrates this point well enough.

The setting up of the Cuajone copper project in southern Peru (Figure 4.7) involved reaching an agreement between the mining company and the government over the division of profits. This took several years to negotiate and was finalised in 1969. A further six years was needed to set up the complex financing of the operation (which thereafter included several metal mining companies, the World Bank and 29 commercial banks) before construction, planned to take seven years to complete, was begun. In the end the cost of development was $US726 million. But shortly after operations commenced, the Peruvian government, quite unexpectedly, imposed a range of export taxes on the output of the mine which lowered the return to capital investors. For potential mining investors this was but one example from many which graphically illustrated the point that hard-pressed governments in less developed countries may attempt to finance their social and other programmes by extracting the revenue they need from mining companies. Mining companies cannot resist such measures and have no means of redress.

This particular message was certainly taken to heart by the transnationals and those from whom they raised capital, causing by the early 1980s a dearth of mining development in the less developed countries. The 1990s, however, have seen a change in the attitude of transnationals towards their investment plans resulting from a combination of factors.

▲ *Figure 4.7*
Cuajone copper mine, southern Peru. The mine, which was difficult to negotiate with the state and to finance in the world capital markets, took much longer than average to develop and construct, and was for some years considered a dubious commercial investment.

First, corporations operating in many of the developed countries now perceive themselves as beleaguered by increasing government restrictions, regulation and taxation. In the Canadian provinces, for example, interventionist administrations now exist whose practices have resulted in a growing list of levies, including those related to licensing fees, fuel taxes and the funding of workers' compensation claims, all of which are unrelated to company profits.

Additionally, and especially in North America, mining transnationals see themselves as being under attack by a well-funded environmental lobby which is sapping public support for exploration and mining. The intrusion into wilderness areas by minerals extraction, the subsequent damage to ecosystems caused by primary processing and, in some cases, smelting, and the problem of site rehabilitation when mines close, have all been subject to the antipathy of such groups, even where the proposed developer has tried to minimise the environmental impact of mining.

At the same time the attitudes of many of the developing nations have undergone a remarkable transformation. In the Latin American countries, often under pressure from the International Monetary Fund which sees mining development as an important tool in promoting economic growth, they have had to become committed to financial and trade liberalisation, increased foreign investment and a reduction in their indebtedness to the developed nations. They are now offering transnational mining corporations preferential tax treatment, which can include repatriation of capital and profits with limited restrictions and total taxes on profits at 30% or less. They are also offering security of mining tenure; government joint ventures guaranteed through third parties, and a prompt government response to, and assistance with, development plans and proposals, including those related to environmental controls.

These location changes, however, increase the apparent dependence of the developed world for minerals on the developing countries. If the European Union alone is considered, its dependence on raw material imports is running at 95–100% for nickel, manganese, antimony, chromium, molybdenum, platinum, titanium, tungsten, vanadium and phosphate. For aluminium, copper, cobalt, tin, lead, zinc, iron-ore and uranium the level of import dependence is between 50% and 80%. It is a dependence further underscored by recent explorations in Latin America and China involving the latest forms of advanced technology which have pushed the reserve of the developing countries towards 60% of the world total (Plate 2).

The whole situation regarding the changing location of minerals extraction raises a number of issues of importance for this chapter. First, it assumes that the less developed countries will continue to have low levels of minerals consumption themselves. A rise in internal demand in these countries could have other consequences, as will become apparent in Section 5.2. Secondly, it takes for granted that the level of the consumption of newly won minerals in the developed world is really necessary. It may well be that real needs can be met by much lower levels of minerals exploitation and that recycling and substitution can play a much larger part than is the case at present in meeting world demand. Again, these are matters addressed in Chapter 5.

Finally, it raises questions regarding the environmental acceptability of minerals extraction both in the developed and developing world. It is to the framework of environmental controls that have been built up in these countries over recent years that attention must now be given, not only to look at how these have developed, but also to illustrate the kind of constraints these offer on the extractive process, and therefore upon their propensity to affect the environment.

4.2 *Statutory production constraints*

That mining companies are now rarely allowed to pursue activities with completely unbridled freedom and without regard for any other interests was made apparent in Section 2.6 of this chapter as well as 4.1. Apart from being constrained by regulations which in developed countries concern the safe working of pits and underground mines, the extractive industries have found themselves increasingly operating within a framework of controls not only relating to rehabilitation of land but also to the use of land and its air and water environs. These are legally enforced and administered by one or more government departments operating nationally or regionally.

The reason for the growth of such controls in most developed countries lies in the view that in highly populated and complex societies land is increasingly seen as a scarce resource for which at any one time there may be competing uses, and that the market place cannot be relied upon in terms of equity to decide what activity is carried out where, when, and to whose benefit (though, of course, in the case of minerals, they can only be extracted where they exist and where conditions allow extraction at a cost that will enable them to be marketed). At the same time, in the wider general interest, the costs of visual damage and pollution (which will be examined in detail in Chapter 5) cannot reasonably be borne by the public at large. They must be dealt with at source by those who create them, although such costs will almost certainly be reflected by an additional charge upon the products of the company concerned. Nor can it be appropriate, it has been argued, for mineral operators, once the material has been extracted, to abandon land, leaving it to other public agencies to undertake and to pay for its rehabilitation.

The development of the framework of controls to deal with these concerns began after the First World War. In the USA, federal legislation was passed in 1922 to regulate land use by a system of zoning ordinances – a set of rules which govern the location of factories, educational institutions, shops, hospitals, housing, etc. and prevent the close juxtapositioning of incompatible forms of land use such as heavy industry and housing. At the time, mineral working was outside its scope, but the zoning of other forms of development was a major factor in ensuring that the production of, say, aggregates, did not occur in urban or semi-urban environments, even though benefits might have accrued to the operator from the close proximity of suitable markets. Similar development controls spread in the inter-war years to Europe, but it was mainly after 1945 that the closely defined regulation of such matters occurred in Sweden, Italy, France, Denmark and the United Kingdom.

In the case of the United Kingdom, a rather more sophisticated approach was adopted and set out in the Town and Country Planning Act of 1947. By the beginning of the 1970s many developed countries had a similar system of development control. This system involved the preparation of broad general plans relating to future land-use needs, which were usually formulated at the provincial or county level with much more detailed programmes set out at the level of the municipality or district. Such arrangements also offered a means by which the desirability of specific development proposals could be considered as and when they were put before the relevant authorities for adjudication (Figure 4.8). A means of appeal by the unsuccessful applicant for a particular land-use change was usually incorporated in the system. Within such broad generalisations there remain, of course, wide variations. Although a number of countries (including the Scandinavian countries and Germany but not the United

▲ *Figure 4.8 Merehead, the largest limestone quarry in the Mendips, UK. Somerset County Council were one of the few counties to use their powers under the 1947 Town and Country Planning Act to draw up a coherent plan to meet the future needs for Mendip stone in a way which would prove to have some degree of environmental acceptability.*

Kingdom) have a comprehensive approach covering all land, it remains the exception rather than the rule. Even in the United Kingdom agricultural activities and forestry remain outside the remit of the 1947 Act (discussed by Blunden in *Sarre & Blunden, 1996*, Chapter 5).

In the USA in the meantime, although such a system of development control had not emerged by the 1970s, more rigorously applied land-use zoning put minerals exploration and development off-limits in ever larger areas of government land. Although this was done in the interests of environmental protection and wildlife conservation, it has not been without protest from the mining industry, unions, local government, and regional economic development groups. But claims that these withdrawals were excessive have largely been ignored. Indeed, constraints on other lands have been increased, as the next section makes clear.

4.3 *Acts and agencies*

The system of national controlling agencies is also variable, and although in most countries minerals come within the purview of ministries of mines, minerals, energy and/or natural resources, in some countries departments of environment, planning, construction and even agriculture may be concerned. For a number of nations the responsibility for looking after minerals may vary according to the mineral in question, particularly if strategic considerations are involved. Belgium and some Australian states have such an approach.

In the case of the United Kingdom, although all minerals are handled together, there is little centralised control at national or provincial level; responsibility is almost completely devolved to counties or, as in Scotland, the regions. Only where mineral development proposals are likely to be of a particularly contentious nature are they 'called in' by the Secretary of State for the Environment, who, after receiving the evidence of a local public inquiry through an inspector's report, will then make a judgement on the issue. The relevant Secretaries of State for Wales and Scotland perform this task for those countries.

The growing environment debate of the late 1960s and early 1970s in the USA introduced a new element into minerals development which was embodied in the Federal National Environmental Policy Act of 1969. This attempts to address the issues of resource conservation and pollution control *together* whenever a new major official land-use development is proposed. These issues are reviewed in each case through the preparation of an **Environmental Impact Statement** which sets out the ways in which the proposed development will affect the environment. In practice it usually also incorporates a cost-benefit analysis of that development. This has offered the opportunity to supplement existing legislation, whether its aim is land-use control or the maintenance of public health, and to provide a means of analysing all the factors for and against the proposal. It is an approach which has been copied by many US states, by many Canadian provinces, by Australia at national and state level, and, more recently, by the European Union.

Although a wide range of other considerations is addressed by Environmental Impact Statements, it should not be supposed that land-use planning controls, by contrast, can perform that role alone. In the United Kingdom context the submission of an application to extract minerals will cause the recipient local authority's planning department to address considerations which cover its interaction with other forms of land use, and therefore, human activity. If the potential extractor wishes to open up new ground not already designated for mineral working, the proposal will be referred to other government departments such as the ministries of agriculture and transport and the Environment Agency, and to bodies such as English Nature in England and the Countryside Commission (or their equivalents in Wales and Scotland) and the Health and Safety Executive. Their respective views as to the desirability of the proposed operation will be made available, as will any conditions they feel should be applied to the working, if it is permitted to go ahead. For example, the Environment Agency, using the Rivers (Prevention of Pollution) Acts of 1951 and 1962, might demand that the disposal of mine or quarry waters should not alter the mineral content of streams or rivers, while the Health and Safety Executive may wish to control dust or other emissions from secondary processing activities under the Clean Air Acts of 1956 and 1968.

Agencies controlling mineral development, in whatever ways they exercise their role in the developed industrial countries, are increasingly seeking to apply conditions to consents which demand a degree of site rehabilitation either as a mining operation progresses or upon its completion. In the United Kingdom such a requirement was first demanded under legislation dating from 1951, but it took another 20 years for this to be upgraded into a demand for a much more sophisticated scheme of aftercare.

Such rehabilitation considerations were first invoked in the USA just before the Second World War in connection with the strip mining of coal in West Virginia (Plate 3), but they have chiefly been embraced by other countries only since the early 1970s. Apart from the United Kingdom, these include Germany, France and other states or provinces of the USA, Australia and Canada. In some instances the government authority has insisted on the deposition of a cash bond before extraction commences, which may be drawn on for rehabilitation purposes. British Columbia in Canada, for example, operates such a scheme. Since its development as an important source of minerals did not begin until the late 1960s, the province was able to impose it on most mines. The virtue of the scheme plainly lies in ensuring that rehabilitation is not at the public expense should the extractor become bankrupt or otherwise cease operations prematurely.

4.4 *Cost of environmental controls*

Chapter 5 discusses the impact on mining enterprises of legislation to control dust, noise, effluent discharges to air and to natural drainage systems, and the disposal of solid wastes, as well as to ensure landscape rehabilitation. Here it is only necessary to note the particular importance of such legislation for mines in the developed countries and that the significance of such measures of control can be seen as an additional cost factor. What were once environmental costs borne by the wider community before the days of control have now become part of the internal cost structure of mining companies. But just how important can this additional financial burden be?

Unfortunately, little analytical work has been carried out into answering such an important question outside that undertaken by the Centre for Resource Studies in Canada. However, that country is one of the world's key mining provinces and the results of the Centre's research are clearly applicable elsewhere, so it is worth citing what their answer would be. In general terms, it would appear that the costs of such controls can indeed be critical in the case of workings operating at the economic margins, whether they are metalliferous mines, aggregate undertakings or those otherwise connected with the construction industry. In a case study reported by the Centre in 1981, for example, an evaluation was made of the effects of water pollution control on the economics of 131 metal deposits discovered in Canada over the previous three decades, as well as all other potential mining developments in that country. The deposits were mainly copper, zinc, lead and molybdenum. It found that the total capital and operating costs (at 1979 prices) of the necessary pollution controls were $Can1200 million. The figure was 3–10% of total mine investment and 2–5% of operating costs. But it rose to $Can2100 million if it was assumed that the best possible water treatment schemes (given known technology) were employed. In the latter case water pollution controls would be 13% of total investment and 8% of operating costs.

As the Centre's study points out, while the impact of these controls varies according to mine size, deposit type and region, the incidence of expenditure to combat pollution is greatest for smaller sulphide deposits because of the need for more sophisticated means of treatment, the costs of which cannot be spread across a large output of ore. However, given the prevailing market conditions for copper and the other main metals at the time of the study and the rate of taxation, five deposits would have been rendered uneconomic by the levels of water pollution control then required. This would have meant forgoing a potential net value to society (defined as the increase in society's real wealth by investing in minerals as opposed to some other activity) of around $Can700 million.

Another study from the Centre for Resource Studies, dating from 1980 and using a sample of 57 Canadian metal mining and processing operations, considered the impact of air pollution in the form of sulphur dioxide resulting from local smelting activities. Although this study was less detailed in its approach compared with that for water pollution, it showed that control costs only marginally diminished the number of economic deposits. This is perhaps because of the location of the sample entirely on the Canadian Shield where deposits with reasonably high metal content can be profitably exploited. However, assumed sulphur dioxide control costs (based on legal requirements at the time of the study) did have a significant effect on the potential value to society of the base metals mining sector. When all other costs except taxation were taken into account, the amount by which pollution controls of this kind diminished their potential value was around $Can271 million.

There are specific tax advantages to be gained, however (especially in Quebec and Ontario), from the use of integrated processing systems in which smelters are presumed to be part of the total operation (as opposed to the use of an entirely separate smelter facility). This arrangement further favours the larger mines which have richer ore bodies to exploit. Overall these controls can be regarded as a disincentive to the development of smaller deposits of poorer quality while bringing extant medium and larger mines working ores of lower metal content closer to the margins of profitability. Clearly, either through reduced mining profits or tax reliefs on integrated smelters, the end result of air pollution controls for governments is a substantial reduction in their revenues from the minerals industry.

While some presumption of satisfaction must be supposed among provincial ministries of environment in Canada, this would plainly not be shared by mining companies. Moreover, Ontario's Ministry of Natural Resources recently agreed in a report that changes in the standards of environmental control had a distinct impact on mining investments in Ontario (which is Canada's leading minerals province and one of the most important in the world), but it also argued cogently that a similar disincentive existed where tax changes were concerned. When these were combined with additional environmental burdens they led to 'investment substantially lower than it would have been had the pre-1972 tax and environmental regimes prevailed'. Indeed, in a study published subsequent to this government report which used a computerised simulation model, research workers were able to show that 'the combined effect of changes (in taxation and environmental regulations) was a reduction in mineral investment of at least 20%' over that which would have occurred in the province had these changes not been introduced.

Although there are obvious environmental implications of this reduction in the number of places at which exploitation occurs, it is equally apparent that additional costs incurred as a result of environmental controls have not, on their own, been sufficient to overcome the resistance of resource developers, usually in the guise of transnational companies, to the less developed countries, where the concept of environmental protection is sometimes not pursued with such rigour. This assumes, of course, that there is a choice to be made – that is, that the desired minerals are available from both the developed and the less developed world.

A less developed country can effectively trade off environmental protection against the benefits of incomes and jobs, by giving the former a much lower priority. But if some less developed countries appear to offer lower marginal costs in this way, many do not. Indeed there is evidence to show that where a less developed country has created its own state-run mining company, transnationals operating there are called upon to take much stronger environmental protection measures than the native company. Thus, in general terms, if any environmental advantage is to be found in locating in a developing country, it is largely through the absence of an effective environmental lobby opposing such operations that the benefit can be identified. As Section 4.1 makes clear, there are now other more telling factors at work in favour of locating new mining actively in developing countries.

4.5 Summary

This section has set out to show that minerals extraction does not occur in a situation dominated by the free interplay of supply and demand, but that constraints are imposed on production which go well beyond the limits of geological circumstance. These additional constraints, which may result in decisions to mine at one location rather than another, may be imposed by the mining companies themselves or by governments. Either way, the geographical incidence of environmental disturbance created by mining activity will be affected.

Even where extractive operations do occur, governments may further impose restrictive covenants detailing the circumstances under which mining and preliminary processing may take place and the treatment of the site once such activities are ended.

Activity 2

To close Section 4, go back through it and list the constraints, other than geological, which influence the *location* of mining activities under the headings of those specifically aimed at limiting environmental damage, and those which may be termed economic or political.

5 *Patterns of consumption*

5.1 *Consumer groupings*

If the supply of minerals is significant in environmental terms, so, too, is their consumption. Here again there are considerable geographic variations. The major industrialised nations consume the greatest quantities. As already noted, iron-ore lies at the heart of an industrialised economy because it is used in so many products. If the way this is consumed around the world in the form of steel is examined, the differences between the industrialised and non-industrialised nations immediately become apparent.

Consider Figure 4.9 which shows changes in the apparent per capita consumption in selected countries for 1938, 1978, 1988 and 1993. Between 1938 and 1978 world steel consumption (excluding the USSR) rose by more than three times. The figure indicates for this period that the highest increases were achieved by the developing countries such as China where consumption increased nearly eight times and in the Latin American countries of Mexico and Venezuela where the increases were around twelve and six times respectively.

Activity 3

Let's work out from Figure 4.9 just how the developed nations have fared over the period 1938 to 1978.

First the countries of Europe which had the greatest amount of ground to make up, compared with their northern neighbours, were those of the Mediterranean. If the former Yugoslavia managed about a 14-fold increase in consumption per head, work out what Spain and Italy managed to achieve. If consumption about doubled in many other industrial countries, how did Japan fare over the period?

USA consumption per head between 1938 and 1978 in absolute terms remained far and away the greatest over the whole period with the United Kingdom approaching it only at the beginning of the period. Identify the countries that were catching up with US per capita consumption or had even overhauled it by 1978.

In the fifteen years since 1978, as Figure 4.9 illustrates, two world economic recessions have affected steel consumption in many of the western industrialised countries, which by now have modest to nil growth and, in some cases, falls in per capita consumption. Latin American utilisation of steel has similarly fallen. Only countries such as China, Japan and India have continued to achieve significant increases in per capita consumption. By considering more comprehensive 1993 data than that selectively used for Figure 4.9, it is possible to divide the world into four main groups of steel consumers. The first group consists of about 1152 million people living in 29 industrialised countries who in 1993 consumed steel at an average rate of 398 kg per head. The second group consists of around 617 million people living in 21 countries, including some of the developed countries peripheral

	1938	1978	1988	1993
Czechoslovakia	90	750	694	466*
USA	320	675	451	400
Canada	130	575	601	450
Poland	30	560	409	174
Japan	80	535	706	647
Sweden	220	465	445	381
France	130	365	304	238
UK	225	360	307	227
Australia	210	360	396	325
Netherlands	135	355	297	285
Italy	50	330	460	378
Spain	16	250	269	243
Yugoslavia	18	245	159	–
Venezuela	35	225	170	96
Mexico	12	140	85	99
Argentina	50	130	93	98
China	6	45	65	108
India	10	16	21	22

*combined Czech and Slovak republics

▲ *Figure 4.9 Steel consumption (kg per capita).*

to the western European industrial heartland and the large developing nations of Latin America. In 1993 the countries in this group had a per capita consumption averaging 176 kg of steel. The third group consists of 7 developing nations with a per capita consumption averaging 60 kg and a total population of 1291 million, of which the people of China made up over 87%. Of the rest of the world's steel users, some 2340 million have a steel consumption even lower, at around 39 kg per head.

5.2 Less developed countries and minerals consumption

The key question to ask here is what would happen if attempts were made by those in the second, third and fourth groups to meet the average per capita consumption of those living in the 29 industrialised countries, let alone emulate the per capita levels of steel consumption of countries such as Canada and Japan. Calculations indicate that steel availability would need to increase from the world total of 736 million tonnes (1993) by a factor of about 2.4. Unrealistic as this is, since it is based on an increase worked out at one point in time when it could only be achieved over a number of years, its implications are manifold. It raises questions regarding the availability of iron-ore from existing sources, the opening up of new mines and smelting facilities and the implications of all of these for the environment. Also, much more serious problems of supply would soon arise for a number of other metallic minerals, as well as hydrocarbons, although not aggregates. It would be a situation exacerbated by a rising world population. Table 4.2 compares the levels of reserve and resource existing for these in 1989 with those postulated for 2030. World stocks will drop perilously low if less developed countries increase their consumption to match the USA. Reserves show quantities that can be profitably extracted using current technology, and resources show the total quantities thought to exist. Estimates of years left to depletion are based on current world consumption, or those for 2030 on the assumption that 10 000 million

Table 4.2 Estimated life of selected minerals (years)

	Current consumption rates		2030 rates	
	Reserves	Resources	Reserves	Resources
aluminium	256	805	124	407
copper	41	277	4	26
cobalt	109	429	10	40
molybdenum	67	256	8	33
nickel	66	163	7	16
platinum group	225	413	21	39
coal	206	3226	29	457
petroleum	35	83	3	7

Source: *Scientific American*, September, 1989, p. 96.

people will consume at current US rates. Assumptions are made regarding the future population but they also need to be made about technological change (including improved extraction techniques), and the ability of shortages to move some minerals from the status of resources to those of reserves because, as prices rise, it becomes profitable to work formerly uneconomic grades. Nevertheless, by 2030, supplies of copper, cobalt, molybdenum, nickel and petroleum could all reach critical levels.

Attempts to resolve such a situation by extracting grades of ore even an order of magnitude less rich than those worked at present could only give rise to substantial environmental problems. For example, although the largest copper mine in the world (Bingham, Colorado) is enormous, it is only one-tenth of the size required in the situation just outlined. Energy requirements would also be a dominant consideration. Look at Figure 4.10, which again uses copper as an example, and consider the energy requirements for the mining, crushing and milling of different grades of a copper sulphide bearing rock extracted by open-pit operations.

Q What do you deduce about the power needed to recover copper moving away from the lower end of currently extracted grades to approach those of the average abundance of the Earth's crust?

A About 80 kW_{th} of energy is needed for the extraction of copper grades of about 0.2–0.3%. This would rise gradually until the copper content fell to around 0.1%. Thereafter the energy requirement would rise steeply as copper content dropped to the average abundance figure of 0.005%. By then it would be around 3000 kW_{th} – an increase by a factor of 37.5 on current requirements.

These increases in both land consumption and energy requirement for copper extraction may appear formidable but they pale into insignificance compared with the gigantic requirements for the extraction of mercury or similar such metals at or near their level of average abundance. In the early 1990s mercury comes from a few deposits with a grade of 0.2–0.5%, whereas the average abundance of mercury is 0.000 04% – four orders of magnitude below the present cut-off grade at which the ore is no longer mined. Thus the working of most ores near their cut-off grades seems most unlikely, given that there is a geometric increase in the tonnage of waste rock to be disposed of for every arithmetic decrease in grade. It also has to

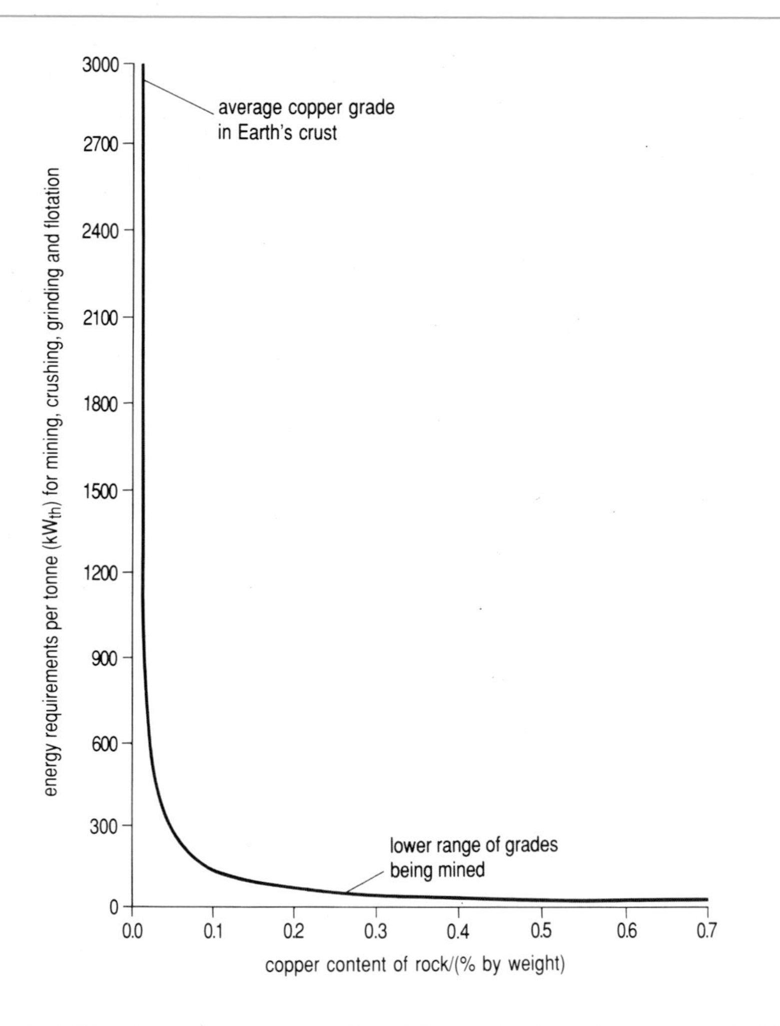

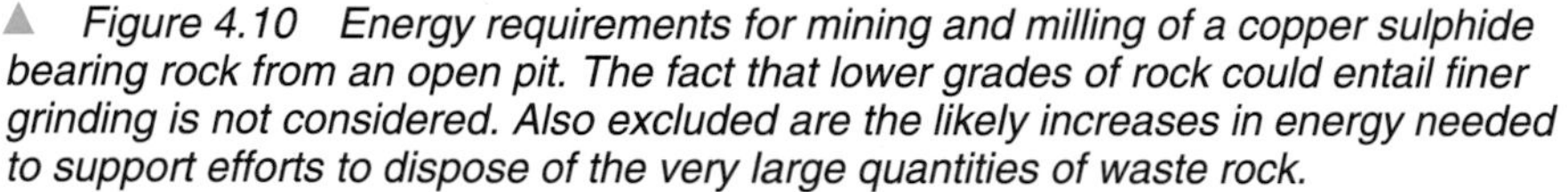

▲ *Figure 4.10 Energy requirements for mining and milling of a copper sulphide bearing rock from an open pit. The fact that lower grades of rock could entail finer grinding is not considered. Also excluded are the likely increases in energy needed to support efforts to dispose of the very large quantities of waste rock.*

be remembered that the volume of every tonne of waste rock removed from the ground will increase by between 20% and 40% after processing, thus further aggravating the waste disposal problem.

Failing the identification of any new reserves (an increasingly less likely situation given the sophisticated state of exploration techniques), solutions to such supply problems can only be met in a limited number of ways. These include substitution between minerals in order to reduce the pressure on those in shortest supply, and the recycling of metals already in fabricated products. More will be said about these two possibilities in Chapter 5. For the moment it is enough to remind you that these spectacular readjustments in the relationship between demand for minerals and possible sources of supply are suggested to point up a particular set of possible environmental implications, even though they represent an extreme in the dynamic of supply/demand relationships unlikely to be realised except in the longer term.

6 Conclusion

This chapter began close to home, using the extraction of china clay in Cornwall as a base from which to examine the processes involved in extracting minerals from the Earth's crust and processing them. It was shown that the environmental impact they have when they are worked must be determined spatially, not only by geological circumstances which dictate where they occur, but also by the propensity of those who invest in mining (largely transnational companies) to choose among possible locations. In meeting a minerals demand that mainly stems from the industrial countries, the location for investment, assuming its availability, was, until recently, largely in the developed countries.

The hypothesis that the levels of demand for minerals made now by industrialised countries might be made in the future by the less developed countries raises questions not only about the availability of some minerals in sufficient quantities, but also about the undesirable levels of environmental damage that could accrue. These are questions to which the next chapter gives further and more detailed attention, and Chapter 5 also looks at the impact of minerals extraction on the environment in terms of the extent to which this impact is manifest at the mine, in the territory immediately beyond, or even across national boundaries.

However, it is important even at this early stage in the discussion of mineral resources to appreciate that their extraction and processing is just the first stage in a production process. Ultimately, in the case of metals, this results in their use in the fabrication of those artefacts that have become important in the lives of most of us, at least in the developed world. (Just think back to the way in which we began this chapter!) Figure 4.11 attempts to establish these forward links by giving a 'cradle to grave' account of the impact of a mineral resource such as a common metal on the environment. In other words, the diagram tries to show the life cycle of a resource and how at each stage it interfaces with the environment.

Figure 4.11 also makes it clear that each stage in the process results in wastes. Sometimes these may be pollutants released to the air or to a water body. But as we recognised earlier in this chapter in our discussion of china clay, they may also be solid wastes. Box 4.5 offers a typology of all forms of such wastes, emphasising the extent to which they can be reused in some form, a theme taken up again in Chapter 5. As the Box suggests, where waste material is in fact a redundant artefact – for example, a disused aluminium can that contained beer or a soft drink – it may be disposed of to a landfill site. A greater environmental advantage would be gained by recycling of such artefacts, and it would also help to conserve the resource itself.

It is with the problem of wastes that we conclude this book, for Chapters 6 and 7 are devoted to the ways in which we cope with the discharge to the environment of, respectively, the gaseous and liquid effluent from the chemical industry, and the spent nuclear materials that are the result of generating power from nuclear reactors. Here the limitations of Figure 4.11 are evident enough. We make no reference in this to fissile materials used in the production of nuclear power, but it is clear that the diagram cannot show the political and economic constraints on the location of extractive activities mentioned in this chapter, or the way in which these constraints may apply to the subsequent processing of

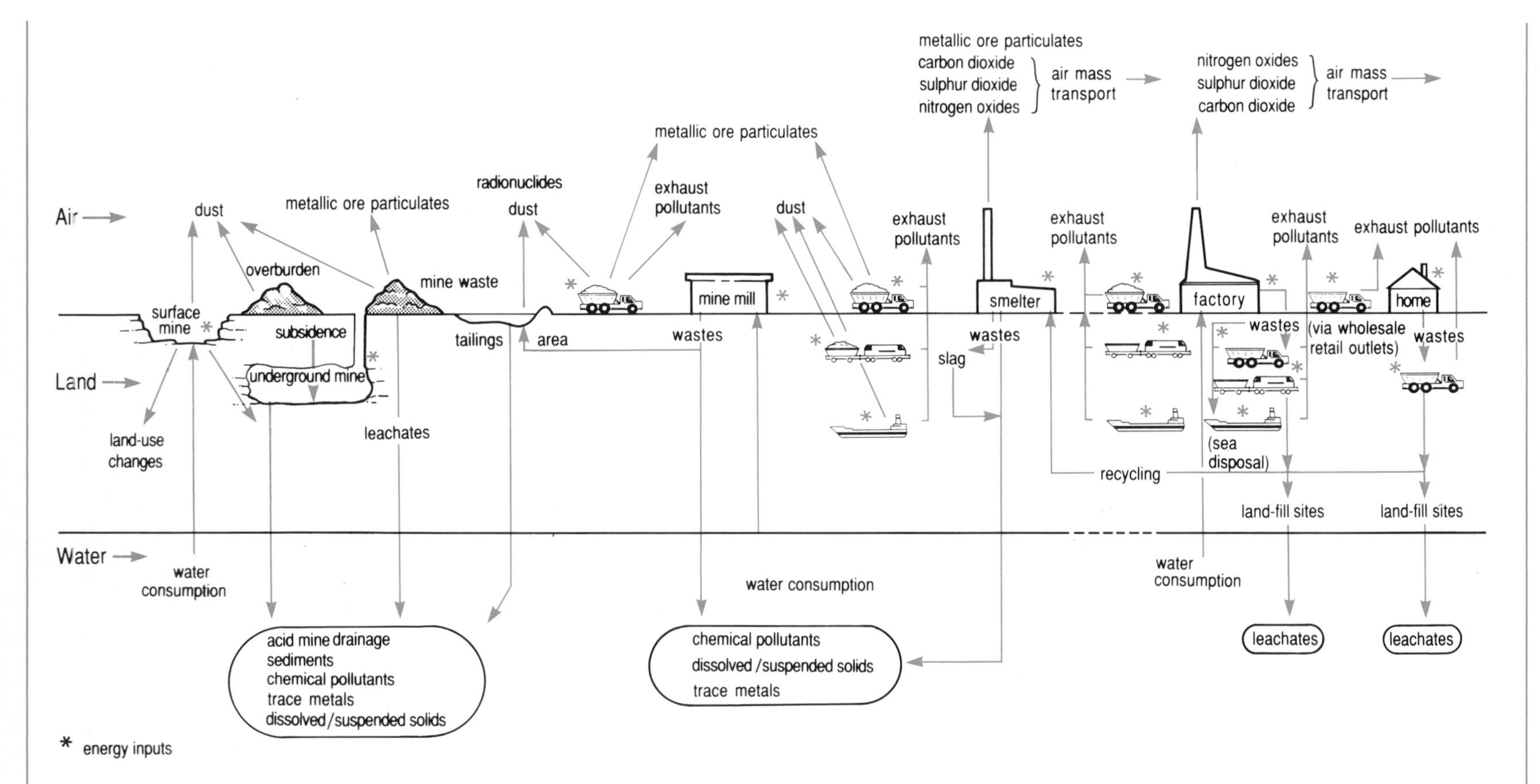

▲ *Figure 4.11 Minerals extraction and use; a summary of environmental linkages.*

Box 4.5 Wastes

The industrial processes outlined in Figure 4.11 and the agricultural, trading and urban systems of the previous volume in this series (*Sarre & Blunden, 1996*) – all aspects of our technological society – produce large quantities of wastes. The term 'waste' reflects a human perception. In natural ecosystems (e.g. *Sarre & Reddish, 1996*, Chapter 7) nothing is wasted: virtually all materials are recycled. Although we are beginning to return to this idea, reinterpreting 'wastes' as 'resources' (see Chapter 5), it is still a long way from realisation.

Not all solid wastes are equally available for reuse, but a broad distinction can be made between the following.

1 Materials which are part of natural ecological cycles – human sewage and animal slurry, plant and food residues, wood products and by extension the paper derived from them. They only present problems in entering the natural cycles because of the localised quantities resulting from concentrations of population; they contain energy which can be recovered, either directly by burning, or as methane from controlled 'digestion' processes.

2 Materials, like metals, which do occur naturally, but not in the purified forms which have to be extracted from natural ores, and otherwise processed, by the expenditure of energy. Recycling avoids the use of more energy, and conserves the primary resource.

3 Chemical compounds which do not occur naturally, and result from chemical operations. Some of these are not degraded readily by natural processes (e.g. some plastics, or the CFCs associated with ozone depletion in the upper atmosphere) or are toxic to humans or other species; their safe disposal is a challenge to the chemical industry to be faced alongside their production and use (see Chapter 6).

4 Chemical elements, in particular radioactive isotopes, which do not occur naturally, and result from nuclear operations. The radiation hazards associated with differing levels of radioactivity and half-life are compounded by the specific ways individual elements enter the food web and particular organs of the body. The minimisation and/or safe disposal of such wastes is a similar challenge to the nuclear industry (see Chapter 7).

minerals, their consumption by industry, their use by individuals and their ultimate disposal. However, both Chapters 6 and 7 see wastes and their disposal as an essential part of a highly politicised process in which decisions made by governments reflect the outcome of complex interactions between different interest groups.

References

DU MAURIER, D. (1972) *Vanishing Cornwall*, Penguin Books, Harmondsworth.

SARRE, P. & BLUNDEN, J. (eds) (1996) *Environment, Population and Development*, Hodder & Stoughton/The Open University, London, second edition (Book Two of this series).

Answer to Activity

Activity 1

Prices on the London Metal Exchange are quoted in sterling and US dollars per tonne. At the time of writing (1995) zinc had recently been trading between $1043 and $1035, lead between £637 and £634, and copper between £2965 and £2942. You will see that rare metals such as platinum, silver and gold are given in $ per ounce.

Chapter 5
The environmental impact of mining and mineral processing

1 Introduction

Bingham Canyon open pit to the south-west of Salt Lake City in Utah, USA (Figure 5.1) is roughly 4 km across and 1 km deep. Into it you could comfortably fit the cities of Bath or Winchester. From it more than 100 000 tonnes of copper ore are extracted every working day, making it one of the largest open-pit mineral workings in the world. Consider too, the metals smelter at Copper Cliff near Sudbury in Ontario, Canada (Figure 5.2). The substantial quantities of gases and minute metal particles (particulates), which up to the early 1990s came out of its 387-m chimney stack each year, could seriously affect environments many hundreds of kilometres away.

These are extreme examples of two different kinds of environmental impact caused by minerals extraction, and by minerals processing, one physical and one chemical. Whether minerals are won from the land's surface by open pit or quarry or by underground mine, and however they are subsequently processed, there will be environmental effects. This chapter sets out to examine these first of all, showing that they are both

As you read this chapter, look out for answers to the following key questions.

- What kinds of impacts are caused by mining and mineral processing at micro, meso and macro scales?
- How could these impacts be reduced by reclamation, recycling and substitution?

Figure 5.1 Bingham Canyon Copper Mine, Utah, which is 48 km south-west of Salt Lake City and reputedly the largest excavation produced by human activity in the world.

◄ *Figure 5.2*
Inco's giant 387-metre smelter stack. Opened in 1972, it was designed to reduce levels of effluent in the city of Sudbury.

physical and chemical and have the ability to affect both amenity and ecosystems. The effects are examined at a local or micro level, where problems of visual disturbance will prevail (although air and water environments may also be affected), at regional or meso level, where effluent gases from smelters can have a major impact on finely balanced ecosystems, and at inter-state or macro level, where air and water environments may be affected by smelter contributions to ambient levels of sulphur dioxide in air masses which may travel over considerable distances. The chapter concludes by considering the means by which environmental damage associated with mineral mining and processing may be ameliorated, as well as ways in which the primary extraction and processing of minerals may be reduced in the first place.

2 Micro environmental aspects of mining – visual intrusion

2.1 Topographical determinants

Mines, pits and quarries can make an environmental impact simply by their appearance. Bingham Canyon is a spectacular example of this. But the extent to which a mineral working intrudes on a landscape is generally local, to be appreciated at close range, and not necessarily related to the area of the working or the volume of material produced by it. Self-evidently, in an underground mine this must be so, since what is visible on the surface is related only to the various aspects of minerals processing (such as the mill or the smelter) and the areas given over to the disposal of waste rock, especially the tailings. But even in the case of surface workings, the operation can be large yet not obtrusive.

One of the biggest limestone quarries in Britain is at Merehead in a designated Area of Outstanding Natural Beauty in the Mendip Hills on the borders of Somerset. Before the recession of the early 1990s it was producing four million tonnes of limestone a year, but is almost wholly invisible from points of public access. This is because the Merehead operation involves extraction from a deep pit located on a broad and gently sloping rock surface of considerable size. An operation which extracted minerals from a valley side or an exposed rock face would be far more obtrusive.

The nature of the topography and the spatial relationships between the excavation and the surrounding land forms are usually the prime factors that determine whether or not an excavation will be visible from public vantage points. This assumes, of course, that there has been no addition to the site of any artificial earthworks (usually known as amenity banks) or specially planted tree screens, so often insisted upon by planning authorities as a condition for consent to operate the extractive enterprise.

Nevertheless, the sheer size of a minerals working can still determine its visual impact. As Chapter 4 of this book indicated, the large-scale china clay workings of the St Austell area of Cornwall in England are a major visual intrusion in a rural landscape otherwise largely exploited by agriculture and tourism. Travellers approaching the area from the north-east along the A30 road can catch their first glimpse of them at least 12 km away. The same is true of the smaller Lee Moor china clay workings abutting the Dartmoor National Park in Devon – these can be seen from Plymouth 16 km away as well as from a number of tors (hills) inside its south-western boundary.

Of considerable size by British standards, the open pits at St Austell are roughly circular, with the larger pits typically around 100 ha (hectares), or about twenty times the size of Wembley football stadium. These are frequently around 60 m deep, roughly two-thirds the width of the grassed area of the same arena. In addition to the open pits there are the extensive ponds in which the fine waste materials from the processing of kaolin are placed. Of the material extracted, seven out of every eight tonnes are waste, which is piled up in immense white tips – this is why the workings are such a landmark. Not surprisingly, the land requirements for the industry in the St Austell area alone are of gigantic proportions, as Table 5.1 demonstrates.

Table 5.1 Land requirement (in hectares) for the china clay industry in the St Austell area

	Land occupied by industry 1969	Land occupied by industry 1984	Land occupied by industry 1991	Anticipated land occupied by industry 2011	Anticipated increase in land occupied by industry 1991/2011
For pits	645	992	1108	1232	124
For tips (quartz sand, overburden)	830	1318	1287	1649	362
For micaceous residue disposal	77	301	295	383	88
For plant	310	378	345	338	7
Total	1862	2989	3035	3602	567

The problems of waste disposal for the china clay industry are particularly problematic. The four-fold increase in land used for the dumping of micaceous residues between 1969 and 1984 reflects the phasing out of marine disposal in favour of tailings ponds (see Chapter 4, Section 3). In the years since 1977 when a reclamation scheme was instituted, 348 ha of the land taken for tipping and 80 ha of that given to tailings has been reclaimed, mainly for amenity purposes (see Section 8). Open pits remained unreclaimed until recently but now 162 ha have been back-filled with waste materials from processing. In the period 1995 to 2011, 34% of tip requirements will come from back-filling. The relatively static requirements for plant over a period of increasing product output are an indicator of improved processing efficiency.

Source: ECC International.

2.2 *The impact of aggregates*

It is neither industrial mineral workings of the kind just discussed, nor metallic ore extractive enterprises that people are likely to be most aware of, or, for that matter, most environmentally concerned about. The widespread quarrying operations undertaken for the extraction of aggregate materials are far more important. As was noted in Chapter 4, their relative abundance is due to the wide range of rock types that can be used for aggregate, their low value, and in market economies, their inability to bear high transport costs. Moreover there has been a remarkable expansion of such enterprises in all industrialised countries to meet the rapid increase in building and road construction in the last 30 years. Production figures from the United Kingdom alone make this point clearly enough (Figure 5.3). Such figures obscure a related feature – that of significantly enhanced output from a declining number of sites. From around 5000 in 1950 the United Kingdom total has fallen by over 500 over a 40-year period.

From what has been said, it would be rightly concluded that the demands for aggregates from largely urban consumers will be met from the closest available sources. The most important of the aggregates group, sand and gravel, are usually to be found close to urban centres, simply because such settlements have tended to occur in lowland areas and frequently in river valleys where such materials are most likely to exist, so their ready availability ought not to be a problem. However, the appetite for sand and gravel has been particularly voracious. In London's green belt alone, in excess of 3300 ha has been actively worked for aggregates, almost wholly sand and gravel, with more than another 3000 ha approved by planning authorities for the same purpose. Nearly all of this is located in areas where there is significant pressure from other land uses which are not at all compatible with aggregate extraction. As sand and gravel workings are

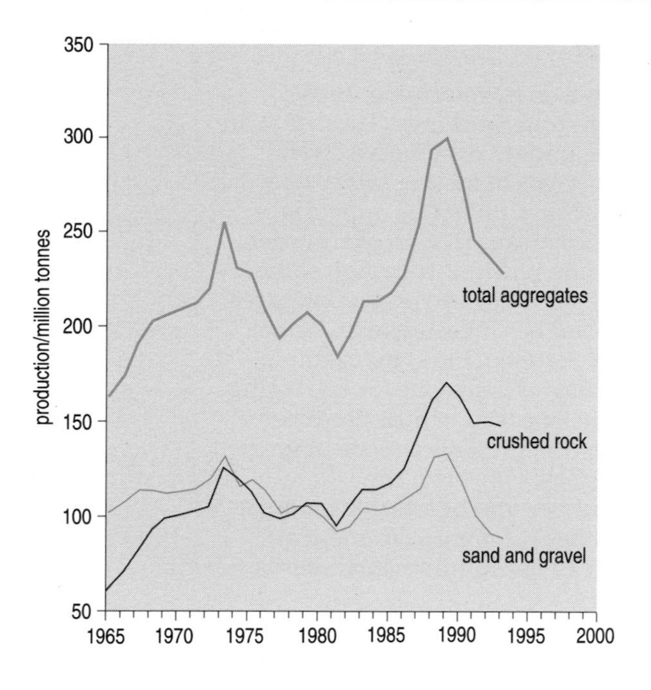

Figure 5.3
Aggregates are sensitive to the economic fortunes of the country, with the recession of the early 1990s particularly apparent. However, the failure of sand and gravel to follow the overall upward trends in production of aggregates and that of crushed rock (limestone, igneous rock and sandstone) reflects increasing problems of extraction (see text).

frequently unsightly, noisy and dusty, it is not surprising that they give rise to complaints about their environmental undesirability.

Public dissatisfaction is all the greater where the workings are associated with further processing. A survey carried out in Britain during the mid-1970s indicated that local authorities received more complaints about sand and gravel workings, even allowing for their greater prevalence, than any other extractive activity. Initially there were grounds for believing that the environmental problems might be reduced by increasing the output of sand and gravel extracted by dredging from the sea-bed (long a common practice off the coasts of south-east England and the eastern seaboard of the USA). However, claims have been made about the interference of such operations with navigation, submarine cables and pipelines. Even more importantly, it is alleged that they damage fishing grounds and cause coastal erosion.

To reduce the production of sand and gravel in areas close to urban centres, the working of hard rock aggregates at greater distances from them has increased; the higher transport costs are offset by bulk handling, specially advantageous transport contracts and by economies of scale at the point of production. In the United Kingdom this has led to greatly increased quarrying activities in places such as the Mendip region of Somerset, the Craven district of North Yorkshire, the Brecon Beacons of South Wales and the Charnwood Forest district of Leicestershire, all of which supply centres of consumption in the south-east of England. Unfortunately, these locations are in places of considerable scenic interest or are in or adjacent to National Parks or Areas of Outstanding Natural Beauty. Thus the amelioration of one problem has raised others, the only mitigating circumstances being that hard rock quarries are not as extensive in area, per tonne of product, as sand and gravel. However, mammoth quarries as far afield as the island of Harris are now being considered as possible sources of such material.

2.3 *Methods of mineral extraction*

Hard rock, like sand and gravel, is generally won by quarrying. In the United Kingdom there are few underground stone workings. Those that are in operation produce under 0.5% of the total amount of stone each year. Nevertheless, there are environmental advantages to mining rather than quarrying. To begin with one major source of environmental impact is absent; that of the excavation itself. Its absence means that no overburden need be tipped and noise and dust nuisance are reduced, although some surface plant is likely to be in evidence. Because of the obvious advantages, the Department of the Environment in England has investigated whether damage to the limestone scenery of the Peak National Park, for example, might be reduced by the underground working of limestone for use by the chemicals industry. As it is, the ICI quarry at Tunstead, which produces about two million tonnes a year for this purpose, presents a scene of great visual disruption to the visitor entering the Park. However, the underground alternative would raise the cost per unit of saleable material by 50% compared with quarrying. Since neither ICI nor government are prepared to meet the additional cost, in spite of its environmental merits, nothing has been done.

In the USA a national committee was set up to study the underground working of stone and recommended a major research effort into trying to reduce its cost and sustain the rate of advance of underground excavation to the benefit of the environment. At present the main areas of progress in this field have been where major conurbations overlie horizontally bedded limestone strata. In such areas opportunities exist to extract this easily worked material and to sell the resulting space for underground car parks, shopping malls and even offices.

In spite of the undesirability of quarry and open-pit methods of minerals extraction, because of the likely damage done to the environment, the trend worldwide in this century has been away from underground production. This has been especially true in the USA where, until the mid-1960s, underground working predominated, with surface extraction being mainly confined to the aggregates industry. By 1980 though, over 80% of that country's minerals by weight was produced by open-pit methods, a trend which has continued in the period since.

Q From what was said in Chapter 4, why has this change occurred?

A This change is due to a widening gap between the respective productivity of open-pit methods and underground extraction. The former technique has been favoured by modern earth-moving equipment. In the 1960s it was estimated that the surface extraction of minerals could achieve a productivity rate of 500 tonnes per worker per shift compared with underground mining where 50 tonnes per worker per shift was common. This change also owes much to the fact that the extraction of metals, other than iron, demands ever greater emphasis on the use of lower grades of ore as the more concentrated veins of metallic ores have been worked out.

Although open-pit workings for metallic ores are much less common than quarry workings, where they do exist much larger extractive operations are needed. The leanness of the grades of ore will need to be offset by the massive scale of open-pit operations and by high levels of output per annum if companies are to minimise costs per tonne extracted and cope with increased demand. Such operations are not always going to be of the

▲ Plate 1
Former china clay pit near St Austell, Cornwall.

▲ Plate 2 Bom Futura in Rondônia, Brazil, is the world's largest tin-producing operation. In 1989 the 50 000 tonnes of tin concentrates it produced represented nearly 60% of Brazil's output.

◀ Plate 3
The aftermath of strip mining for coal near the city of Gillette, Wyoming. Much land dereliction of this kind remains unrehabilitated in the USA, especially in Appalachia.

▲ Plate 4 Open-cast coal excavation at Shipley Lake, Derbyshire. Bucket excavators dig out the coal, the overburden and soil having been separately tipped adjacent to the site for subsequent reinstatement.

◄ *Plate 5*
Salt mining 'flashes' at Winsford, Cheshire. Surface hollows caused by natural brine pumping from the underlying strata in the eighteenth and nineteenth centuries have turned into shallow lakes.

▲ *Plate 6 The same view of Shipley Lake as in Plate 4 after mineral extraction had ended. Open-cast working usually results in a high standard of rehabilitation achieved in a relatively short time.*

▲ Plate 7 Acid drainage from a strip mine in New York state. Surface water drains into the local river system from workings which exploited an acidic material, garnet, used as an abrasive. This has produced an environment alien to aquatic life. Strip mining for coal which has been associated with sulphides and exposed to the weather is a more common cause of acid drainage problems in the USA.

▲ Plate 8 Baled-up drink containers awaiting recycling. Both aluminium and plastic can be reprocessed. In the USA, in states where there are mandatory deposit laws, 68 million kg of bottles made from polyethylene terephthalate were collected for this purpose in 1988.

◀ *Plate 9*
Damage to trees in the Erzgebirge mountains of central Europe.

◀ *Plate 10*
Cleaning up after an oil spill on the south Devon coast.

▲ *Plate 11 Agroforestry on farmland in China. Intercropping with* Paulownia *trees improves the yield of crop plants such as tea and wheat. The trees are fast growing and help to alleviate the shortage of timber, fuelwood and animal fodder.*

◄ *Plate 12 Vertical axis wind turbines.*

▲ *Plate 13*
A passive solar house in the Milton Keynes Energy Park, Buckinghamshire. Large areas of glazing on the south side favour heat capture in winter.

▲ *Plate 14 An earth-sheltered house in the UK. These houses are built partly underground with concrete walls to take advantage of the almost constant temperature 3 metres below ground, reducing the need for heating in winter and cooling in summer.*

scale of Bingham Canyon open-pit copper mine; nevertheless, they could have considerable environmental impact.

The hypothetical proposals for an open-pit metal mine in the highland zone of Britain and its comparison with an underground mine illustrates this point (Figure 5.4). The proposed working area of 325 ha for the open pit within a site of 800 ha would have an extractive operation covering some 81 ha. This underlines the need in the case of open-pit metal mining for large areas to be given over for the disposal of overburden as well as tailings, and for the buildings required.

The usual arrangement is for most of the minerals processing plant to be housed within a single building or small group of buildings. Where metallic ores are concerned, this will be the mill. The crushing and grinding of the rock and the removal from it of its metallic content is a complex activity compared with the processing associated with aggregate materials, so mills tend to be much larger than their quarry counterparts.

Regardless of the particular features of the fixed plant, whether associated with quarrying or with ore processing, its visual impact will depend upon location. In the past, plant sites were chosen only to fulfil primary operating requirements; these inevitably involved minimising haulage distances and avoiding sites beneath which valuable minerals remained to be exploited. As operations proceed, it is of course possible to move the processing plant inside the worked-out space, thereby decreasing its obtrusiveness. However, where the plant has some years of life remaining in it, or where the visual damage is not too great, then there are few incentives for such action.

▲ *Figure 5.4 A hypothetical open-pit copper mine in a highland zone of Britain. Overburden is tipped to the left of the open pit. To the centre rear of the picture, the tailings area can be seen behind the dam wall. The processing plant is in the foreground.*

2.4 Wastes from minerals extractions

The main solid waste produced from the development of a quarry or an open-pit operation is the overburden, which is usually dumped on the surface adjacent to the working site. The amount of waste rock resulting from the processing of aggregates is normally relatively small (compared, for example, with the wastes produced from the processing of metallic ores) and the ratio of excavated waste to saleable product does not usually exceed 1:1 and may be zero. Other wastes, such as silts from stone washing and dust, are generally small in volume and do not cause serious visual impact when they are surface dumped.

The considerable problem of waste materials from china clay working has already been fully described, but the quantities produced have been exceeded at slate quarries in Britain. This industry, largely centred on North Wales, has considerably reduced its output as a result of the competition from substitute materials for roofing. Nevertheless, waste from slate workings in the major producing area is thought to amount to a total of 400–500 million tonnes. The processes of splitting and dressing the individual slates can result in the production of waste up to twenty times that of the actual saleable product so that even modest production results in large waste dumps. Even today large tips of broken slate and slate chippings are frequently found disfiguring parts of the otherwise outstandingly attractive scenery of Snowdonia (Figure 5.5). These are only slowly eroded by the action of wind and rain so that natural vegetation finds it difficult to establish itself on such dumps.

In the case of underground coal extraction such levels of waste production are never reached. For deep mined coal in the United Kingdom, the average waste to marketable fuel ratio was, in the early 1970s, 1:1.9. This meant that, by the end of that decade, about 67 million tonnes of waste had to be disposed of annually. Three million tonnes of this were retained underground; 61 million tonnes were dumped on the surface in tips and the rest was disposed of into the sea. But whereas the winning of coal has always involved the extraction of some worthless shales along with the coal, increasing mechanisation led to a waste to coal ratio approaching 1:1 at fully automated coal faces.

National figures for waste produced by coal mining mask the considerable geographical concentration of the disposal problem. Calculations made in 1978–79 showed that the Yorkshire/Derbyshire/Nottinghamshire coalfield accounted for 63% of waste output. The old coal mining areas of South Wales and Scotland had 26% and 23% of all spoil tips in the United Kingdom but over 80% of these were no longer in use. (**Spoil** refers to mining wastes tipped on the surface of the ground near the mineral working.) This contrasted strongly with a county such as Nottinghamshire, where up to the early 1990s many remained active. In this county alone over 900 ha are covered in spoil heaps (something like the combined areas of Hampstead Heath and Hyde and Regent's Parks) and the collieries produce about 13 million tonnes a year of spoil. The county council had estimated that, by the year 2000, using present tipping methods and assuming production constant at 1980 levels, 5000 ha of the county will be covered by waste tips (something approaching the size of the county town itself), with all that this means in terms of sterilised land and severe visual disturbance. Such forecasts, however, could not foresee changes in the UK coal industry that were to make them totally invalid, for while the total UK coal pit provision stood at 170 in 1984–85, producing more than 91 million tonnes, it had fallen to a mere 16 underground workings at the

Figure 5.5 Slate waste at Blaenau Ffestiniog. An area understandably left outside the Snowdonia National Park at its designation, but surrounded by it.

privatisation of the industry in 1994, with output at half the level of ten years earlier. This small core of pits represented the few that could be run as efficient automated enterprises capable of surviving in the world market place unaided. Just before the demise of British Coal, it had established that using modern extraction technology, every 50 million tonnes of mined product would require 200 ha of new land for waste dumping. Since present circumstances are unlikely to push annual demand beyond that figure given above, then it is possible to estimate year on year the likely needs for waste dumping, at least in the short term. This is not to imply that some waste cannot be suitably disposed of in an acceptable manner which could have some impact on such speculative calculations, or that in situ reclamation techniques that can improve the quality of local environments do not exist; these matters are explored in Section 8. Nevertheless, the environmental issue raised by the need to handle such large quantities of waste must give cause for concern. Indeed, underground operations strongly contrast with the winning of coal by open cast methods.

Although the adverse environmental impacts created by open-cast coal workings can last as long as 15 years, they are more usually short term because the area concerned is soon rehabilitated. However, given the present scale of operations worldwide, with 50% of all coal derived by such methods, the short-term environmental problems can be of considerable importance, especially in the more densely urbanised and populated zones of Europe. In spite of land rehabilitation charges, open-cast mining is highly cost effective. In order to sustain the output levels demanded, the land requirement is high. In the UK during the 1980s it was around 10 000 ha (Plate 4). In the Ruhr the figure is 19 000 ha and about 25 000 people have been temporarily displaced by these excavations. In the two to five years that an open-cast pit is in operation, apart from the displacement of people and other forms of human activity, the nuisances of noise, dust and traffic can be a problem. Visual disturbance is a considerable drawback and is always difficult to deal with in the short term. The only exceptions are those workings which exist far from populated areas in countries such as Australia, the USA and Canada.

2.5 *Wastes from minerals processing*

The preparation of coal for sale, whether mined or produced by open-cast methods, inevitably gives rise to the need to dispose of the fine material into tailings ponds. These are not a significant element environmentally. Even at collieries which produce coking coal, where the largest quantities of fine waste materials are produced, tailings do not amount to more than 15–20% of total waste produced. This contrasts strongly with quantities of waste produced in the winning of metalliferous ores.

Although high-yielding veins of concentrated ore can be selectively extracted by underground mining techniques, thus considerably reducing the quantities of tailings produced in the milling process, the ore is frequently disseminated in the rock. Then, the valuable mineral content seldom exceeds a few per cent and may be a fraction of 1%. This means that the recovery of the metallic minerals in the milling process leaves tailings which are substantially the same mass as the ore treated, but which occupy a far greater volume than that material did when it was in the ground. Tailings ponds can be superficially similar in appearance to water reservoirs, but without their redeeming features of neatness and cleanliness. Both have a dam wall, but a tailings pond often consists of an extensive pool of discoloured water with a silt or sand beach which is usually dotted with the remains of dead trees and shrubs. The dimensions of a tailings pond inevitably relate to the output of the mill, in the case of metal mines, and the life of the mining operation. Self-evidently the disposal needs of a mine with a five-year life and a mill capacity of 2000 tonnes a day are quite different from one that has the same throughput but a life of 25 years. Operations which fall between these two examples would not be untypical for many non-ferrous metal mines, although a large enterprise could have a mill throughput of more than 9000 tonnes a day. The gigantic Bougainville copper mine in Papua New Guinea, until the premature closure in the late 1980s because of political unrest, was perhaps quite exceptional. Its 78 000 tonnes per day of tailings were disposed of into the sea. Such a means of tailings disposal, together with their discharge into lakes, has the advantage of getting rid of a visual problem, but they can and do create water pollution and damage to the aquatic ecology.

It is difficult to generalise about the surface area of tailings ponds. In Ontario most cover less than 245 ha, but some are over 1000 ha in extent (see Figure 4.2). The problem is not just one of mill output, but the nature of the terrain. This may determine whether the mine operator opts for a shallow, but really extensive pond, or a deeper one of limited size. If there is any choice in the matter, the former has advantages since the larger surface area will allow greater opportunities for oxygenation, thus improving water quality. But the latter may be visually more acceptable, a matter of consequence where new operators may be required by government agencies to seek to diminish the impact of this environmentally disturbing aspect of their operation.

Without regulatory controls the operators would inevitably minimise costs by transferring wastes by pipelines to a site closest to the mill which fulfilled their disposal requirements. In areas of considerable scenic attraction where a mine has been allowed, it has been known for the regulatory agency to insist on the dumping of tailings many kilometres away in an area of less environmental significance. In Japan one mining operation is more than 70 km from the tailings pond. In Britain one planning authority was insistent that any further development of an underground fluorspar mine in the Peak National Park would have to depend on the dumping of tailings in such a way as to prevent any interference with the environment. A feasibility study was therefore undertaken into the possibility of transporting these by pipe to a site outside the boundaries of the Park.

Eventually, when the tailings area is full, there will probably be some attempt at site rehabilitation in order to stabilise it and improve its visual appearance. This frequently involves establishing some form of vegetation cover although this may not be easy. These are matters which will be discussed in Section 8. In the meantime, during the life of the tailings area, which may be a long one, it may be necessary in the interests of near-by communities to do as much as possible to improve the appearance of the site. This is usually achieved by maintaining a cover of water over as much of the tailings as possible.

2.6 Summary

This section, which has concentrated on the visual obtrusiveness of mineral extraction, began by showing that although sheer size can be determinant, topographical considerations are important. Aggregate production may also pose problems, but not because of size – rather, the frequency with which they occur. Indeed, all kinds of open-pit mining operations are increasing at the expense of the less obtrusive underground workings. The quantitative aspects of waste production were then considered as a visual hazard with particular reference to the winning of a number of specific minerals, and the section concluded with an analysis of the particular difficulties created by the disposal of tailings.

Activity 1

Go back over the discussion of tailings disposal, listing the advantages and disadvantages of tailings disposal in (a) the sea;
(b) a narrow but deep valley; (c) a shallow broad land depression.

Although the brief section on subsidence which follows shows that it obviously has a strong visual outcome, its propensity to cause damage to the interests of other property owners and land users justifies its separate treatment and the exploration of the particular problems that can be caused by the extraction of coal, salt and oil and natural gas.

3 Micro environmental aspects of mining – subsidence

3.1 Subsidence and coal mining

Subsidence can occur as a result of the underground working of minerals, but its impact on the surface may be considerable and will undoubtedly affect more than just the eye. Perhaps the most frequent cause of subsidence is the working of coal. The removal of whole seams of material, frequently of considerable width and many metres thick, inevitably results in the collapse of the overlying strata. Mining a vein of metallic ore does not usually have this result. Coal mining subsidence mainly causes damage to buildings, roads, underground pipes, sewers and cables, and to agriculture through the disruption of drainage systems. In the USA, the Bureau of Mines reckons that over 35 000 km^2 have been affected by subsidence, nearly all of which can be attributed to the underground working of coal. This figure is expected to rise to 42 000 km^2 by the end of this century. Of the areas especially affected, the Illinois Basin is outstanding with 10% of the nation's total subsidence problem, followed by western Pensylvania and northern West Virginia, where around 45 major subsidence incidents are reported each year, mostly involving abandoned mines.

In most other countries where underground mining is important, no overall estimates of subsidence are available, although there are individual case studies. One of the best documented is of a ten-year period in the English East Midlands and included the Nottinghamshire, Derbyshire and Leicestershire coalfields. This showed that one in three houses had to some extent been damaged by subsidence, although of the 62 500 properties involved, only in 4% was the problem appreciable. In the rural context about 728 ha were sufficiently affected for the National Farmers Union to seek compensation. This, however, represented less than 2% of the total agricultural land inside the area influenced by mining.

Another study in the Nord Pas-de-Calais coalfield of France has concentrated on measuring subsidence itself. In the early 1990s nearly 17 000 ha of land, both urban and rural, are directly affected. In one district alone, around Lens, the ground has sunk about 5 m over an area of 20 km^2, but there have been changes in level in four of the basins which have formed within this zone of between 9 m and 14 m.

Although subsidence, along with waste disposal, has always been the major environmental impact of coal mining, its worst effects can still only be controlled rather than totally mitigated. Complete avoidance of subsidence would involve the leaving of unacceptably large quantities of coal in situ, destroying economic viability. However, there have been great advances in the understanding of subsidence since the late 1940s. These derive from work carried out in western Europe, Poland and the USSR. Although results do not relate to the prevention of subsidence, they enable predictions to be made of the degree of physical damage on the surface so that surface development and coal exploitation may be co-ordinated.

Three examples of co-operation between local authorities and the British coal industry to effect planned subsidence may be cited. When the plans for Peterlee, a new town in the county of Durham, were first prepared in 1947, about 30 million tonnes of coal remained under its 948 ha. After taking the precaution of carefully aligning surface constructions, it was found to be possible to build on land below which as many as three seams of coal remained to be worked. The phasing of coal extraction with motorway development took place when that section of the M1 which crosses the Nottinghamshire/Derbyshire coalfield was constructed. In places the motorway was built to allow for subsidence.

More recently, when British Coal (responsible for the nationalised industry), was planning a new mining development at Selby in Yorkshire, it acknowledged the need to prevent damage to buildings of historical importance. It therefore left coal in place in the seams both beneath and well beyond the area covered by such structures. Available evidence seemed to indicate that to protect a circular area on the surface with a diameter of 200 m, a 900-m diameter circle needed to be left intact in the coal deposits at 500 m, and a circle of over 1300 m diameter at 800 m. Estimates of the tonnages thus sterilised can be made; for example, if a seam one metre thick is abandoned at a depth of 500 m in a 900-m diameter circle, something of the order of 500 000 tonnes can be lost.

3.2 *Other subsidence problems*

The situation with regard to salt mining compares unfavourably with that of coal. A method is now in use which ensures the removal of salt in solution through the drilling of boreholes, the pumping in of water and the extraction of the resulting brine. This leaves behind large caves, but sufficient salt to support the overlying rock. This process, invented by ICI, is known as controlled brine pumping and largely eliminates subsidence. However, in the United Kingdom a great deal of salt has been won since the eighteenth century by the random pumping of salt dissolved in the groundwater from beneath the Earth's surface. This completely uncontrolled method of extraction has caused considerable subsidence, which has resulted in the permanent flooding of agricultural land in parts of Cheshire (Plate 5). In 1969 court action was needed to restrain further natural brine pumping activity. The problems created in the past, however, remain. In Cheshire the continuing unpredictability of subsidence is particularly serious since the areas of dissolved salt are not clearly established. A county council survey dating from the mid-1970s acknowledged just this problem. It suggested that at that time the total subsidence resulting from natural brine extraction was equivalent to a drop of 0.5 m over 119 ha in any one year. In reality, falls of greater magnitude, but varying widely, have been experienced over a much larger area.

The most notable cases of subsidence caused by the extraction of crude oil and natural gas are where they have occurred near sea-level. Flooding has been the highly unfortunate consequence. Evidence of this problem has come from an oil field on the shores of Lake Maracaibo (Venezuela), from coastal locations in Japan and the USA, and from the Po delta in Italy.

Although they will have occurred elsewhere, the depression of the land surface will very likely have gone unnoticed.

Q Why is the extraction of oil and gas less likely to create a surface problem than the underground extraction of coal and salt?

A Subsidence resulting from oil and gas extraction does not derive from the subterranean *removal* of solid material such as with coal and salt, leaving cavities that can suddenly collapse, but from the reduction of fluid pressure in the underground oil and gas zones. This allows the strata above to settle slowly and evenly at a lower level.

The subsidence at Long Beach (Los Angeles), California, an industrial township lying 2–3 m above sea-level, provides a well-documented case study. As a result of the removal from 1936 onwards of oil from the Wilmington and Signal Hill fields, a substantial part of the central urban area of Long Beach began to sink. By the time the fall had reached 8 m, it was decided to take action to prevent the area being inundated by the sea. Retaining walls and levees were constructed and repairs to damaged oil well boreholes caused by earth movements were carried out. The cost of these projects was over $100 million. In the late 1950s attempts were made to repressure the oil zone by injecting water. This resulted in a 15% rise in the level of the land surface.

4 *Micro environmental aspects of mining – aquatic damage*

4.1 *Water pollutants – the general background*

One of the environmental hazards of mining that has received attention over recent years and is a local or micro problem is the contamination of surface waters with materials that can prove toxic. Such contamination may be by particulates of heavy metal (such as those of copper, lead, zinc, nickel and uranium) or by **sulphides**, or by other materials which may have been used to extract the ore. These problems are the direct result of the processes of mining and milling which break what was solid rock into fragments, so allowing air and water to react chemically with it. As these reactions depend on contact between air, water and rock surfaces, they are most prevalent where the material is finely crushed or ground. The finer the materials the greater the potential surface area is exposed within the body of material. Under the circumstances it is not surprising that tailings can offer the most serious problems, although waste tips or exposed stockpiles of untreated ore have been known to cause difficulties.

Even greater dangers of environmental pollution are likely to occur where the constituents of the rock fragments lead to an increase in acidity. Acid production, or **acidification**, is usually associated with the action of oxygen on sulphide minerals, most particularly pyrite and pyrrhotite, both iron sulphides, which often accompany the ores of lead, zinc, nickel and copper. In theory, this oxidation process can occur chemically or biologically. Although differences of opinion exist about the relative importance of the two pathways, evidence largely supports the latter. Initially, the reaction (whether chemical or biological) needs a supply of oxygen – that is, access to air. Then, the overall effect is to convert sulphide ions (electrically charged atoms) into sulphuric acid. This affects any waters that drain through the minerals by lowering their pH value, i.e. making them more acidic. This is important because acidic solutions react with other metallic compounds present in the tailings, releasing metal ions (some of which may be toxic: for example, those of heavy metals) into solution and so into the groundwater. In waters where higher pH values prevail, most metals form insoluble oxides or hydroxides. Overall, then, acid generation can be considered as bad news – leading to enhanced levels of 'dissolved' metals and greater potential toxicity.

Where ores are not associated with these iron sulphides, acid drainage is less of a problem. It can also be counteracted by the presence of other rock materials. In the world's major lead mining area, that of south-eastern Missouri, for example, the ore is not associated with pyrites, but with dolomite, the base rock through which the natural drainage flows. Since dolomite is a limestone, the waters become alkaline and at alkaline pHs insoluble lead compounds are formed which precipitate out of solution. Thus, where it is necessary to engage in minor discharges of excess water (the situation at three of the five mines), a nine-day period of retention allows traces of lead, zinc and copper to settle out. This, plus some elementary sedimentation removal techniques, is all that is required before discharge into the drainage system. The pH levels in local streams are between 7.8 and 8.3, so the dissolved metal content in these is low.

A method of avoiding the problems that arise with pyrite discharges to the tailings area is to remove them in the milling process by flotation (as explained in Chapter 4). Unfortunately, this often reduces pyrite discharges by only 95%, their complete removal being impossible. Moreover, it would add other undesirable chemicals to the tailings, while still leaving 5% pyrite to form acid. At the same time, if the pyrite concentrate was to be dumped, careful isolation and monitoring would be essential.

Q If the pyrite is not removed from the tailings, how could acid drainage be prevented?

A Since acid formation involves the reaction of oxygen with pyrites, one approach would be to exclude air from the tailings. Another would be to use limestone to raise the pH of the drainage waters.

The exclusion of air from the tailings might be achieved through encapsulation, or inundation (i.e. keeping the tailings permanently under water), and, once the area is no longer in use, through the employment of physical covers, chemical 'fixation' or by creating a vegetation cover over the surface of the tailings, thus limiting the amount of oxygen that can enter them. Currently, vegetation offers the most practical long-term approach for reducing the problem when the tailings area is complete, a matter taken up in Section 8.

Once the abandoned areas of tailings have been flushed of process waters and vegetated, effluent treatment should no longer be needed. But while the operations continue, effluent waters may remain to be disposed of. A lead–zinc–copper mine at Sturgeon Lake in Ontario not untypically treats its contaminated water by adding lime slurry to bring the pH to 10. Large tanks with mixers are used to precipitate out the heavy metals as hydroxides, a practice which can be, and often is, pursued in the tailings pond itself at other mines. Flocculents (agents which encourage any remaining metallic compounds to settle out) are added before the fluid is passed to clarifiers. The settled-out sludge, once filtered and dried, is disposed of as a zinc concentrate.

The problems of working metallic ores that contain pyrites can also arise as the result of the mining of coal. The oxidation of the sulphur associated with the extraction of coal, particularly where local aquifers or streams are disrupted, leads to acid drainage problems. Areas of surface tipped waste can also be affected as rainwater drains through them (Plate 7). In regions where a considerable amount of land is given over to strip mining, such as Pennsylvania, Ohio and West Virginia in the USA, coal extracting companies have had to build drainage treatment plants to cope with acidic waters that can run as high as 5400 p.p.m. of acid. In compliance with the federal water pollution control standards, these plants use a relatively simple treatment of acid drainage water using hydrated lime. In the United Kingdom since 1995, the regionally organised coal mining companies have become similarly legally liable to control acid drainage from their active mines and tips. Where these have been long abandoned, the Coal Authority, a new government agency, is responsible.

Where metal mining is concerned, a report from Sturgeon Lake, Ontario, serves as a reminder that the water available from the active tailings area, and the water pumped from the mine or other parts of the processing plant, is not all disposed of. At this particular location, which is not untypical of lead–zinc–copper mining and processing undertakings, about half the water is reclaimed and retained to meet the requirements of the milling operation. Indeed, it provides a typical example of the substantial advances in recycling and in the standards of water treatment that have been achieved in base metal mines over the past 30 years.

Some water pollution can be generated by materials used in mining or processing. Mine drainage waters can contain both ammonia and nitrate, usually present as a result of the use of ammonium nitrate in explosives used for rock blasting. This chemical makes the explosives safe to handle, but because it is very soluble, excess quantities and spillage may be washed away. Ammonia in its molecular form is toxic and especially so in water where the pH is high. (In more acidic conditions, those of lower pH, the ammonia is converted to ammonium which is less toxic.) Discharges containing molecular ammonia into river water of pH 7 and above can cause severe problems for fish.

Q What problems can arise from low pH values:
(a) in the waters of the tailings area of a metal mine;
(b) in the drainage waters of a mine site where ammonium nitrate explosives have been used?

A (a) These can lead to enhanced levels of 'dissolved' metals and a greater potential for toxicity. If these waters are discharged into the natural drainage system of the area problems can arise for aquatic life.
(b) There should be none. Ammonia used in explosives for rock blasting only becomes toxic where pH values are high.

4.2 Radioactivity – a special problem

The operations at Elliot Lake, Ontario, raise the question of another pollutant primarily associated with uranium mining, that of radioactive waste. This arises from the presence of an isotope of radium which is a major waste product deriving from the milling operation and remaining undissolved in the process used to extract the valuable ore. Since 99% of this waste product passes through to the tailings, any disposal of surplus water from it into the drainage system can cause problems. Although a possible hazard of this kind was not recognised when uranium mining began in earnest in the 1940s, by the end of the 1950s it became clear that, for example, the ingestion of drinking water or the consumption of fish contaminated by the radium isotope could cause bone cancer. The first attempts at abatement procedures involving tailings were tried at the Durango mill on the Animas River in the USA. Since these were unsuccessful, subsequent and exhaustive research concentrated on techniques which, at reasonable cost, might remove the radium isotope at the mineral processing stage, but again without result. However, one short-term remedy is to treat the effluent from the mill with barium chloride which reacts with sulphuric acid to form a precipitate of barium sulphate and this carries the radium out of the solution. The longer term answer must be the stabilisation of the tailings, their vegetation and the prevention of the passage of further water through them that might leach out into the drainage system, a matter to which further reference is made in Section 8.4.

An even safer alternative method to dispose of radium waste might be to backfill the mine or the open-pit excavation with tailings. This process, which has been strongly recommended by the Australian government as a preferred solution, would consist of mixing the wastes with water and pumping them into place. The fine material contained in the wastes would then need to be removed so that the wastes could dry out after placement. Unfortunately, this fine material, which makes up 50% of the total, would still have to be dumped. One company, Anaconda, working in New Mexico, USA, has found a means of disposing of its separated fine material by injecting it deep under the Earth's crust via boreholes. This approach is also practised in Texas, where some land owners have demanded it in order to restore the landscape following the termination of open-pit uranium mining.

Whatever long-term measures are currently recommended to contain this problem of radioactivity, evidence from Elliot Lake has shown that mining prior to the late 1970s was certainly less than environmentally considerate. Indeed, two or three branches of the Serpent River system had been much affected by uranium tailings. These had been dumped in all four lakes on the central arm of the system, which, together with three in the eastern arm, all showed high levels of an isotope of radium released by the milling process (Figure 5.6). In the late 1970s a scheme for the expansion of mining in the area was proposed, but a provincial inquiry into environmental issues received evidence that all of these water bodies were quite unsuited in their present state for fishing or swimming. Since the area was in demand for recreational purposes, the inquiry concluded that the complete district would have to conform to specific standards of water quality before further development could be contemplated. This would involve, amongst other things, reducing the dangerous radium isotope to levels of 1.11×10^{-11} Bq/l (becquerels per litre) or below.

As a longer term aim, both the Atomic Energy Control Board and the Ontario Ministry of the Environment agreed that all radium exposures

▲ *Figure 5.6 Uranium tailings at Elliot Lake, Ontario. More than 100 million tonnes of these were deposited directly into the environment, including the Serpent River system, before the problems of radioactivity in uranium tailings were properly appreciated. Present day legislation imposes strict controls on the disposal and proper containment of such wastes.*

must be the minimum that it is possible to achieve in the light of prevailing social and economic factors. The ultimate objective for the district at large had to be 'the restoration of the aquatic habitat to support a viable self-sustaining healthy and wholesome fishery'.

4.3 *Water pollution beyond the mine*

In most circumstances it would appear that problems regarding water contamination no longer have a deleterious impact in drainage systems at any great distance from mine sites, particularly if treatment schemes are undertaken. Where minerals extraction other than for metals or coal is concerned, there are also many examples of water quality improvements away from the immediate neighbourhood of the extractive operation. In the china clay working area of St Austell, Cornwall, over 600 000 tonnes per annum of micaceous residues were being discharged into St Austell river in the early 1970s. At the same time a further 300 000 tonnes per annum were being poured into the Luxulyan river. This waste material was transported downstream and deposited as white slimes in St Austell Bay. The unpleasant appearance of the river and the beaches of the coastal area led the responsible company, now ECC International, to assess a wide range of disposal alternatives. These included carrying the material 800 m out to sea by pipeline, a solution not favoured by fishing interests because of its propensity to have an adverse effect on the habitat of shellfish. Eventually, technical progress in the better separation of kaolin from these fine residues, thus reducing the amount of micaceous material, was achieved. As a result, it was decided that this waste should be settled out behind land-based dams, or disposed of into used pits, a solution that has been practised since 1975.

In spite of such improvements, some sites with long histories of extractive activity still remain polluted. For example, the Rio Tinto mining district of Spain, north-west of Seville in the foothills of the Sierra de Aracena, has been a site of the working of copper ore for 3000 years. But it was not until 1873, when the rapid development of its ore body had made it one of the largest copper producers in the world, that water pollution reached major proportions. The ore body is pyritic and contains high levels of iron, so acidic waters, stained red by iron ions, drain from old waste tips as a result of natural leaching and from the foot of disused, as well as working, tailings areas. Because these waters are not contained, they continue to do much damage to local river ecosystems.

In Cornwall, the designation of some rivers as industrial has allowed their continued use for disposal of effluents from tin mining operations. For example, the South Crofty mine near Camborne, the UK's one remaining mine following the collapse of world tin prices in 1986, discharges the bulk of its mill waste water in the Red River. Heavily discoloured with haematite, 32 million litres of water a week, containing between 60 and 90 g/l of particulates, are carried to the sea some 20 km away. In West Penwith, Cornwall, similar problems arose at the now defunct Geevor tin mine, but here the effluent of 62 000 tonnes a year of solids suspended in more than 700 million litres a year of process water, passed immediately into the sea, causing heavy haematitic discoloration over a wide section of the coastline of outstanding scenic value near Land's End.

4.4 Summary

This section has shown that mine sites offer considerable possibilities for the contamination of rain or groundwater, and steps need to be taken to ensure that these are treated before they pass into the natural drainage system. However, by far the greatest problems arise from tailings. In a more detailed consideration of the hazards to water environments imposed both by uranium tailings and those containing sulphides, which have a propensity to generate acids, methods of reducing or eliminating these have been discussed. It is shown that while substantial advances in the standards of water treatment have been made in the last 30 years, including the recycling of much of the water used in the mill and collected on site, it is still possible to find river systems in developed industrial countries which have largely been abandoned to the discharge of mining effluents.

5 Micro environmental aspects of mining – air pollution

5.1 Problems of dusts

Drilling, blasting, waste tips, stockpiles and the loading and transport of ores from the quarry or open-pit face can all create dust. The dust can be dealt with by spraying water over the working surfaces and the watering of on-site roads or sweeping them if they are surfaced. The first stages of crushing the extracted rock can also create a dust problem. But here enclosure of the working plant and the use of dust collectors can both play an important part in minimising the nuisance. Metal concentrates may be dried before shipment to the nearest smelter. Where this occurs, the concentrates are more likely to release particulates rich in heavy metals into the atmosphere during transport unless steps are taken to prevent it.

Tailings ponds, apart from the problems which can arise from their appearance and the waters that can drain from them, can also expose finely ground and easily blown materials to the wind. Although there are circumstances in which the tailings may be completely covered by water, more often than not they lie above water level forming a sizeable, gently sloping beach. Substantial areas of the pond may also dry out at certain times of the year, or on closure, and if the coarser fraction of the tailings has been used to help build the dam wall, the remaining finer material may be even more susceptible to wind action.

If in exceptional cases the tailings area is very large, as at the Bingham Canyon Copper mine in Utah, USA, where the pond is more than 2100 ha in extent, remoteness means that dust is not a problem. However, at the Elliot Lake uranium workings, where one of the four major tailings areas in use covers more than 280 ha, the situation is quite different because of nearby settlements. Here, two of the tailings areas no longer in use are located on predominantly flat, high surfaces where some natural barriers and a number of dams contain the deposited tailings. Efforts have been made to reduce dust blown from these through the use of vegetation.

The Environmental Assessment Board, investigating the expansion of uranium mining in the late 1970s, recognised the hazard caused by dust with a radionuclide content to adjacent townships, especially Elliot Lake and Nordic, but also Denison and Quirke town sites. Its findings were echoed in a series of radiation surveys undertaken in 1980 at 20 uranium processing areas in the western USA which emphasised the serious hazards posed by such dust, both because it delivered a dose of radiation to the lungs and other internal organs of humans and other animals, and because it contaminated grass and became concentrated in local milk supplies. The province of Ontario's criteria for suspended particles were certainly being exceeded in the Elliot Lake area and it would appear that levels at Nordic in particular were well above 153 g/m^2 identifiable as tailings with a radionuclide component.

Radioactive dust is not the only problem in the vicinity of uranium tailings ponds. These impoundments tend to give off radon gas as radioactive materials decay. The breathing in of this gas can result in radiation doses to the respiratory tract that can lead ultimately to deaths from cancer. Evidence seems to suggest that levels above the natural background level may be detected from uncovered tailings to distances of 2.5–3.3 km or more. A further hazard derives from the fact that radon can build up inside buildings near tailings to quite dangerous levels.

Although levels of radon and radioactive dust in the Elliot Lake area have not caused immediate alarm to the provincial environment authority or to the Atomic Energy Control Board, they have been perceived as both an inconvenience and a possible danger to people and crops.

5.2 *Smelting and the local environment*

The processing of metalliferous ores beyond the stage achieved in the mine mill to the point at which the refined metal becomes available is carried out at the smelter. This is often located at the mine itself, although smelters which handle a wide range of metals are sometimes to be found close to the industrial complexes that use their end products.

The processes involved in smelting give off considerable quantities of waste gases and metal particulates which are vented to the atmosphere via chimneys. In more recent years, as concern about possible environmental damage and health hazards has risen, the waste gases vented from primary iron and steel, aluminium and iron-alloy production have been subjected to some cleaning, using methods which involve electrostatic precipitation or bag filtration. The smelting of sulphide ores has also been controlled. Sulphur dioxide (by reaction with oxygen and water) is now largely converted to sulphuric acid within the plant rather than in the atmosphere. The effectiveness of these containment procedures is no longer a technical problem; instead, it depends on what smelter operators are forced to spend on them. As it is, gases and metal particulates are still vented via stacks to the atmosphere in a compromise which not only reflects wider community needs, but also the capacity of the smelter operator to bear the expense of improved 'house-keeping'.

Although the use of very tall stacks as a means of achieving high-level dispersal of effluents has moved the impact of air pollution well beyond the immediate vicinity of the smelting operation and the community in which it is situated, this technique is not universally used and in many instances the effects of smelter discharges mostly remain a local problem.

This is particularly true of the former Communist countries of eastern Europe. Evidence from Poland, for example, indicates the grave damage done both to people and crops in Upper Silesia by smelting. However, although steps have been taken to improve the situation in Poland, Russia's chief smelting complex remains a major hazard, as Box 5.1 makes clear.

Itai-itai disease can result from the inhalation of dust containing cadmium but is more likely to be from the eating of foods which have at some stage been subject to air pollution from smelters that process ores associated with this metal. The disease (which leads to bone porosity, followed by skeletal deformation, fracture and ultimate collapse) was first recognised in the vicinity of some of the 25 smelters in Japan. To cite one example: in 1960 amounts of cadmium in the atmosphere around the Annaka city smelter in Japan were as high as 18 p.p.m. in plants at a distance of 0.4 km, falling to 10 p.p.m. at 1 km away. Soil analyses showed cadmium had built up to levels of 33.5 p.p.m. 0.5 km away from the plant and to 9 p.p.m. at 1.5 km. In fields in the Usui river basin immediately beyond, cadmium in rice crops averaged 0.39 p.p.m., but in samples taken from areas where there was no known pollution, cadmium levels averaged around 0.07 p.p.m. By the end of the decade average levels in rice crops had risen to 0.49 p.p.m. while in the soils it averaged 22.3 p.p.m. By then the wider spread of the problem had been acknowledged by the setting up in Japan of a Committee on Cadmium Poisoning and the Itai-Itai Disease. Its deliberations eventually led to the banning of the use of cadmium in that country.

There are several other studies of cadmium smelting operations in England, the USA, Canada and Australia. All show that where the levels of cadmium in soils and vegetation near smelter chimneys are high, this rapidly falls as distance from the plant increases, just as it did in the Japanese examples. Only at Avonmouth (England) and Belledune (Canada) is there evidence of high levels of cadmium uptake by crops within 0.8 km of the smelters. Fortunately, in neither case are the crops concerned a staple part of the diet in the way that rice is in Japan.

One other smelter effluent also has a relatively local impact on plants as well as people and animals. Fluoride can be vented into the atmosphere in the process of making aluminium, but it may also be given off as a result of the processing of clays in the manufacture of bricks and ceramics, and the production of iron and phosphate fertilisers. Gaseous fluorides can be taken into the human respiratory tract and ingested through the intestines. Fluorosis, a form of osteosclerosis, has been known to result, but only in those exposed to high levels of fluoride on a continuing basis, as happened in the early European aluminium smelter plants. A more notable problem has been the ingestion of forage containing fluoride concentrations in excess of 40 mg/kg by animals, especially dairy cattle. This has been known to damage teeth and cause bone degeneration and lameness. Tests on cattle at 43 farms close by a source of gaseous fluoride near Peterborough in England were carried out by the Ministry of Agriculture, Fisheries and

Box 5.1 Russia's pollution capital

The small river which runs just to the north of Norilsk has water the colour of blood. The river falls from the hills far to the west of the city as clean as you would expect in this distant Siberian wilderness, but once it passes through Russia's largest mining metropolis, it is choked with poisons.

The stray chemicals which give the river its outlandish tint come from the Norilsk Nickel metalworks. The factory produces more nickel and copper than any other in the world. Because of it the city of Norilsk – population 250 000 – is the most polluted in Russia, possibly in the world.

The metalworks employs 90% of the local workforce and is a city in itself: an industrial nightmare, surrounded by mounds of earth displaced from the mines, the only mountains for miles. Sulphur dioxide gas billows from the chimneys, the pipes, the cracks in the furnaces; the tops of the tallest chimneys are hidden in a dense acidic haze. The air around it stinks and the snow is covered in a black crust. When it rains the sulphur falls to earth as sulphuric acid and contaminates everything. Most of the trees are dead or dying.

There is, however, a tendency among the city authorities to ignore the problems (including the severe health problems of those who work in the factory) and revel instead in the prestige which comes with being part of Russia's most important mining and smelting centre. 'We are not specialists on ecological sustainability' admitted Alexander Krivula of the Norilsk Planning Institute, with remarkable understatement. 'But…we are supplying Russia with hard currency.' He suggests that it is impossible to cut pollution significantly from the nickel factory because the place is too large for any existing cleaning technologies to deal with. He also appears to take pride in the management's ability to escape Russia's increasingly strict but ineffective environmental laws. 'We all know how to get around the rules', he said.

Norilsk Nickel which owns the metalworks, accounts for over 20% of the world's nickel output (and 80% of Russia's), 19% of the world's cobalt (70% of Russia's), 42% of world platinum (100% of Russia's) and 3% of the world's copper (40% of Russia's). (Michael Shaw Bond, *Financial Times*, 2 September 1995)

Food in the 1970s and revealed fluorosis at 19 of them. As to plants themselves, fluorides are absorbed through the leaves and concentrated in the tops, causing a burnt appearance. Among the most sensitive species are apricots and gladioli and among the least is celery. The evidence is that a concentration of 0.5 $\mu g/m^3$ (ten times ambient levels) to 1.0 $\mu g/m^3$ for 30 days in air could give rise to the 40 mg/kg level in vegetation, at which problems occur. A variety of techniques for the control of the gaseous fluorides from a specific source exists and can be relatively easily exercised up to 95% effectiveness. These are usually brought into play once a problem has been identified or, in the case of new plant, to meet the relevant environmental protection standards.

Activity 2

Look back over material in Section 5.2 and comment on the most important differences in the attitudes of various governments towards smelter pollution.

5.3 Summary

Mining operations can create a series of problems at the local level, including dust from working surfaces, waste tips and stock piles, all of which may be rich in heavy metals unless control systems are used. The major problems have really arisen from the smelting of metal ores, where the discharge of particulates from smelter chimneys create a major health problem for the area in the immediate vicinity – so much so that the emission of cadmium (which is readily taken up by field crops) from Japanese smelters has been banned. As with water pollution, the story is one of increasing awareness of the damage that can be done by the discharge of effluents, combined with ever stricter controls, although as in the case of Russia, they can be ignored!

6 Meso environmental impacts of mining

6.1 Sudbury – a classic case

Several studies from the 1970s have shown that following the expulsion of effluents from very tall stacks, under certain circumstances mineral particulates, and especially sulphur dioxide (if a sulphide ore has been smelted), can travel substantial distances from their sites of release. Only their removal from the air by rain is likely to hinder this process. The history of Sudbury in Ontario, Canada, illustrates the main trends in air pollution that derive from such a source and its changing role, first as a polluter of Sudbury and then of the region around Sudbury, but later as a major contributor to transnational problems.

The early years of mining and smelting at Sudbury (summarised in Box 5.2) provide the background to the more recent developments of the 1970s and 80s.

The key to the spread of air pollution problems into the meso environment came in 1972 with the construction by Inco of a 387-m tall chimney at Copper Cliff. This 'superstack' was designed to achieve a high level of effluent dilution, well away from the Sudbury area. Given that this was by then the largest single source of pollution in the world, such an approach seemed a reasonable one. In effect what happened was that

Box 5.2 The Sudbury story – mining and smelting from the 1880s to 1980

It is said that it was a blacksmith, Tom Flanagan, who in 1883 first noticed a rusty stain on a rock near what is now Murray Mine. There was some disappointment when it was discovered that it was an ore which contained nickel as well as copper, since the market for nickel was poor, and inadequate technology existed for the separation of the two metals. In the 1880s, however, a separation process was discovered giving rise to a tremendous expansion in the mining industry in Sudbury.

In 1886, S. J. Ritchie, an American, started the Canadian Copper Company and mining began in earnest. At first the ore was sent for smelting to Britain and Bayonne, New Jersey, but in 1888 the first roast yard and production-oriented smelter were set up in Copper Cliff. In the roasting of copper-nickel ore, sulphur is removed mainly from its associated pyrrhotite by heating in the presence of air to a sufficient temperature to begin a process of self-maintaining oxidation. In the early days the crushed ore was piled on huge beds of cordwood several feet deep, covered to prevent open flames, and ignited. The beds of ore would burn for two or more months before being loaded into cars with steam shovels and smelted in the furnaces.

The discovery of nickel-steel for armourplate and the Spanish–American war of 1898 boosted the demand for nickel, and by 1901 there were 80 roast heaps and 9 furnaces in the area. In 1902 the International Nickel Company of New York (Inco) was organized through the merging of many existing companies, and in the years that followed it purchased a number of other companies, including Mond in 1928. The only company remaining independent of Inco to this day is Falconbridge Nickel Mines, which was incorporated in 1928.

In 1916 the Canadian public demanded that the assets of Inco be placed under Canadian control. Also, in 1916, its open air roasting activities were moved from Copper Cliff to O'Donnell, where the remains of its main roast bed can be seen today. This roast bed operated until 1929, when government insistence resulted in the building of a new Copper Cliff smelter which had roasting furnaces and a 510 foot stack, the highest in the British Empire at that time. Although a number of new stacks were built at the various smelters over the years, the 510 foot stack was only exceeded in height by the 637 foot stack at the Copper Cliff Iron Ore Plant (1956) and the Inco 'superstack', which became operational as an anti-pollution measure in 1972.

In many ways, from an environmental improvement point of view, 1972 was a 'red letter' year, since it was also the year in which the Coniston smelter and a pyrrhotite roasting plant at Falconbridge were closed, the Inco iron ore plant cut back on emissions and both a strike and a holiday shut-down occurred. Probably the period most damaging to the environment, with the exception of the early roast bed days, was in the early years of the second world war, when production of copper and nickel were doubled in response to wartime demands, and pollution control had to take second place in company priorities.

A pleasing aspect of pollution control which has developed over the years, though not reaching the level hoped for by many environmentalists, has been the use of waste gases to manufacture sulphuric acid. The first acid plant was built at Coniston by Mond Nickel in 1925. Although Inco's first plant was built in 1930, its latest plant opened in 1967, producing 1400 tonnes of acid per day. In 1978, Falconbridge Nickel opened a sulphuric acid plant along with its new smelter. The cutback on sulphur dioxide emissions in 1972 (from more than 2 million to less than 1.3 million tonnes per annum for Inco, and from 344 000 to 244 000 tonnes per annum for Falconbridge), together with the better dispersion achieved by the 'superstack', have combined to give a noticeable improvement in plant growth in the Sudbury area and a marked reduction in what the Ministry of the Environment calls 'potentially injurious fumigations'.
(Keith Winterhalder, 1978)

Sudbury, through the superstack, managed to export most of its problems. These have been recognised in a belt extending some 60 km downwind of the plant in Sudbury, where not only some of the metal particulates in the plume, but also a good deal of its sulphur dioxide has tended to fall back to ground level in this zone, having been converted by the action of the oxygen and moisture in the atmosphere to dilute sulphuric acid, popularly known as 'acid rain'.

Studies undertaken in the 1970s of some of the lakes in the affected area showed that in many instances the chemical balance of their waters had altered. The uncontaminated lakes were nearly neutral (neither acid nor alkaline) and, having a granite bedrock rather than a limestone or clay base layer, lacked resistance to acidification. So acid rain appears to have tipped the pH of a number of these lakes to values below 5.5, thus providing an environment in which fish and the organisms on which they feed could not survive.

Not unexpectedly, many of these water bodies also contained in solution abnormally high levels of heavy metal ions, particularly nickel and copper discharged from the plume. Although there was a lack of evidence from those lakes that had a low pH and were already biologically dead, it is known that metals of this sort can enter aquatic ecosystems and become concentrated in animal food chains with devastating effect. Not only are copper and nickel toxic, evidence shows that if both are taken into the food chain at the same time they enhance each other's toxicity.

Of the lakes closest to Sudbury, within 20 km of the major smelter, about 20 were part of a survey reported in the early 1980s. Because their bedrocks varied widely in chemical and physical characteristics, the degree of stress shown as a result of acid rain varied considerably, with about a quarter having unacceptedly low pH values. These lakes also varied in their metal content according to their pH but especially according to intensity of atmospheric input and thus distance from Sudbury. If the global mean value of 10 $\mu g/l$ of nickel is taken as a reference point, only four lakes at the outer edge of the 20-km radial were remotely close to or under this figure. Most were far above it, with one reaching the highest level recorded anywhere in the world – 512 $\mu g/l$. The highest levels of copper concentration in lake waters (102 $\mu g/l$) were to be found at this same location. Indeed, copper concentrations in the lakes in question followed a similar pattern to those for nickel.

Between 40 and 60 km from Sudbury, further studies concentrated on the presence of nickel and copper and the impact of sulphur dioxide upon lakes in the Killarney Provincial Park, relating these to the Inco stack at Copper Cliff. The Provincial Park is important for recreation, particularly fishing and canoeing. In the lakes largely located on granite, sulphur dioxide fall-out as dilute sulphuric acid resulted in their waters falling below pH values of 5.5 (Figure 5.7) with a resulting loss of aquatic life. Although nickel concentrations at this distance (Figure 5.8) fell to levels which were acceptable in terms of the standards generally applied to domestic drinking water (at not more than about 25 $\mu g/l$), the presence of copper in many of the lakes remained above that level (Figure 5.9).

The effect of the situation created by the transport of effluents from Sudbury on a provincial park, once celebrated for its fishing, was not a matter of minor environmental significance. In the early 1980s water-based recreational activities, such as those common in Killarney, directly generated annual expenditures of $900 million and indirectly a further $750 million in Canada. Because of their importance in terms of wealth creation it was perhaps not surprising that an attack was begun on the dual

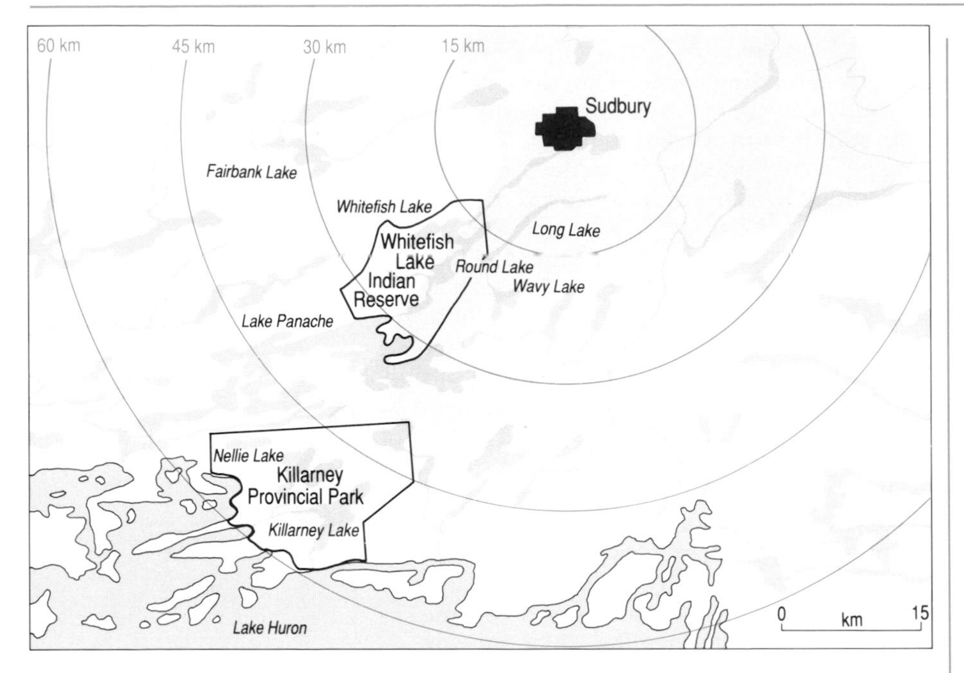

▲ *Figure 5.7 Sulphur dioxide emissions after the construction of the superstack at Sudbury. The shaded area may contain lakes with waters of pH 5.5 or less, depending on their bedrock.*

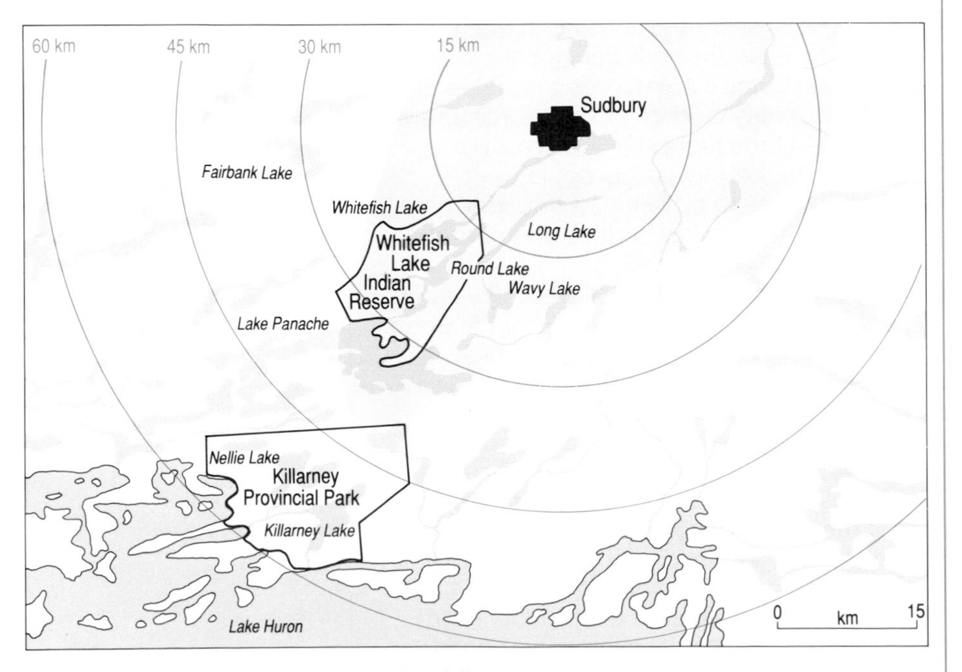

▲ *Figure 5.8 Nickel emissions after the construction of the superstack at Sudbury. Nickel can be used as a reasonable definer of metal emissions from the Sudbury area smelting complex since it is naturally low in lake waters. A global mean is 10 µg/l, while lakes in the shaded area can be as high as 300 µg/l, even where there are no direct discharges to these lakes of liquid effluents containing nickel.*

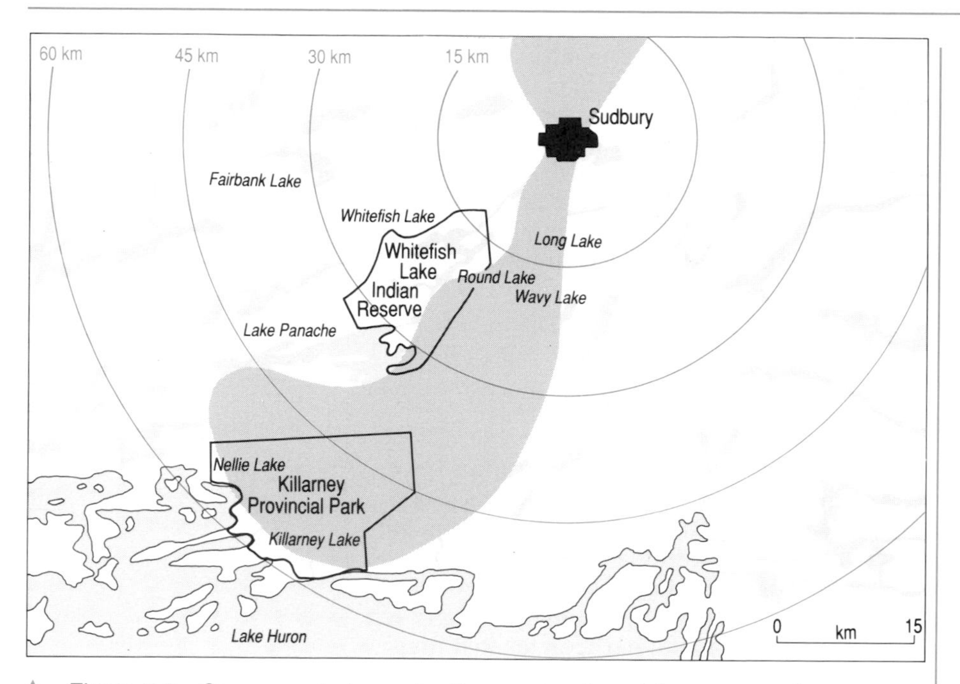

▲ *Figure 5.9 Copper emissions after the construction of the superstack at Sudbury. The shaded area may contain lakes with over 25 μg/l.*

problems of low pH and elevated metals levels through the addition of lime to the lake waters. Such efforts to raise the level of the former and lower the level of the latter were seen as a prelude to restocking the lakes with fish. A similar programme was applied to lakes less than the 20 km from Sudbury. The results coming from all these attempts at aquatic rehabilitation were far from entirely satisfactory. For example, Lohi Lake, some 6.5 km from Sudbury and one of the most adversely affected water bodies, had a first liming six years earlier, after which its pH rose from 4.3 to 4.8, whilst nickel concentrations dropped from 264 μg/l to 223 μg/l and copper concentration dropped from 84 μg/l to 61 μg/l.

6.2 Regional improvements

Notwithstanding such limited efforts to improve lake water quality, the situation with respect to the impact of Sudbury emissions on water bodies in the region has not remained static. Initial levels of discharge from the smelters were reduced by process improvements at Falconbridge in 1978 and Inco in 1980. The latter were also coupled with a regulatory control placed on the superstack in that same year limiting its sulphur dioxide emissions to an average of 2500 tonnes per day. At the same time, industrial problems, leading to an eight-month shutdown at Inco in 1978 and 1979 and the more recent world-wide industrial recession, resulted in a decline in the quantities of ore smelted and also, therefore, in stack discharges. Thus a comparison of data for the quality of 209 lakes collected in the period 1974 to 1976 with that for 1981 to 1983 shows signs of an improvement unrelated to any rehabilitation programme undertaken at individual lakes.

The key to a real and continued improvement in the water quality of these lakes has been the addressing of the problem at source. In 1984 the Ontario Ministry of the Environment started a ten-year programme of sulphur dioxide reductions applied to all major sources of this gaseous effluent. This 'Countdown Acid Rain' scheme set for the Inco smelter a series of target falls in stack emissions, starting at 728 000 tonnes a year, reducing to 685 000 tonnes by 1986, and then to 265 000 tonnes by 1994. Actual emissions in 1994 were below the negotiated level, largely as a result of the use of new milling and smelting technology introduced in 1993.

6.3 *Other meso studies*

Although the Sudbury region remains an extreme case, other studies have been undertaken at a meso level of the impact of smelter discharges. For example, attention has been drawn to the propensity for not only the major components of smelter effluents to be transported long distances, but also trace elements as well. Indeed, in work carried out on five copper smelters in south-east Arizona, it has been shown by trace element analysis that each smelter has its own distinctive 'fingerprint' of metal particulates that may be identified at a considerable distance from a point source of pollution.

The world's largest lead smelter at Port Pirie on the Spencer Gulf in south Australia contributes not only lead particulates to the atmosphere but also those of silver, zinc and cadmium. Indeed, its impact on soils in the area has been traced up to 65 km from the operation, with the highest concentrations to the north and south in the direction of the prevailing winds. The overall spatial distribution patterns of contamination recognised in this initial study implied that stack out-fall would especially affect the Gulf area of the north-west and south-west quadrants. Further research in the 1980s established just such a situation with over 600 km^2 of sea-bed sediments affected by these metal particulates. The probable ambient background levels of metallics in the region as a whole are for silver, cadmium, lead and zinc respectively: 0.3, <0.03 (less than 0.03), 2.9 and 10 μg/l. These may be compared with the sediment measurements shown in Figure 5.10. Analysis of the flora and fauna of the area has shown that almost all organisms contain elevated concentrations of cadmium, lead and zinc. Twenty common species of fish appear to have been eliminated or reduced in numbers by the presence of metals. Indeed, concentrations of 0.7 μg/g and above of cadmium and 92 μg/g of zinc seem to have a clear impact, though they cannot be assumed to be acutely toxic to all species. As far as the less mobile members of the fauna are concerned, the range of species could be observed to increase away from the most contaminated zones.

Of these other investigations, only the situation at Flin Flon, Manitoba, Canada, provides a strong parallel with Sudbury in that the discharge of sulphur dioxide and metal particulates from the smelting of sulphide ores containing zinc, copper and cadmium has also created environmental problems. Like the Sudbury area, Flin Flon lies on the granitic Precambrian Shield. But a few kilometres to the south of the city, this gives way to the limestones common in the central part of the American continent. Studies which began in the mid-1970s not unexpectedly indicated that as a result of the discharges of its own 281-m superstack, many lakes in the area had become contaminated with airborne metals and had also been affected by sulphur dioxide fall-out.

Q Given the two kinds of bedrock in the Flin Flon area and the knowledge you have gained in this section so far, what broad conclusions do you expect the study to show?

A It will underline the potential for ecological disturbance in those lakes on the granitic Shield with their poor ability to neutralise acidity. This is important not least because of their use for tourism and sport. They may be contrasted with those water bodies located on limestone, albeit more distant from the smelter, which remain largely unaffected. This is because the limestone is chemically basic and so able to neutralise added acidity.

One of the more recent assessments of the impact of the smelter on its surrounding area is summarised in Table 5.2, which shows the annual deposition of metals at selected sites at different distances from the smelter. You can use Table 5.2 to compare deposition levels and their effects in two contrasting lakes, Nesootao and Hook. Both are considerably influenced by sulphur dioxide fall-out from the smelter (as acid rain). Although this has not been measured on a site-by-site basis, discharges from the chimney, at the time the table was compiled, were in the range 232 000 to 306 000 tonnes per year, accounting for 4.5% of Canadian emissions. Because of the rapid decay factor for metal particulates with distance from a point source of emission, Nesootao is less contaminated than Hook. However, since Hook

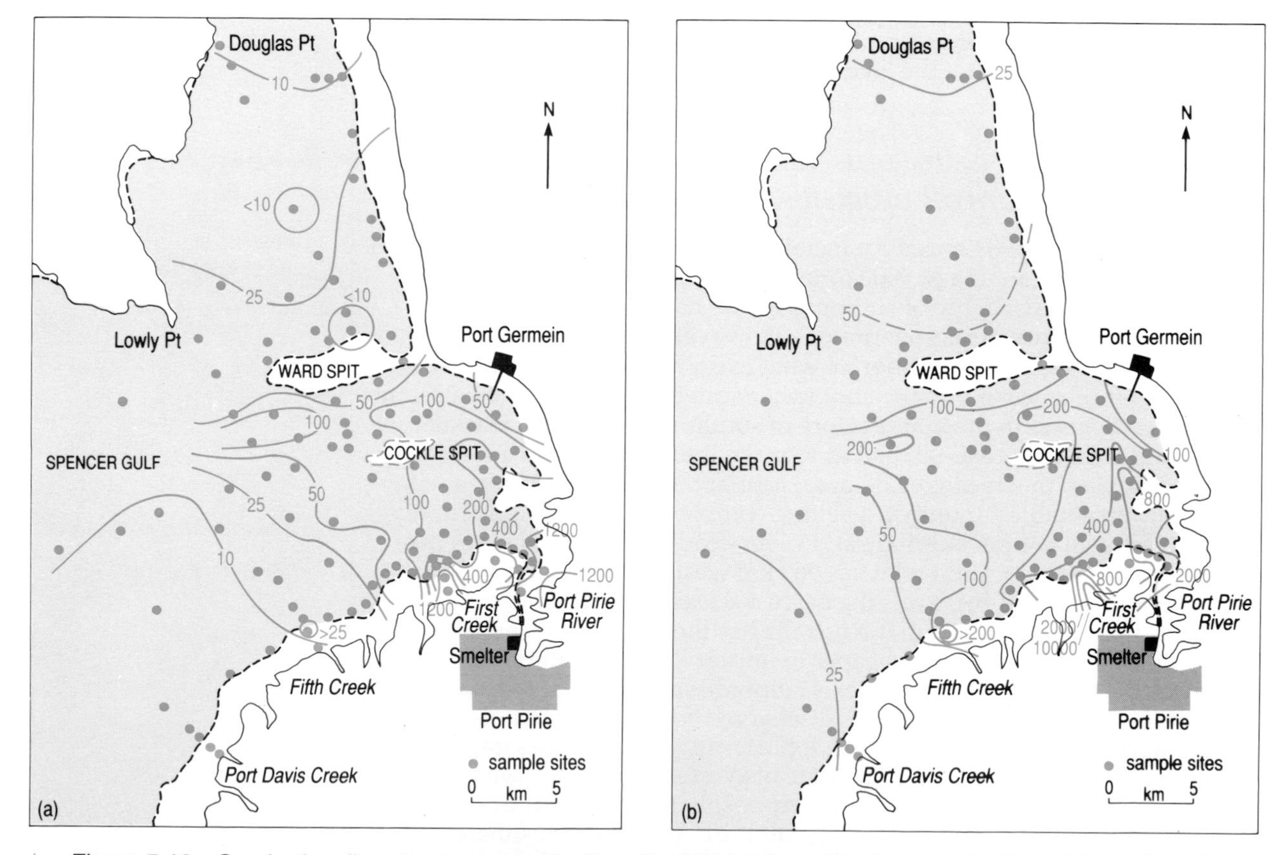

▲ *Figure 5.10 Sea-bed sediment measurements (in µg/l) of (a) lead and (b) zinc near the Port Pirie lead smelter, New South Wales, which contaminates areas with metallic particulates up to 65 km from its stack.*

Table 5.2 Deposition and distribution of smelter metals in six lakes near the smelter at Flin Flon, Manitoba

Lake	Distance from smelter/km	Deposition/(mg/m²/yr)			
		Zinc	Copper	Cadmium	Lead
Cliff	4.0	1400	62	3	32
Hamell	4.2	1287	58	3	31
Lake 6	5.2	889	42	2	24
Hook	9.5	313	27	2	27
Nesootao	12.5	195	11	0.7	9
Thompson	20.2	87	5	0.4	5

Source: Franzin, W. G. (1984) Aquatic contaminants in the vicinity of the base metal smelter at Flin Flon, Manitoba, Canada, pp. 523–78 in *Environmental Impact of Smelters*, ed. J. O. Nriagu, Advances in Environmental Science and Technology, vol. 15, Wiley, New York.

Lake is well to the south of Flin Flon and on a limestone bedrock which assists in the process of maintaining alkaline pH levels in its waters, its biota were found to contain relatively low metal concentrations. This contrasts with Nesootao to the north-west of the smelter. Its granitic base cannot modify the acidity of the lake waters arising from acid rain. This has led to the considerable bioaccumulation of metal particulates, especially by fish, in spite of the fact that the lake is less subject to metal particulate fall-out than Hook.

6.4 *Pollution control problems*

Steps were taken in 1983 to reduce metal particulate emissions from the Flin Flon smelter, then running at 9600 tonnes a year, by more than 80%. Reduced levels of discharge of sulphur dioxide have subsequently been called for by the provincial government. However, at extremely remote locations such as this, the question of what to do with the sulphur dioxide removed before discharge to the atmosphere can be a difficult one. Sulphur dioxide can be used in the manufacture of sulphuric acid, but if markets are distant and transport costs so high as to make it uneconomic as a saleable commodity, then this means of disposal cannot be used. This applies with particular force to the Mount Isa smelting complex in Queensland. There the processing of copper sulphide and silver–lead–zinc sulphide ores take place at a semi-arid location which is 960 km west of Townsville and 1560 km north-west of Brisbane, the nearest sizeable settlements.

Thus, in spite of the fact that Australia has the means to meet its own sulphuric acid needs, it is still cheaper to import the raw material for its manufacture, sulphur. Only the nickel sulphide smelting plant at Kalgoorlie, which in the early 1990s discharged some 330 000 tonnes per year of sulphur dioxide, offers a real opportunity to reduce this import bill. This is because of its location in the state of Western Australia, which consumes one-third of the nation's superphosphate, an agricultural fertiliser which is manufactured using sulphuric acid and phosphate.

If the only prospect open to such remote operations involved with sulphides is the release of sulphur dioxide to the atmosphere, at least the barren nature of the terrain for many hundreds of square kilometres around Mount Isa means that environmental damage is unlikely to be great. Nevertheless, such effluents still need to be vented away from the 30 000 inhabitants of Mount Isa township itself. These smelters produce between them about 400 000 tonnes per annum of sulphur dioxide, derived from the processing of 120 000 tonnes of copper and 120 000 tonnes per annum of lead. Certainly, readings taken around the settlement prior to 1977 and the implementation of the Queensland Clean Air Act were sufficiently unacceptable for the National Air Pollution Council to call for a programme of high-level sulphur dioxide dispersal from new chimneys (300 m and 270 m) for each of the smelters. More than this, discharges were only to be permitted when the prevailing wind blew stack emissions away from the community.

As for the particulates of copper, lead, arsenic, antimony, cadmium and mercury given off by the Mount Isa smelting operations, agreed pollution control standards after 1977 of a maximum total of 23 $\mu g/m^3$ effluent levels in the stacks were a substantial reduction on levels prior to that year. Formerly the lead smelter alone often discharged at values of 70 $\mu g/m^3$, and the copper smelter was putting out 530 $\mu g/m^3$ made up of lead, arsenic and copper. The improvements carried out at the behest of the National Air Pollution Council resulted from modifications or additions to existing gas-cleaning hardware and the installation of major new items of bag and filter equipment.

Activity 3

Describe the similarities and differences between the control measures imposed by governments on Mount Isa and Sudbury and the reasons for them. Your response will help you to summarise some of the key issues in this section of Chapter 5.

7 *Macro environmental aspects of mining*

7.1 *Transfrontier air pollution*

One definition of the macro impact of mining pollutants might be where these cross national frontiers. Such a definition could stretch credibility in the case of the Trail zinc smelting complex in British Columbia. The largest complex of its kind anywhere in the world, it has a long history in terms of air pollution.

Before the invention in 1929 of a practical method of producing sulphuric acid from the sulphur dioxide in the smelter effluent, emissions ran at around 300 tonnes a day of sulphur dioxide. The mountainous terrain in the vicinity of the smelter made it difficult for the gas to escape, and the area surrounding the operation had been denuded of vegetation. Today, however, 95% sulphur fixation is achieved, the spin-off product is used by the local agricultural industry, and the area has long been revegetated with acacia and conifers. Measurement of sulphur dioxide in the ambient air in a 7-km radius around the smelter indicates an average of 0.02 p.p.m. The air standards in the region beyond, because of the close proximity of the US border and the dispersal achieved from the release of effluents from the smelter stack, are controlled by an International Joint Commission, along with the provincial government. Critical emission levels are varied according to wind direction and the state of the vegetation growing season, but in all cases values fall within clearly defined limits.

Accepting that Trail must be a special case of transfrontier pollution, the smelter itself, like others discussed above and many more around the world, contribute their share of sulphur dioxide to the atmosphere. Depending on location and country they feed into the overall levels of atmospheric sulphur dioxide, along with power stations (especially those that are coal fired) and heavy industry. For example, in Canada, where emissions of sulphur dioxide prior to the recession of the early 1990s reached a peak of about 6.5 million tonnes a year, smelters contributed two-thirds of the overall output. These Canadian figures compare with the US total for sulphur dioxide emissions of 3800 million tonnes per year, but the contribution to this by smelters was much lower at around one-third.

Nevertheless, the fact remains that sulphur dioxide from a wide range of sources, including smelters, can persist in the atmosphere and travel considerable distances across continents. Thus the out-wash of sulphur dioxide, which falls as dilute sulphuric acid, together with anthropogenically derived oxides of nitrogen, which are mostly created through the combustion of hydrocarbon fuels, can affect areas remote from their points of origin. The impact of these effluents can be adverse where the environment is one of peculiar sensitivity, in much the same way as they were in Killarney Provincial Park. Evidence now available shows that sensitive environments located on granite bedrocks such as those in eastern Canada, in the north-eastern USA, in Germany, in The Netherlands and in Scandinavia, are increasingly succumbing to damage from acid rain. These matters were introduced in Chapter 2 and are pursued further in Chapter 6.

7.2 *Summary*

In the discussion in this chapter so far, consideration has been given to the impact of minerals extraction and processing at micro, meso and macro levels. Throughout, some attention has been given to ameliorating the environmental effects that such activities may have on the environment during the operation of the mine or the smelter. These include the working of some minerals underground instead of by open pit, where visual damage can be a particularly sensitive issue; the extraction of minerals such as coal and salt in such a way as to avoid subsidence; the disposal of tailings away from areas of scenic quality and their placement underground if they are radioactive; the avoidance or reduction of acid drainage problems along with other attempts to diminish the exposure of tailings to the atmosphere during their operative stage; and the containment and reduction of smelter effluents.

In the next section the argument is taken a step further by considering ways to deal with those environmental problems which occur at the conclusion of minerals extraction and processing or at the end of a phase in such operations, for example, as the completion of a tailings pond.

8 Reclamation of mined areas

8.1 Some underlying principles

When mining operations at any site end, it becomes necessary to decide what is to be done to the area to mitigate the impacts of the operation on the environment. Several options are available.

- First of all, the site can be neglected, leaving natural processes to do what they will with it. In spite of legislation, this can still occur where sites are remote or where reclamation may be particularly difficult. However, the result is not always to be deplored. For example, disused chalk quarries on the downlands of southern England have been neglected because of the difficulties of rehabilitating vertical faces, but they have ultimately become valuable as nature reserves.

- More often than not, schemes will be undertaken for **reclamation** of the area as quickly as possible in order to improve its appearance, return it to some productive capacity or contain toxic materials that might otherwise damage adjacent habitats or, indeed, human beings. This may involve attempts to achieve **restoration** of the area by recreating the original topography and perhaps re-establishing the land use prior to mineral extraction. For all these purposes, a major aim is to establish soil and vegetation which will stabilise the surface, exclude air, reduce water penetration and so prevent acidification.

- **Rehabilitation** may also be considered in which the return of the land to some productive use involves the development of new land forms and/or ecosystems. In Kansas, for example, a grassland ecosystem that has been surface mined has been replaced with a lake because the area lacked standing surface waters. A well-managed lake might be regarded as more desirable than the original and certainly more desirable than the area in its unrehabilitated state.

- Other forms of 'enhancement' have involved such considerations as the rehabilitation of surface mined sites at slopes shallower than the original. In west Virginia, for example, the argument has been that reductions in the pre-existing slope of at least 20 degrees diminish the potential for surface erosion in an area prone to such problems.

Whatever the complex of factors that may be involved in the decision-making process relating to how to treat worked-out mine sites, the fact remains that the characteristics of the waste materials and land surfaces after extractive operations are important factors affecting the ability of such sites to undergo effective reclamation or restoration. These may be considered under the key headings of physical and chemical.

Physical properties

As was stated in Section 4.1, the physical and chemical properties of mining-derived materials are commonly associated with their particle size distribution, which may vary from the large person-sized blocks sometimes obtained from stone quarrying to the finest clay-sized particles resulting from the processing of metallic ores. As particle size decreases there is a corresponding increase in surface area per unit of volume. This determines moisture retention by particles, drainage and availability of nutrients to plants.

In the course of soil formation, physical processes tend to reduce particle size. Similarly, residual materials from mining activities may be reduced in size by environmental processes, although for tailings and hard resistant materials this may not be significant within a human generation.

Chemical properties

Plants require a number of elements which they absorb mainly through their roots. Some plants specifically require more of one element than another, but in general those elements needed in fairly large quantities are nitrogen, phosphorus, potassium, calcium, magnesium and sulphur. Those required in only small quantities include chlorine, iron, manganese, boron, copper and molybdenum. The precise requirements of vegetation for each element are difficult to define and depend on a wide variety of factors such as type of vegetation and on the level of production required. Some elements may be available in such low concentrations in soils and waste materials that plants show deficiency symptoms and reduced size. On the other hand, some elements may be available in unnaturally high concentrations and produce toxic conditions. Spoil and tailings may contain elements which are directly toxic to plants even in quite low concentrations: cadmium and arsenic are examples.

The availability of a nutrient element is usually strongly influenced by pH conditions, as shown in Figure 5.11. In the favourable range 5.5 to 7.5, most plant nutrients are available, usually in the form of ions in the soil solution. These include both simple metal cations (like K^+, Ca^{2+}, Mg^{2+} and Fe^{2+}) and polyatomic anions of non-metals (like sulphate, SO_4^{2-}, and nitrate, NO_3^-). Above pH 7.5 the availability of iron and manganese diminishes, because they begin to form insoluble hydroxides, and the same happens to copper and zinc above pH 8.

The original sources of nutrient input to the soil are rainfall, natural weathering and the fixation of atmospheric nitrogen by bacteria in the soil. Once a soil is established, nutrients are also available from decaying organic matter. Precipitation inputs may be significant but are usually insufficient for plant growth in most climatic regions. The natural weathering of rocks reduces average particle size and provides most of the plant nutrients with the exception of nitrogen. This can be taken through the leaves as well as the roots. Soil organic material accumulates large quantities of plant nutrients which on decomposition again become available for absorption by plants. A particular type of organic material, humus, also plays a key role, along with clay particles, in holding mineral elements in an available form.

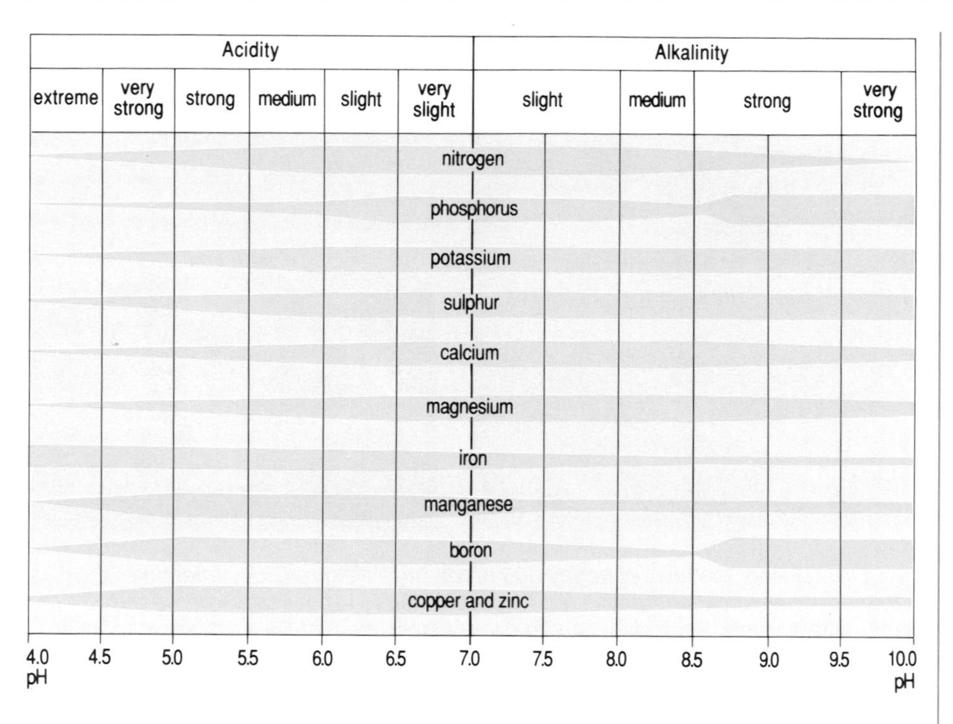

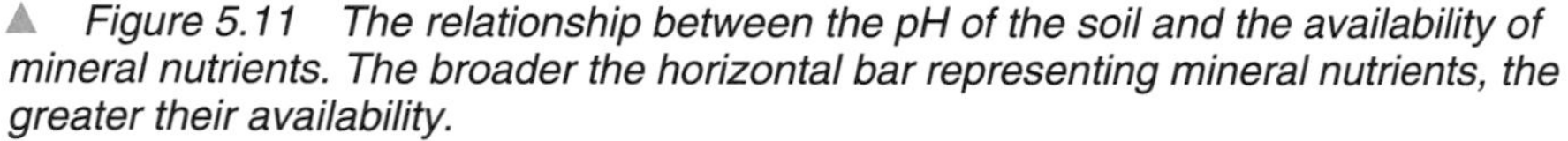

▲ *Figure 5.11 The relationship between the pH of the soil and the availability of mineral nutrients. The broader the horizontal bar representing mineral nutrients, the greater their availability.*

Against this background discussion of the key factors regarding the properties of materials that will play a major part in site reclamation, consideration can now be given to the processes involved. This may best be undertaken by grouping the sites into three: voids or open pits and quarries; overburden and waste tips; and tailings ponds.

8.2 The reclamation of voids

The commonest of all open-pit operations is that for sand and gravel. More often than not, workings have taken place below the water table. Even so, the voids concerned may be readily incorporated into a general reclamation plan. Any waste sand or gravel poses no threat to the environment and may be readily landscaped by using earth-moving machinery, often to grade gently into a central lake. In the United Kingdom the best known examples are along the River Thames, both in its lower reaches west of London and near its headwaters on the Wiltshire/Gloucestershire border; near Olney in Buckinghamshire; and at Holme Pierrepoint in Nottinghamshire. Most of these areas have been reclaimed as recreational facilities in the form of water parks, catering primarily for watersports such as sailing, water skiing and fishing (Figure 5.12). Such pits have also long been considered as suitable fill sites for domestic, industrial and other wastes by many planning authorities. Several hundred hectares of wet sand and gravel pits have been filled in the Greater London area where the demand for land for development has been strong. Where these wet pits are concerned, the main type of material used is rubble or inert waste, for in spite of encouraging

▲ *Figure 5.12 (Left) The Cotswold Water Park, an area of nearly 600 ha on the Wiltshire/Gloucestershire border in the valley of the upper Thames, where local authorities have collaborated to rehabilitate worked-out sand and gravel pits as country parks. Alternatively, such sites can be given a new lease of life as a venue for water sports, as at Ringwood, Hampshire (right).*

research into disposing of domestic refuse into wet pits, no satisfactory economic method of eliminating the risk of polluting groundwater supplies has been found.

Where open-pit operations have involved the extraction of rock aggregate, clays or metallic ores, considerable volumes of rock overburden may have been removed. Pits located in dense rocks or clays will likely fill with water and their fate will largely depend on their degree of isolation or the population density of the surrounding countryside. In the case of the large derelict brick clay pits of Bedfordshire and the Peterborough area, both in the highly populated south-east of England, their impermeability makes them ideal for the disposal of wastes such as garbage, low-toxic material and fly ash from coal-burning power stations. Indeed, research undertaken in Illinois on brick clay pits shows that toxic elements in wastes deposited in them have not migrated more than 0.5 m in more than 30 years. Nevertheless, waste containing a relatively high organic content can allow the formation of methane gas with the risk of explosion unless special precautions are taken. Deep pits of the dimensions which commonly remain after metal mining in most instances are too far from urban centres to be of use for waste disposal. Pits in limestone, or where there was some limestone overburden, may sometimes be useful as disposal sites for certain toxic wastes. The high pH of water percolating through the rocks around the site and into the pit, frequently around 8.4, will serve to hold many toxic elements out of solution. The soil and overburden removed from the pit will have been placed away from, but close to, the excavation. With the progress of time, forest vegetation will probably colonise these materials, as has happened, for example, at the iron mines near Star Lake in New York state.

Prior to excavation of the pit, however, sensible practice often dictates that the soil be carefully stored simply because it is an ideal medium for the establishment of the vegetation cover after mining. As soil consists of a number of distinctive layers called horizons, it is preferable to remove and retain the uppermost horizon on its own as this is the site of plant growth and organic activity such as that created by worms and microbes. This adds to the cost of rehabilitation so it is common practice to keep all the soil horizons together. There is evidence, though, to show that in soil which is stored for long periods of time the microbial and earthworm populations are reduced, soil structure deteriorates, and fertility declines.

In spite of efforts to restore voids to their original surface levels, this remains the exception rather than the rule, especially where hard rock quarries are concerned. In the United Kingdom, of the annual take of land for limestone, only 5% is completely restored. In such situations attempts at rehabilitation are confined to trying to mellow as rapidly as possible the disused vertical working bluffs. Here the natural or artificial regeneration of flora can be assisted by creating screes at their bases and by layering their faces with ledges on which plant life can take a hold.

8.3 Dealing with overburden and waste tips

Physically, these materials are usually very coarse but can consist of boulder-sized to clay-sized particles. The finer particles can fall into the voids between the larger fraction, leaving an uneven surface. Their physical nature makes them an unsympathetic site for plant growth, since they can be prone to drought. But of even greater concern in land reclamation is the increase in volume of the overburden once extracted. This poses certain problems when attempts are made to backfill the material into the original void. Where this is practicable it can often result in the creation of a surface higher than the original one.

Overburden and waste materials may range from relatively chemically inert rocks, such as quarzites, to soluble limestone and gypsum. Others may contain amounts of sulphide ores. The main indirect effect of these materials on the environment will be their influence on the quality of water draining through and from them. As was made clear earlier, calcareous materials such as limestone generally improve the quality of drainage waters. Sulphides, on the other hand, lead to acid generation. The best known example of the problems caused by this comes from Appalachia in the USA where acid drainage following the strip mining of coal has also caused the chemical release of heavy metals. Since no attempts have been made to rehabilitate these areas, great environmental damage has resulted along with much human suffering. At sites where fine quartz particles accumulate, these will provide sandy and infertile sites for plant growth, whereas accumulations of silt contain more clay and are generally more fertile. There are exceptions, however. Overburden consisting of heavy clay particles impregnated with salt overlying coal deposits near Edmonton, Alberta, has made restoration of the area very difficult.

The disposition of the overburden has a considerable influence on the logistics of land reclamation since its final resting place may vary from steeply inclined mountain slopes to reasonably flat surfaces. Among the most straightforward reclamation exercises are those where coal has been extracted from gently sloping or flat-lying sedimentary rocks. Here the overburden may simply be backfilled and its surface contoured to that of the original land surface as the operation progresses. Soils are first removed and stored in their separate horizons for ultimate replacement in the correct order. This makes the revegetation of the surface a relatively simple task once ditches, fencing and walls, etc. have been put in place. Perhaps one of the most spectacular as well as successful examples of restoration from open-cast coal operations is that at Westfalen in Germany. In the United Kingdom restoration has been equally effective, with the Ministry of Agriculture, the Agricultural Land Service and the Agricultural Development and Advisory Service supervising a five-year course of treatment once the land surface has been restored (Plate 6). Only then is a permanent drainage system installed. However, the re-establishment of something like the pre-existing ecology takes considerably longer and may never be achieved.

In some instances in the process of open-cast coal working, land restoration has been attempted without separating the soil from the rest of the overburden. In southern Illinois the re-established surface materials contain metals such as iron, aluminium and manganese and are deficient in potash and phosphorus. The restoration of the land to agriculture has only been possible by the offsetting of these difficulties with high and continuing applications of fertilisers.

Of all the wastes tipped as a result of mining, the spoil from coal extraction is among the most problematic. If the coal contains sulphides it will not only be toxic, but also liable to spontaneous combustion as a result of reactions between finely divided iron pyrites and oxygen. Even though such spoil heaps may lose much of their toxicity as a result of combustion, they can remain physically unaccommodating for plant growth. Nevertheless, most colliery spoils become colonised in time by local grass and tree species.

Apart from coal, as far as the United Kingdom is concerned, two other materials can pose a reclamation problem when tipped. As was noted earlier, the slate quarries of North Wales have produced waste on a gigantic scale following their rapid expansion of output in the nineteenth century. The reclamation of these tips is desirable because of their close proximity to a national park. It is, however, hampered by the long, broad fragments of variable size involved and the absence of fine material. Thus the only solution is to crush and compact the waste and cover the tips with imported topsoil, but this is rarely adopted because of the expense involved. Such tips remain, therefore, largely unvegetated.

In the case of china clay, it is the physical qualities of the huge tips of quartz waste that hamper reclamation. Quartz is geochemically resistant and weathers very slowly. As silicon has no value as nutrient material, these are utterly sterile growth sites. Plant life is further restricted by the steepness of the sides of the tips and their inability to hold water. The material itself is poorly consolidated so that if trees are eventually established there is a danger that they will be toppled in high winds. But before this can happen, attempts have to be made to produce a vegetational cover of grass by **hydroseeding**. This involves spraying onto their surface a mixture of seed, a rich mulch and fertiliser in a chemical binder. Once the grass takes root, the sites require sustained additions of fertiliser for up to ten years; by then the soil and plants will have accumulated sufficient nitrogen and the vegetational cover will be self-sustaining. Many quartz tips have now been treated but only after these conically shaped dumps have been recontoured into more acceptable land forms.

Undoubtedly, hydroseeding techniques of the kind used for quartz tips or the use of suitable fine materials (if not soils) on deposits of overburden and other waste materials are the best and quickest means of providing adequate growth sites for plants and thus of most readily achieving a good standard of reclamation.

8.4 *The vegetating of tailings areas*

It must be apparent from the discussion earlier in this chapter that the nature and composition of tailings can vary greatly. Although tailings can in many respects be regarded as a primitive form of soil, they may also have a much higher potential to develop toxicity and are more physically homogeneous than many soil parent materials. In recent years some tailings have been deposited into basins which have been lined with a clay or plastic material to prevent acids and dissolved heavy metals leaking into the outside environment. Where such liners are not used it will be necessary during the active life of the tailings area to ensure that all waters emerging from it are collected and treated with lime to reduce acidity and precipitate the dissolved toxic elements. As even the most efficient lining material will have a limited life, its greatest value will have been to check problems of leakage at least while the tailings area is in use. At the closure of the site, the establishment of vegetation on its surface not only diminishes acid generation by limiting the amount of oxygen entering the tailings, but also regulates the amount of water entering the tailings, most of which should be accounted for by evaporation and plant transpiration.

After a number of years of experimentation with the vegetation of tailings areas, a relatively simple approach to its establishment and maintenance has been adopted. The area in question will require dressing with amounts of agricultural lime according to the chemical nature of the tailings, as well as the addition of considerable quantities of fertiliser, i.e. a mixture of nitrate, phosphate and potash. However, other problems may occur. For example, coarse textured dark tailings are strongly warmed by the sun and tend to dry out and blow quite readily. Drought and wind action may also retard seed establishment. Sharp sand-sized grains will shear off emerging seedlings. Very fine materials such as the micaceous china clay residues rapidly become compacted and as a result are only slowly permeable and resist root penetration. But once limed and fertilised, even the most unpromising tailings, such as coarse-textured sulphide tailings of Inco's Sudbury nickel workings, will usually support a dense covering of vegetation.

Tailings from non-sulphide ores are much more sympathetic growth sites as they are not potential sources for acid production. Tailings derived from some igneous rocks such as granite may provide sufficient supplies of most of the chemical elements necessary for reasonable plant growth. In this case only nitrogen fertiliser would have to be added.

In some tailings areas, the limiting factor for plant establishment may be toxicity derived from the chemicals used in the milling process. The production of gold involves the use of cyanide in the milling process which then appears in the tailings. In due course even this may be washed out of the surface of the tailings and vegetation can be established. However, at a gold mining site near Long Lake, Sudbury, Canada, the unvegetated medium to coarse sand tailings remain sharply divided from the boreal forest even after 45 years.

A factor which complicates tailings reclamation in some areas is the spontaneous combustion of waste materials. Although this phenomenon is well known in coalfields where spoil may burn for many years, it is also found in metal mining. One example of this involves pyrrhotite which occurs with the nickel ores of Sudbury. Here, the fine particles of the mineral burn on contact with the atmosphere producing a crust-like material. One mining company, Inco, separates out the pyrrhotite and stores it under water in a pond in the expectation of smelting it as an iron-ore. In the same area another mining company does not remove the pyrrhotite and disposes of it with the rest of the tailings. As the pyrrhotite comes into contact with the atmosphere, it hardens to a concrete-like mass so that all areas of the tailings pond above water resemble an uneven tarmac surface. In rehabilitating such areas, that company has covered the ponds with layers of gravel and fine sands available from glacial deposits on their property. These have been successfully sown with a grass/legume mixture.

The ultimate fate of a tailings body will mainly depend on the population density of its associated region, the climate of that region, the chemical nature of the body itself, and the existence of reclamation legislation imposed by the state. In order to provide some picture of the range of possible circumstances which can prevail three examples can be considered.

1 *Tailings close to an urban area*, e.g. those belonging to Inco at Sudbury – the company which separates out its pyrrhotite. Because their tailings lie adjacent to a city of over 100 000 people, legislation was invoked for their reclamation. Completed ponds were first treated in the 1960s. In the initial years following the establishment of a range of grasses, regular fertilisation and liming was needed, but after ten years a substantial organic horizon developed, forming an embryo soil, and treatment is now no longer required (Figure 5.13). The use of aircraft to spread lime, fertiliser and seed was initiated in 1990. This led to a considerable acceleration in the revegetation programme; in the ensuing five-year period, around 650 ha were reclaimed. Although there are considerable areas of native forest and

Figure 5.13 Tailings at Copper Cliff, Sudbury, first rehabilitated in the early 1960s. After the successful planting with a mixture of grasses, eventually the voluntary invasion of indigenous deciduous trees occurred. The conifers were subsequently added as part of the management plan for the area.

lakes in the region, these reclaimed areas present a convenient domain near the city of Sudbury for recreational purposes, and thus will be ultimately designated as public parks.

2 *Uranium tailings close to urban areas*, e.g. at Elliot Lake in Ontario. The need to contain radioactive dust and reduce radon emissions to safe levels has meant that completed ponds have had to be drained and covered with imported earth. This can be successfully achieved if the soil is added to a depth of between 2.4 and 3.7 m, the thickness depending on its characteristics.

3 *Tailings in isolated areas*. These are by far the most common with the major extractive complexes to be found in the North West Territories of Canada, the Scandinavian Shield, the western United States and Australia. In all these places stringent reclamation demands are placed on mining companies before mining permits are issued. Once the tailings body has been vegetated, the area is allowed to revert to variations of the natural surrounding ecosystem. Most commonly, both boreal and mixed forest will invade the area, which in a relatively short time will be recognised only by its unusually flat aspect. However, the oil crisis of the 1970s precipitated a search for new sources of oil supply that has, for a remote area of north-west Canada, resulted in a reclamation problem that is not only of enormous proportions but virtually intractable. The techniques used to remove petroliferous tar from the extensive deposits of tar sands in Alberta employ hot caustic soda, which results in a fine quartz residue waste suspended in a liquid. This has to be deposited in huge tailing ponds in which the settlement of this sand can occur. Estimates of the time this may take vary from hundreds to thousands of years, thus making any attempts at reclamation a very long-term goal indeed. In the meantime, these great lakes, consisting of a mixture of sticky bitumen, oil, caustic soda and sand, present a serious hazard to wildlife, especially birds.

4 *Tailings in arid environments*. Problems may occur over water availability, influencing both seed germination and the maintenance of vegetation. Indeed, there are many examples of newly seeded areas that have been abandoned for lack of enough water to establish a plant cover. However, in the semi-arid regions of Arizona in the USA, immediately south of Tucson, tailings from copper mining have been reclaimed. With judicious selection of plant species and seeding to exploit times of maximum available soil moisture, the tailings have been successfully stabilised. This was done by using barley as an initial crop in order to establish organic matter in the tailings and then by planting local shrubs, trees and perennial grasses. No additional irrigation was used and evidence suggests that the natural beauty of the desert was maintained. This example emphasises the need to use local vegetation species, particularly in stressful environments.

8.5 The positive utilisation of mining wastes

So far in this chapter it has been shown that waste materials produced by mineral extraction and processing may be used in the reclamation of voids, or cosmetically treated on site to blend as far as possible with the existing local landscape. Indeed, they may also form the basis of screens (amenity banks) designed to obscure from general view the least attractive aspects of the extractive activities. However, another possibility exists and that is to treat wastes as saleable products.

The opportunities for making use of wastes in this way are closely related to the quantities of waste available. Depending on the mineral being worked, these quantities at any one extractive site tend to be either very large in relation to the valuable product, or very small. Look at Table 5.3, briefly refer to Section 2.4, and then answer the following question.

Q Which minerals wastes come into which category: (a) large or (b) small producers of waste in relation to the valuable product?

A (a) Coal mining is a large producer of waste wherever it is worked. In the United Kingdom kaolin and (in the past) slate quarrying are notable contributors. Copper accounts for 50% of wastes in the USA. Other large quantities of waste, taken in a world context, arise from the working of iron-ore and taconite (a low-grade form of iron-ore), uranium, phosphate, gold, gypsum, lead and zinc.

(b) The small producers of waste include hard rock aggregate enterprises and those of sand and gravel, although here, as Table 5.3 indicates, remarkably little is known about exact quantities.

Table 5.3 also provides a picture of the purposes for which wastes are positively used, sometimes in their raw state and in others upgraded as major inputs to the production of bricks, artificial aggregates or cement. Although the range of uses may appear impressive, it remains a fact that there are few alternative uses of upgraded wastes that are economically viable in competition with natural products, particularly if significant transport costs are incurred in reaching markets. This is even more the case with respect to the use of wastes as a form of fill away from their sources of production. Over short distances (a radius of 15 to 25 km) it is quite possible that abandoned excavations may be advantageously filled with wastes from mines or quarries in production. This course of action has the environmental advantage of eliminating the need to take virgin land for tipping and, if the land surface of the abandoned excavation can be restored by filling with waste, of producing useful new land. If the distances are short, transport costs can be offset by the avoidance of land purchase for tipping and the value of the reclaimed land.

Computer data banks listing voids and wastes have been produced and programmes which associate one with the other, according to suitability and least cost, have been run successfully in at least one region in the United Kingdom (the West Yorkshire conurbation), as well as in Canada. However, the problem of moving waste materials to some more remote location for use in a major project remains a problem.

Feasibility studies have certainly looked at the possibility of transporting colliery wastes to reclaim sites on the Firth of Forth in Scotland and Humberside in England, in spite of objections that such actions destroy local marshland habitats. In the 1970s when a third London airport plan would have involved the reclamation of a coastal site at Maplin in Essex, the notion was also examined of moving 400 million tonnes of fill from the china clay wastes of Devon and Cornwall, the slate wastes of North Wales, or the colliery wastes from the Yorkshire/Nottinghamshire coalfields as well as those in the north-east of England and South Wales. In the case of coal spoil alone, about 12 000 ha of serious dereliction could have been removed, but all the schemes failed to be adopted because of transport costs.

Table 5.3 Production and utilisation of mines and quarry wastes

	Production/ (Mt per annum)				
	Waste rock (inc. overburden)	Waste rock and tailings	Tailings	Stockpiles/ Mt	Utilisation
Australia					
Lowgrade ilmenite	0.2		–	?	–
Lead/zinc ore	–		0.5	4.75	–
Coal		*c.*60		3000	Small quantities for road construction and fill material (burnt spoil).
Belgium					
Coal		4		–	Lightweight aggregate manufacture.[a]
Canada					
Iron ore		45		–	Aggregate fill; roofing granules.
Coal			5	?	–
Finland					
Misc. operations	3.1		5.1	94.8	Road construction; aggregate for concrete.
France					
Coal		20		700	Approx. 35% production used for: (i) road construction – fill material; (ii) lightweight aggregate manufacture; (iii) brick making.
Germany					
Coal		63		100s?	Approx. 35% production used for: (i) road construction – fill and embankments; (ii) land fill.
Holland					
Coal		0.3		–	40% production used for block making (from fired mixture of colliery waste, flyash and sawdust).
India					
Laterite		?		–	Clay pozzolana[b]; lime substitute in mortar.
Gypsum	2		–	–	Building plaster.
China clay	0.2		–	–	Clay pozzolana; brick making.
Mica	–		0.005	–	Mica insulation bricks.
Coal		8–10			Mine backfilling (large %); road construction; fill.
New Zealand					
Coal	?		?	?	Small quantities used for land reclamation and fill.
South Africa					
Gold ore	4		*c.*40	–	50% waste rock used for road construction and aggregate (for concrete); silicate bricks (minor usage).
Misc. quarrying					Waste rock used for road construction and fill; tailings in silicate bricks.
Coal		9		150	Very little used.
Sweden					
Iron ore	25		9.6	289	Land fill (limited use); road construction and brick making (small quantities).
Sulphite ore	–		11.7	57	
UK					
China clay		26		400	Less than 5% production used in: (i) aggregate for concrete; (ii) silicate bricks; (iii) road construction; (iv) fill; (v) amenity banks.
Slate		6		400–500	Inert fillers and roofing felt (small % of total used, i.e. 0.5 mt/pa).
Tin ore		0.5		*c.*0.30	Aggregate for concrete (minor usage).
Fluorspar		0.23		–	Road construction and aggregate for concrete (minor usage).
Misc. quarrying		?		–	Road construction; silicate bricks; amenity banks.
Coal		50–60		3000	11–13% production used in: (i) road construction (fill and building); (ii) brick making; (iii) cement manufacture; (iv) lightweight aggregate manufacture.
USA					
Copper	624		234	7700[c]	Road construction; bitumen filler.
Taconite	100		109	3600[c]	Aggregate-skid resistant.
Phosphate ore	230		54	907[c]	–
Iron ore	27		27	730[c]	Road construction.
Gold ore	15		5	450[c]	Road construction; aggregate for concrete.
Uranium ore	15.6		5.8	110[c]	Aggregate for bituminous concrete.
Lead ore	0.5		8	180[c]	–
Zinc ore	0.9		7.2	180[c]	–
Misc. quarrying	68		–	?	Amenity banks.
Gypsum	14.2		2.7	?	–
Asbestos	0.6		2	14[c]	–
Barite	1.9		3.1	25[c]	–
Fluorspar	0.1		0.4	?	–
Feldspar	0.2		0.8	?	–
Coal (bituminous)		100+		2000	Small quantities used in: (i) road construction – base and subbase materials; surfacing aggregates; (ii) cement manufacture; (iii) mineral wool manufacture.
Coal (anthracite)		1		700	Anthracite waste used in: (i) concrete block making; (ii) brick making; (iii) manufacture of lightweight aggregates.

[a] Lightweight aggregate is used for building purposes where structural strength needs to be combined with lightness of construction material.

[b] A form of hydraulic cement.

[c] Stockpiles for USA are tailings only.

Source: Blunden, J. (1985) *Mineral Resources and their Management*, Longman, London.

Also in the 1970s a major study was undertaken by Devon County Council into the disposal of 28 million tonnes of waste quartz from the Lee Moor china clay works on the boundary of Dartmoor National Park, since only about 8% of each year's production of waste was used locally. The idea was to use it in the manufacture of an artificial aggregate and sell it in south-east England where demand was high. Not only did the costs of this run out at four times those for natural aggregate available in the south-east, but it was also unable to compete on transport costs compared with its nearest rival, the Mendips, some 140 km closer to the market.

When considered in conventional economic terms, the long-distance transport of large quantities of waste is seldom likely to be attractive. For this reason it is unlikely that mineral operators will ever voluntarily undertake such movements. Nor is the market for upgraded wastes a particularly viable one, given the competition from natural products and the fact that these also have to be despatched to centres of demand. Conventional economics does not take account of unquantifiable benefits such as visual improvement, the lessening of nuisance or the removal of danger to the public. But it is when these factors appear important that there is scope for government intervention on the basis of a policy that could decide whether the benefits of waste removal are worth the apparent financial 'loss' in doing so.

It seems, therefore, that in the case of small-scale waste production (as at quarries) and because many opportunities for environmental improvement work at extraction sites still exist, current and future waste production will largely be devoted to on-site amenity works. Certainly, for major schemes of waste transport and utilisation, government support would have to be forthcoming.

The formulation by individual countries of any policy for the positive utilisation of mineral waste is hampered by the lack of quantification of existing waste stocks, and of current and future production, as well as insufficient information on the properties of the wastes and their suitability for different purposes. Nor does any national picture exist within those countries of the relative locations of waste heaps and the excavations to which they might be returned. The task of making such studies in a country such as the United Kingdom is feasible because of its small size and is important because of the shortage of land arising from a population density of 234 persons per square kilometre, as compared with 27 persons/km^2 in the USA. Even so, within North America and elsewhere there are areas that may be considered to be heavily urbanised where such studies could be equally valid. Until they are undertaken so that the relevant data may be obtained, proper planning for waste utilisation will not be possible.

8.6 Summary

In order to approach the question of the reclamation of mined areas, this section has tackled the subject under three main headings according to type of site. In a consideration of excavations both the problems and the opportunities have been explored, especially the use of such voids for the disposal of waste materials already occupying otherwise valuable land. Overburden is, of course, material which can be returned to an excavated area. However, conflation, the size or the size range of the materials in question, their chemical composition and their location may be difficulties that require to be circumvented in any successful attempt to rehabilitate the overburden. Many of the same considerations can also apply to waste tips and to tailings. If tailings have the advantage of being level and having a uniform relatively fine texture, they often offer a hostile environment for plants. Adjustment may need to be made to pH values and plant nutrients added before anything resembling a medium conducive to plant growth can be created.

This section concludes by taking a more positive attitude to wastes from mining by considering them not just as a problem requiring on-site treatment, but as materials to be used positively, even if this may require initiatives which are not forthcoming from the workings of the market-place.

While it is self-evident that any positive use of mining wastes must be beneficial to the environment, the need to undertake such a task at all is based on the assumption that the market demand for minerals is satisfied largely by the winning of fresh supplies from the ground. But need this necessarily be the case? An answer to this question is attempted in the following section.

9 *Recycling and substitution*

9.1 *Technical problems and policy needs*

Any review of mineral **recycling**, especially for the metals, would show that in the industrial nations between 25% and 30% of output is now made up from materials previously fabricated into consumer goods, industrial plant, etc., or wastes from their manufacture. This falls far short of what *might* be accomplished, in spite of what may seem the relatively high percentages of re-use already being achieved, especially for some of the precious metals (Box 5.3).

Steel, however, has gone somewhat against the trend, and the use of scrap, which had been rising in the 1960s and 1970s, actually fell during the 1980s. There were corresponding increases in scrap availability: in the USA alone this rose from 610 million tonnes in 1980 to more than 700 million tonnes by the end of the decade. This failure has to be seen, however, against the background of changes in production methods designed to reduce the amount of environmental pollution caused by the old open hearth technique. In this method, pig iron, steel scrap and limestone were melted in a furnace to allow the oxygen in the air to combine with carbon impurities, while the limestone formed a basic slag to remove phosphorus and silicon in the iron. Unfortunately, the new basic oxygen method, in which air is replaced by pure oxygen, can utilise only 15–30 tonnes of scrap steel mixed with every 100 tonnes of pig iron, as opposed to a nearly equal mix for the open hearth approach – hence the decline in scrap utilisation.

Some compensation for this state of affairs has been achieved by the development of a new low-volume production process. The minimill, as it is called, relies on an electric arc furnace and consumes almost entirely scrap steel. Although minimills are increasing their share of steel production, especially in the USA, so far it has not been enough to compensate for the lost demand for scrap to feed open hearth furnaces.

Box 5.3 Recycling precious metals

High value and relative ease of recovery in the practical sense make possible the recycling of most precious metals. For example, nearly all of the mercury in cells, boilers, instruments and electrical apparatus is retrieved when such items are scrapped. The same is true of gold and silver in jewellery, optical frames, photographic laboratories and chemicals.

However, with the platinum group of minerals, only in their use as metals and in the chemicals industry have they so far been retrieved in large quantities. And as the flow chart shows, little has been achieved for one major use, that of an agent in vehicle exhaust systems which controls the output of noxious gases, so nine-tenths of the mineral is currently being wasted. Such a situation could be resolved by the imposition of a sales tax on vehicles, returnable when the customer's redundant exhaust unit containing platinum is returned to a collecting centre for recycling.

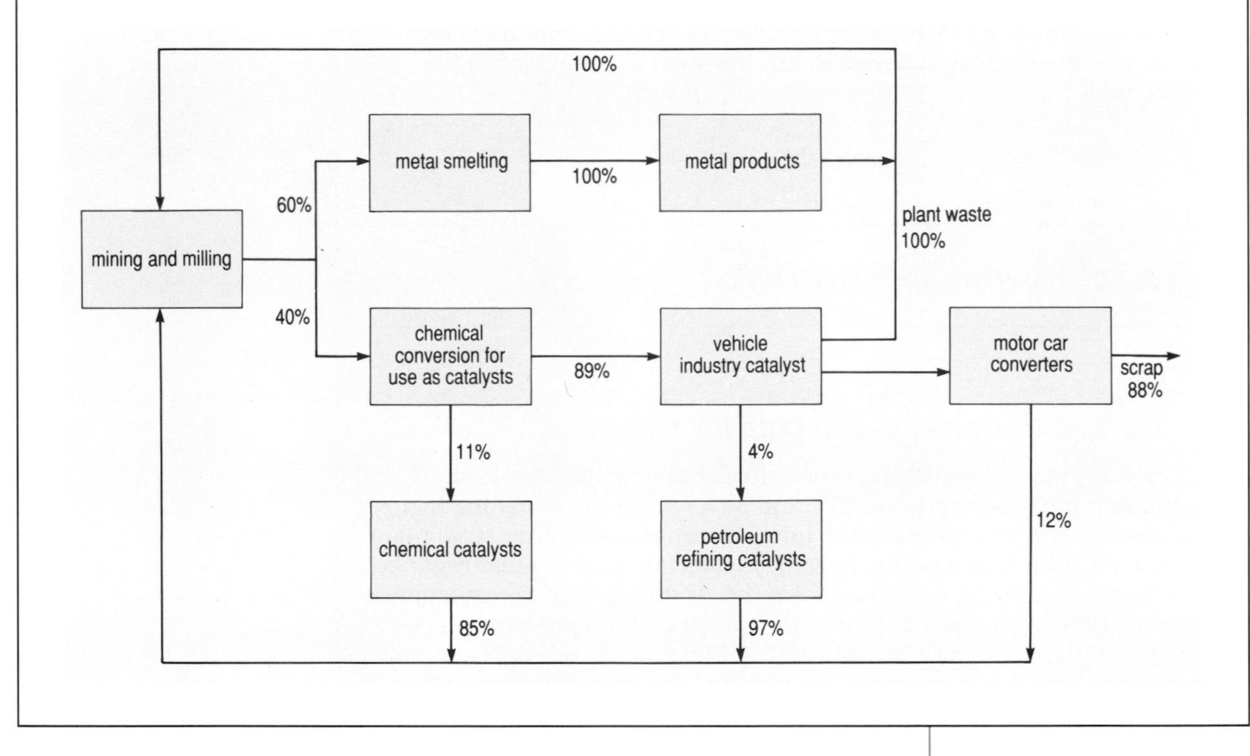

Moreover, the range of steels that may be produced from minimills is limited and these can only be made from scrap having low levels of impurity. This is a distinct problem when it comes to recycling special purpose steel scrap which will contain a high percentage of chromium, cobalt, molybdenum, nickel, tungsten and vanadium. At present these have to be remelted without the desirable removal of the alloy metals, if they are to be re-used at all.

A somewhat similar overriding constraint can be found in the recycling of aluminium. Scrap is only suitable for use in castings because this sector requires lower metallurgical standards than, for example, rolling and extrusion processes. For this reason, as matters stand at present, any growth in the use of recycled materials must come either from greater penetration of the castings market, or a faster rate of growth in the castings market itself. Indications are that although average annual growth in this sector has been healthy, it has not kept pace with the increasing use of aluminium as a whole. It would therefore seem that the extent to which increased recycling may be achieved is constrained by the demand side of the market. If there is no upsurge in the use of castings, increases in the use of recycled aluminium, as with special scrap steel, must therefore await new technological developments which will enable it to be upgraded to compete with the primary material.

Another recycling difficulty lies in the metallurgical complexity of many pieces of redundant household and business equipment. For example, separate copper and high silica steel components are frequently in close association in items such as motor armatures, field cores and transformers. The United States Bureau of Mines has been working with some success on methods of disassociating these, using preferential melting in molten salt baths, a technique which may also assist in the separation of the constituent metallics of ferro-alloys.

Technological developments fostered by the same organisation may also lead to the recovery of lead and zinc, which are known to be given off as oxide effluents in steel-making furnaces. It considers that 100 000 tonnes of zinc and 10 000 tonnes of lead could be retrieved annually in the USA from furnace dusts, once an appropriate methodology is found.

Whatever technical solutions may be produced to assist in recycling, it is also clear that governments can play a much greater part in the conservation of minerals and therefore the environment. For example, the tipping of wastes containing metallic materials at domestic waste sites might be forbidden, while the mandatory sorting of waste at the household or factory end of the disposal chain could assist in the reduction of recycling costs. Moreover, there are a number of practical steps that they might take to lower the consumption of new minerals and to encourage recycling. These include the levying of a tax on the use of 'virgin' minerals for certain manufacturing processes. A number of variants of such a scheme have been suggested but the common view is that the tax would be imposed according to the weight or value of new mineral resources used in manufacture and distribution. Others also favour the inclusion of a negative tax or subsidy, according to the proportion of recycled materials used in them. Proponents of this general approach, though, divide into two camps over the imposition of the tax, with some calling for a uniform rate across the range of exhaustible resources used from new, while others favour a system which penalises the users of the scarcest materials. Yet another proposal suggests the inclusion in the selling price of a consumer durable, such as a car or refrigerator, of a sum which would be refundable when the product is no longer usable and is returned to a centralised collecting centre for recycling.

Whatever the merits of the schemes to encourage recycling, each would at least spread the social costs of environmental and natural resource use throughout the community. Indeed, recycling has the further advantage in that in general it may be regarded as involving a far lower level of energy consumption than that involved in the extraction, processing and smelting of newly won metallic ores. For example, the energy needed to melt down old aluminium is only one-twentieth of that required to produce virgin metal from the bauxite ore (Plate 8). Moreover, remelt furnaces are comparatively inexpensive to construct.

Other savings should also not be ignored. In the case of aluminium, for every tonne recycled, savings are made not only of 4 tonnes of bauxite, but also of more than 450 kg of soda and about 113 kg of lime which would be required to refine the bauxite, together with 680 kg of oil derivatives, 27 kg of cryolite and 36 kg of aluminium fluoride. The use of recycled aluminium also achieves a remarkable reduction in environmental impacts because a smaller quantity of new metal then needs to be produced through a process which, in its first stages of production of alumina from bauxite, presents problems. Gaseous emissions include large volumes of carbon monoxide and dioxide, smaller amounts of hydrogen fluoride, sulphur dioxide, carbon disulphide, silicon tetrafluoride, carbon tetrafluoride, hexafluorethane and water vapour. Of these, hydrogen fluoride emissions cause the greatest immediate concern. In addition, solid emissions of alumina, cryolite, aluminium, fluoride, calcium fluoride, carbon and iron oxide are generated. The environmental impacts of these when vented from a stack will have been ameliorated by the use of equipment called scrubbers, but this produces a highly toxic liquid which ultimately must be disposed of.

A significant problem for the environment is the residual material left over from processing the bauxite itself. As noted in Chapter 4, this takes the form of a red mud which contains in suspension caustic soda and sand and has to be discharged into a tailings pond. Between 0.3 and 1.4 tonnes of this is produced for every tonne of alumina and in the 1980s production averaged about 40 million tonnes a year. The problems ultimately created by the tailings used to derive from the capacity of the solid content not to settle out from the fluid except in the relatively long term. This made rehabilitation a problem. However, this was overcome recently by a new technique which removes the liquid which can then be evaporated. Although the solids remaining can then be contoured, covered with soil and vegetated, the problems of the old tailings areas remain, along with the other environmental hazards created by the processing of bauxite. Thus, the winning of new aluminium compares unfavourably with the quite minor environmental effects of recycling it.

However, the problems attendant on the recycling of aluminium and some other metals cannot be ignored. Besides which, recycling always involves some loss of material and there is uncertainty as to how far its potential may be pursued. In an examination of the immediate possibilities for copper recycling in the USA, recent evidence suggests that an additional 10% could be readily obtained. This is in line with research findings, which also suggest that a further 25% of the aluminium requirement in the USA could be obtained by recycling. While these findings are not optimistic about iron, it is considered that an additional 18% for zinc, and 3–4% for lead are easily achievable objectives, even without draconian legislation.

9.2 *Substitution – old and new materials*

There are two forms of **substitution**. One involves the replacement of one metallic mineral by another in a given usage. The other involves the substitution of a metallic mineral with an entirely different material.

Most metals have one or more metal substitutes for a given end use, as the matrix in Figure 5.14 shows. It also indicates that there are occasions when a metal has no metal substitute at all. However, the matrix cannot indicate that substitution is often only to be made at the price of losing some unique characteristic in a specific application, or, if it were to occur, it could only be at additional cost because the substitute metal is scarcer, its processing more complex, or the amount of energy used in its fabrication much greater. Cobalt and vanadium are, for example, technically substitutable in the production of a number of specialised steels, but the former is six times more expensive than the latter, which makes interchangeability a one-way process. While aluminium is an excellent substitute for copper in its electrical applications, it is less conductive. Thus it can be used in overhead power cables, but not where they are housed in underground ducts; overheating then becomes a problem. Moreover, it should also be remembered that aluminium smelting requires a much heavier input of energy per tonne than that required for iron and many other metals.

If there were a scheme for the taxation of the use of newly won minerals, especially the scarcer minerals, the incentives would be strong to

	Iron	Manganese	Chromium	Nickel	Molybdenum	Tungsten	Vanadium	Cobalt	Aluminium	Magnesium	Titanium	Copper	Tin	Lead	Zinc	Gold	Silver	Platinum	Antimony	Mercury	Cadmium
Iron									■												
Manganese																					
Chromium		■		■	■	■	■	■	■	■	■				■						■
Nickel		■	■		■		■	■													
Molybdenum		■	■	■		■	■		■		■										
Tungsten					■					■	■			■							
Vanadium		■			■	■			■									■			
Cobalt				■	■	■	■														
Aluminium	■	■	■	■						■	■	■	■	■	■						■
Magnesium	■		■						■		■				■						
Titanium	■								■						■						
Copper	■		■	■					■												
Tin			■						■			■		■	■				■		■
Lead	■			■					■		■		■		■						■
Zinc	■								■	■	■										■
Gold			■	■				■									■	■			
Silver	■			■					■			■				■		■			
Platinum				■		■	■									■	■				
Antimony			■								■		■		■					■	
Mercury		■										■	■		■	■					■
Cadmium		■	■	■									■	■							

▲ *Figure 5.14 Metallic minerals substitution matrix.*

speed up substitution wherever this was possible. But for metals, such as tungsten and molybdenum, which are rather imperfect substitutes for each other in some uses, the incentive might be rather to encourage their more sparing use in the short term, and in the longer run to encourage even more resource-saving technological innovation. Where there has already been a move to substitute a more plentiful for a scarcer metallic resource, such as aluminium for copper and zinc, this would probably be further increased, especially in the automobile industry where there are opportunities for using aluminium castings made from recycled metal.

Even without such a tax, technological advances have made it possible to supplant the use of metals with other materials. Much lead and copper piping has been replaced by plastic equivalents, not least because it is light, easy to handle and non-toxic. Also, in the manufacture of vehicles the substitution of a metal by plastic materials made from tough polymers is now increasingly making its mark. While it is difficult to quantify the exact impact of such a substitution, there is evidence from the automotive industry that over the last two decades more stringent safety standards and the desire to produce a lighter vehicle to improve performance in relation to fuel consumption led to the replacement in each unit of 453 kg of metals by 204–217 kg of plastics. The possibilities here are by no means exhausted. It should be added that although plastics, wherever they replace metals, generally require less energy in their manufacture and are readily recycled, they do require hydrocarbons for their production.

Although new forms of substitution from time to time are introduced as part of technological change, such as the use of video-tape in movie cameras instead of photographic film, or the replacement of copper telecommunications cables with those containing glass fibre optics derived from common sands, even more advanced forms of technology could make many more conventional metallic minerals redundant. Much research is being undertaken into the possiblity of using common materials such as sands and clays as parent materials for new products and as a source of aluminium. Indeed, several automobile manufacturing companies are investigating the possibilities of using clays for ceramic engines.

The advantage of using clay and similar minerals as source materials in industry is that they represent a virtually limitless resource and are readily available in all parts of the world to meet whatever demand that might emerge. The rehabilitation of the areas from which these materials are won should prove much less of an environmental problem than those posed by conventional metal-ore mining. But there are several development problems that have yet to be resolved here, not least the high energy demands required to process such materials.

Activity 4

Review the above section, especially Figure 5.14, and look back at Box 4.2 in the previous chapter. Then try to respond to the following questions.

Q Which mineral of those listed in Figure 5.14 appears not to have a substitute?

A Manganese, in its primary role in the manufacture of steel.

Q Which metal is most frequently substituted for copper in electrical applications?

A Aluminium.

Q What disadvantages accrue to the use of aluminium in this context?

A Aluminium is a much less efficient conductor of electricity than copper. Aluminium smelting requires a much heavier input of energy per tonne of metal than that required for copper. Nevertheless, in France, India and the CIS, the use of copper is forbidden where aluminium can act as an adequate substitute.

Q What are the main uses of silver?

A In the making of jewellery, as a soldering agent and in photographic film.

Q Why do you think that its use in the last named context is beginning to decline?

A Because of the increasing use of video-tape in the recording of moving images. Silver forms a vital part of the light-sensitive make-up of traditional film, but this is being replaced as the recording medium by plastic tape coated with iron and chromium oxide.

10 *In conclusion*

In Section 9 consideration has been given to the extent to which the recycling of key minerals (especially those that are scarcer and more expensive to work such as some of the metallic ores) may be undertaken and what opportunities exist to substitute the more abundant minerals for the less abundant or for those that are more environmentally disruptive to extract. More environmentally friendly methods of winning minerals have been touched upon earlier, but here for the first time in the chapter full recognition is given to the fact that there may be alternative strategies for meeting societal needs for minerals.

Some tentative steps down this road may already have been taken because waste minimisation activities are now being practised to a limited extent. The banning in some Canadian provinces of the use of non-returnable containers, such as aluminium cans used for soft drinks and beer, is one example of positive action. Another is the cash incentives given for the return of beverage containers in some US states. In some countries initiatives have taken other directions at a national level. In Japan and Sweden centres already exist for the collection of spent items such as batteries which contain heavy metals and may create environmental problems if just dumped. Indeed, market forces are also playing some part. The problems of a throw-away society in the USA are at last showing themselves in the growing scarcity of land fill sites. This has led to increases in the costs of waste disposal, rising from less than \$20 a tonne in the early 1980s to more than \$200 a tonne in the early 1990s. This has had the effect of making the re-use alternative to disposal more attractive to the point where some communities have initiated waste sorting and recycling programmes.

If in the short term the recycling of metals is to become an important alternative to the mining of virgin resources, it will undoubtedly require, apart from the need for further research, an effective infrastructure put in place by government and backed up by some of the initiatives mentioned in Section 9.2 above in order to subvert the conventional economic approach which takes into account only the immediate effects of production decisions. This revisionist approach to normal economic theory has been elaborated by Professor David Pearce in his report for the United Kingdom's Department of the Environment *Blueprint for a Green Economy* (1989). This is discussed briefly in Chapter 2 of this book in relation to energy and the environment.

A concerted move to what may be described as sustainable industrial development aimed at reducing both unnecessary primary production and consumption wastes may ultimately only be driven by increasing populations and overall rises in living standards in the less developed countries. Clearly, in such a situation, if reliance continued to be made on the securing of mineral resources by conventional extractive methods alone, some shortages would begin to arise. Moreover, as this chapter has demonstrated, ways do exist to overcome many of the environmental impacts of mining and minerals processing, but whether they occur in the vicinity of the mine of smelter or well beyond, the costs of carrying through such schemes are not inconsiderable. Indeed, they will rise sharply as the need to work leaner ores becomes a greater imperative.

Whether or not such a situation does come to pass, the move to a form of sustainable industrial development must be considered as highly desirable from an environmental perspective. Its effect would mean a considerable reduction in environmental problems created by the winning of new materials, especially the non-aggregate materials, and a reduction in ameliorative arrangements that must follow both during minerals production and processing and when these activities cease. Meanwhile, government control of emissions and the regulation of land use must continue to play some small part in what many hope will be a transition to the industrial-ecosystem approach. Fundamentally, probably most of us would agree that if people can raise their living standards without incurring further penalties for the environment that will ultimately degrade the quality of life for all, it is a transition that should be accomplished.

Finally, you will recall that Chapter 4 was concluded with a diagram (Figure 4.11). Now is the time to look back at this since it will act as a reminder of much of the discussion that has taken place in this chapter, and, of course, that which precedes it. It will be apparent from this that although the emphasis so far has been upon minerals extraction and processing and the possibilities of their more effective re-use, the environmental impacts of the industries which employ these minerals in the manufacture of goods has yet to be considered more fully. This is the subject of Chapter 6.

References

BLACKMORE, R. & REDDISH, A. (eds) (1996) *Global Environmental Issues*, Hodder & Stoughton/The Open University, London, second edition (Book Four of this series).

PEARCE, D. *et al.* (1989) *Blueprint for a Green Economy*, Earthscan, London.

WINTERHALDER, K. (1978) A historical perspective of mining and reclamation in Sudbury, pp. 1–13 in *Proceedings Third Annual Meeting of Canadian Land Reclamation Association.*

Answers to Activities

Activity 2

In your response you should certainly contrast the respective attitudes of governments in western industrial economies and those, which, until recently were part of COMECON. In the former the recognition of pollution from smelters has meant the imposition of controls on effluents, almost certainly pushed by public opinion. In Eastern Europe the problem has been suppressed in the interests of production and whatever the recognition of the damage to health from smelter discharge, these were not to be articulated – at least not until the era of *glasnost*! Even now environmental controls may be disregarded.

Activity 3

Emission control programmes were initiated at both locations in order to protect their communities from smelter pollutants. These included, first, the building of tall stacks to achieve the better dispersal of sulphur dioxide and metal particulates, followed by reductions in effluent discharges. Concern at Sudbury through the 1980s has, however, concentrated on the damage done regionally to water bodies by sulphur dioxide, particularly at the Killarney Provincial Park, which has followed the building of Inco's superstack.

The 'Countdown Acid Rain' programme, the first phase of which was completed in 1994, was aimed at achieving substantial reductions in sulphur dioxide discharges and especially from the superstack.

Although local weather conditions which might bring superstack discharges down into Sudbury are monitored so that remedial action can be taken if required, programmes of effluent control are not essentially geared to this objective. At Mount Isa, however, the need to avoid polluting the local community remains the key objective and levels of emission and the circumstances under which these may be vented are entirely aimed at achieving this end. Since the area around Mount Isa is semi-desert with few and well-scattered small communities, the effects of pollutants in the region beyond the township are not seen to be of primary importance. This diminishes the need to convert sulphur dioxide discharges to sulphuric acid in a situation where it is likely to be unmarketable.

Chapter 6
Pollution and hazard: the environmental impacts of chemical manufacture

1 Introduction

Within the manufacturing sector of a modern industrial economy, the use and manufacture of chemicals pose the greatest threats to the environment. Any account of the environmental impact of chemical manufacture must begin by discussing the two principal threats – pollution and hazard.

One of the most frequently quoted definitions of pollution originates from a document published by the Organisation for Economic Co-operation and Development in 1976:

> ... the introduction by man, directly or indirectly, of substances or energy into the environment resulting in deleterious effects of such a nature as to endanger human health, harm living resources and ecosystems, and impair or interfere with amenities and other legitimate uses of the environment. (OECD, 1976)

The term 'hazardous' is applied to substances and processes which, because of their toxicity, flammability or explosiveness, are capable of causing damage of such magnitude (consequence) that the frequency (probability) must be made as small as possible. The allied term 'risk' is now used to represent the product of probability and consequence, and it follows that the control of hazardous substances can be achieved by reducing the probability of their release into the environment and/or by mitigating the consequences of such a release.

Clearly the terms 'polluting' and 'hazardous' are far from mutually exclusive; the differences, which Table 6.1 attempts to summarise, relate to the circumstances of the release to the environment and to the controls over those releases. By way of an illustration, imagine a plant making chemical fertiliser and ammonia as feedstock. If a tank holding this compound as a liquid under pressure were to fail catastrophically, a large release of this dense and highly toxic gas into a populated area could have fatal consequences. Reducing the probability of such an occurrence is primarily a technical issue: quality assurance and inspection procedures – on, for example, the welds in the construction of the tank, the valves and the monitoring equipment – are the responsibility of the Health and Safety Executive (HSE). The manufacture of ammonium nitrate-based fertiliser inevitably entails the continual leakage of some of this material to the atmosphere. This emission is labelled 'pollution' not because its overall environmental consequences are of a much lower order than those associated with ammonia release, but because it is a routine, anticipated and continuous phenomenon. Its limitation (but not its total elimination) by the application of appropriate techniques is the statutory responsibility of **HM Inspectorate of Pollution** (HMIP).

As you read this chapter, look out for answers to the following key questions.

- What are the chief characterisitcs of pollution and hazard as posed by the manufacture of chemicals?
- What factors determined the emergence of controls on the environmental impact of the chemical industry in the nineteenth century?
- What have been the strengths and weaknesses of the UK approach to regulating pollution compared with those of European countries?
- What are the key features of risk analysis in relation to hazards posed by the chemicals industry?
- To what extent is the role of science and technical expertise central to risk assessment in the processes leading to the determination of a planning application for plant that poses a hazard?

Table 6.1 Pollution and hazard: comparison of characteristics

Characteristic	Pollution	Hazard
emission	anticipated	accidental
	permitted	risk minimised
	dispersed	localised
source	mobile	static
	multiple	individual
health effects	latent	acute
	epidemiologically related to source	clinically frank or obvious
regulatory agency	HM Inspectorate of Pollution	Health and Safety Executive

This chapter is intended to illustrate the environmental impacts of chemical manufacturing with particular reference to air pollution and hazard and to analyse ways in which these impacts may be regulated. It draws primarily on past UK practice, but highlights also other principles of regulation used elsewhere or theoretically available. The chapter is based on an account of industrialisation in Britain in the nineteenth and twentieth centuries and of its environmental impact. Special attention is paid to the location and concentration of industry, and to the production processes that have given rise to atmospheric pollution and industrial hazard.

The changes to the UK regulatory system which have occurred following membership in 1973 of the European Economic Community (subsequently the European Community and now the European Union) are considered in detail. Throughout this chapter, we learn that a wide range of disciplines – chemistry, engineering, law, sociology – can be called upon to illuminate society's response to industrial pollution, but none can claim a dominant role. Moreover, there are no 'ideal' solutions which can satisfy everyone. As with other fundamental social issues, the role of state is to search for that compromise in which no single group feels its interests have been ignored. Similar problems arise in industrial areas throughout the world. The development of control mechanisms in the UK and the EU provides an introduction to the global framework that must emerge.

Q What problems are posed by (a) routine discharges of pollutants to air and water and (b) major accidental discharges of pollutants?

A The two sets of problems are not mutually exclusive, but are generally distinguished by the acuteness level of any single pollution event. In the case of routine discharges, the origins of observed pollution are numerous and substances discharged are widely distributed. Environmental effects are cumulative, but can be controlled by methodical interventions at source. Discharges to the air have potentially a very wide effect on human health and on the environment generally. Precise sources of air pollution may, however, be difficult to detect. Water pollution poses less of a problem since few people come into direct contact with polluted water; drinking water, for example, is usually extracted well upstream of industry. Sources of water pollution are nevertheless relatively easy to detect.

Major accidental discharges by definition cannot be precisely anticipated. Most incidences arise from plant failure or operators' error. The consequences for the environment and for people in and around the plant are potentially catastrophic, but the incidence is low.

2 *Pollution and industrial hazards in nineteenth century towns*

2.1 *Industrialisation and urbanisation*

In the nineteenth century, the twin forces of industrialisation and mass urbanisation transformed Britain's landscape. The entrepreneurs and inventors of the new economic order sought ever more rapid and reliable ways to supply power, labour and raw materials to the expanding population and production centres. Two engravings show the striking contrast between the old and the new. The first (Figure 6.1) shows a water-driven grinding wheel in Endcliffe Wood near Sheffield and the second (Figure 6.2) a steam-powered cutlery works in the same city. They illustrate the decline of small scattered rural workshops on the hill streams typical of the eighteenth century and the subsequent concentration of industries in cities, in a mass of furnaces, chimneys and congested workers' dwellings.

The steam engine was the driving force of this industrial and urban expansion. Exploiting technological advances begun in the previous century, steam power both drove and demanded advances in the iron and engineering industries, in mechanisation, and in the exploitation of coal which provided the fuel for furnaces and boilers. James Watt's improved 'fire engines' were already at work in textile factories, ironworks,

Figure 6.1
A water-driven grinding wheel near Sheffield in the nineteenth century.

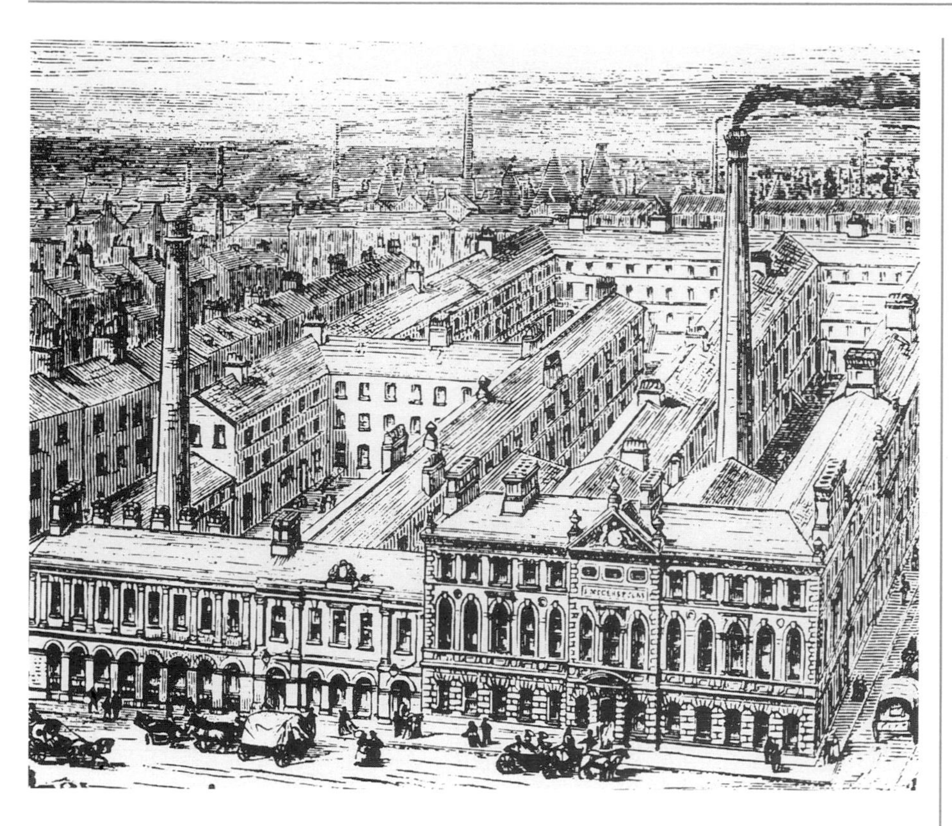

Figure 6.2 Steam-powered cutlery works in Sheffield in the nineteenth century.

coalmines, breweries, flour mills, potteries, waterworks and other industries at the end of the eighteenth century. Increasing reliance on steam power through the nineteenth century produced an urban concentration on Britain's coalfields with factories emitting smoke, dust and various gases and stimulating the beginnings of concern about industrial pollution and hazards.

As important sectors of British industry came to be typified by large-scale organisation so there was a tendency for location to become more specialised. Particular areas developed their own specific industrial character and this was to have important consequences for many aspects of their social circumstances, not least those associated with their environment. Cotton spinning mills and weaving sheds concentrated in the towns of Lancashire and combined with associated engineering works and finishing factories to form vast industrial complexes in cities like Manchester (Figure 6.3). So monstrous were the clouds of smoke 'vomited forth from the numberless chimneys' that in the 1840s the French traveller Leon Faucher compared Manchester to an active volcano.

The application of steam power to the chemistry and art of pottery manufacture made the concentration of the industry in North Staffordshire more pronounced. Inside the factories, workers inhaled dust from clay, flint, plaster of Paris, calcinated bones and feldspar. Outside they were exposed to 'an abundant share of coal dust and coal smoke ... far too much of which is thrown off from the ovens in the shape of dense black smoke, intermingled with gases of combustion' (Figure 6.4).

The glass industry was akin to pottery manufacture in its reliance on minerals (sand, alkali and lime) as raw materials and on coal for firing its furnaces. Here, too, the industry became increasingly mechanised and

▲ *Figure 6.3* Steam-powered cotton spinning mills in Manchester.

▲ *Figure 6.4* North Staffordshire potteries.

▲ *Figure 6.5 Pilkington's glassworks in St Helens.*

▲ *Figure 6.6 A clear view of London and Westminster, 1820.*

concentrated in the factories of a few large firms. One of these, Pilkingtons, employed about 3000 workers in St Helens in Lancashire around 1900 and was one of the largest glass-manufacturing works in the world. Its complex of sheet and rolled plate glassworks, collieries and brickworks are illustrated in a panorama of 1879 (Figure 6.5).

While our primary attention must be concentrated upon industrial sources of air pollution, we must not lose sight of the fact that until relatively recently the domestic hearth was a major contributor to urban air pollution. A city like Glasgow could burn up to 1 million tons of coal a year, most of it on domestic fires though some was used in the production of gas. It was estimated that 95% of London's 'black canopy of smoke' was produced by open fires which discharged 200 tons of soot every day. Small wonder that to Dickens fog had become the 'London particular' producing 'night in the daytime' and blotting out every familiar landmark (Figures 6.6 and 6.7).

◄ *Figure 6.7*
A scene near Temple Bar during a fog in 1844.

2.2 Pollution and the development of the 'alkali' industry and other chemical manufacture

Somewhat different in its impact both on the environment and on environmental regulation in Britain was the heavy chemical industry. Many of the industries already mentioned became increasingly reliant on the chemical industry in the nineteenth century. Its products – soda, sulphuric acid, bleaching powder – were required by the expanding textile, glass and soap making industries. Its large and very specific raw material requirements of salt and coal led to its concentration on Tyneside, Clydeside and Merseyside in close association with the industrial consumers of its products. It was an industry that was highly localised: in 1864, of 84 soda makers, 38 were in south Lancashire and north Cheshire and 19 on Tyneside. Such concentrated mass production of chemicals such as chlorine, sodium carbonate (soda ash), sulphuric acid and hydrochloric acid produced damaging waste gases, and the industrial complexes producing them were dominated by chimneys of great height in an effort to disperse the effluent (Figure 6.8). Because of the significance of this group of industries for the development of pollution control in Britain, it is important to describe the industrial processes and the emissions from these 'alkali industries'.

Originally the detergents, bleaches, mordants and alkalis required by the textile and other industries were supplied from natural products using very slow processes. Many of these products were difficult to obtain and in limited supply. Sustained industrial growth would have been impossible without the development of synthetic substitutes. There was not enough cheap meadowland or sour milk in all of Britain to whiten the cloth of the mechanised Lancashire industry and it would have taken unattainable quantities of human urine to dissolve the grease adhering to the raw wool used in the steam-powered mills of West Yorkshire. It was the heavy chemical industry producing acids and alkalis that provided the products that were required. Its significance to the national economy may be judged from the fact that by 1862 the alkali industry was employing 20,000 workers and producing finished goods worth £2 500 000 annually.

The 'alkalis' – potash (potassium carbonate) and soda (sodium carbonate) – were originally extracted from ash produced by burning vegetable matter (especially seaweed and other marine plants). At the end of the eighteenth century, methods of manufacturing synthetic soda from salt (sodium chloride) were developed. The most successful was the (French) Leblanc process that formed the basis of the British alkali industry. (Later, Ludwig Mond and John Brunner developed the Belgian Solvay or ammonia process to manufacture synthetic soda. The chemical works that they established at Winnington, Cheshire, in 1884 later became the headquarters of ICI Alkali Division.)

There were three major pollutants from alkali works – the hydrochloric acid gas emitted in the initial reaction of salt and sulphuric acid, the hydrogen sulphide from the sulphide waste heaps and the smoke from the large quantities of coal burned. Their combined environmental impact was noxious and destructive, but the mix of pollutants and their wide dispersion made it difficult to attribute blame to any particular works. Owners of land damaged by these emissions had little success in seeking redress through actions in nuisance in the civil courts. The first attempt to apply controls by statute, the Alkali Act of 1863, only related to hydrogen chloride gas, 95% of which had to be condensed. The technology to achieve that reduction in gaseous emissions, essentially a tower in which the hydrogen chloride came in contact with water, had been available for more than thirty years.

▲ *Figure 6.8*
The St Rollox Works in Glasgow manufacturing chlorine bleaching powder.

2.3 Industrial hazards

Impressive as the concentrations of industry became, many nineteenth-century chemical plants remained small scale and scattered, located in residential districts and among commercial property such as shops and warehouses. These situations proved particularly hazardous in case of fire and other conflagrations. Figure 6.9 shows the location of one such works in Manchester as late as the 1890s – a chemical factory shares party walls with tenement housing and faces a wood store at one end and a picture frame manufacturer at the other. The danger of fire was constant, and fire fighting and fire prevention were rudimentary (Figure 6.10). The ubiquitous steam engine itself brought particular dangers and damage-limitation strategies. A Steam Users Association was set up in Manchester in 1854, and after thirty years it was reported that its officers had investigated no less than 1500 explosions in the city – an average of one every week. 'Bad boilers' rather than 'bad firemen' were said to be to blame, and members of the Association welcomed the move to make periodic boiler inspections compulsory. The statistics convinced them that self-regulation had given insufficient protection and they resolved to assist the government in dealing with the dangers.

Less frequent, but more calamitous in their impact, were fires and explosions associated with chemical processes themselves. Some of these

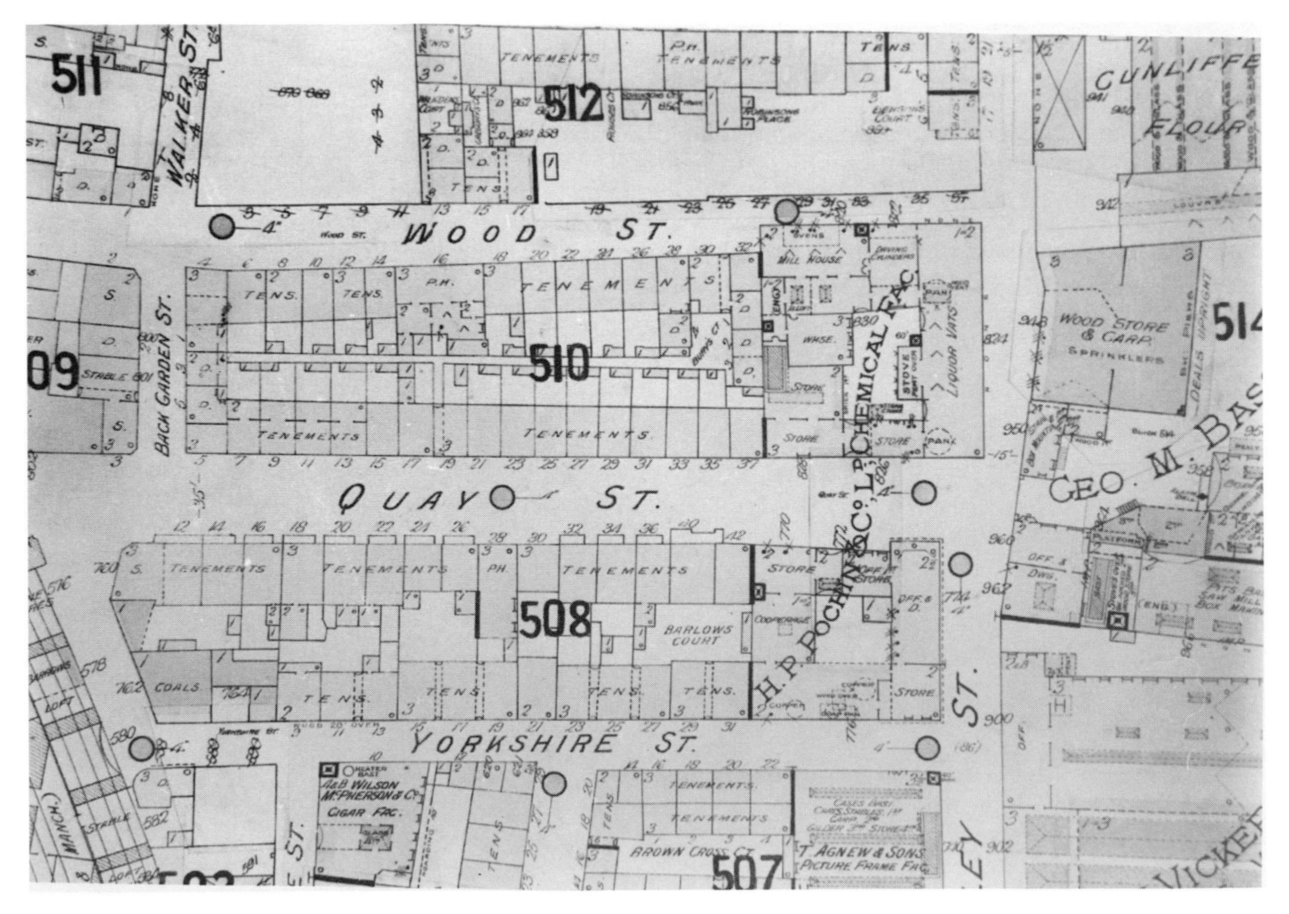

▲ *Figure 6.9 Hazards of industrial location in the nineteenth century. A map drawn for fire insurance purposes.*

▲ *Figure 6.10 Price's Oil Refinery, Blackfriars Bridge, September 1845.*

incidents could be anticipated and precautions of a kind put in place. Others were totally unexpected, with spectacular and disastrous consequences. An example of the former type is the explosion that occurred in 1887 at gunpowder mills near Hounslow, Middlesex. Damage to the works was severe, but its isolated location meant that the nearest houses were merely shaken. Only one worker was killed, while several hundred other men were said to have saved themselves because 'they, in accordance with the rules of the establishment, ran from their work and jumped in the river'.

No such rudimentary rules of self-help existed in the case of a much more extensive and totally unexpected explosion that occurred at a Manchester soap and dyestuffs factory. This incident is important because it illustrates how fast chemical technology was developing at the end of the nineteenth century, making seemingly innocent processes, in apparently safe locations, hazardous in the extreme. The fire and explosion occurred on midsummer's day of 1887 at the works of Roberts, Dale & Co. So extensive was the damage and so terrible were the results that at first it was suspected that the factory had indeed been secretly manufacturing explosives. The works itself was totally demolished, along with adjacent factories and boatyards (Figure 6.11). The nearby Pomona Hall was wrecked and a foreman's home alongside suffered extensive external and

◄ *Figure 6.11 Explosion at Roberts, Dale & Co., Manchester, 1887.*

internal damage. A workman and a fireman were killed. Her Majesty's Chief Inspector of Explosives conducted an enquiry, but Richard Dale, partner in the firm and a chemist of considerable renown, staunchly maintained that he knew of nothing in the factory that could have caused such an explosion. The Inspector's investigations and experiments finally revealed that two substances, picric acid and litharge (an oxide of lead), had been the culprits. They were being used in a novel process invented by Dale to produce new dye colours, and were not explosive in themselves. When mixed and exposed to heat, however, as they accidentally were in this case, they were found to explode readily. At the inquest, the jury recommended that 'the manufacture of such materials should be put under Government control and carried on in an isolated place.' Clearly, scientific knowledge and technological skill were insufficient to secure safe production processes in the rapidly developing chemical industry.

The scale and ferocity of such incidents, though rare, made a dramatic impact, generating vivid images, fear for public safety, and calls for immediate action. However, even in the early part of the twentieth century such major incidents still occurred. During 1917 a 'vibrating tremor' ran across London when a Ministry of Munitions plant, purifying TNT, exploded in Silvertown. This explosion, the largest ever recorded in the capital, completely obliterated the purification plant and every building within a 400 yard radius, killing 73 people and injuring over 1000. After the explosion, 'people stood outside as though transfixed. Some fleeing, some lay flat on their faces on the pavement, some prayed against the walls of the street'. According to a contemporary observer this grave disaster lit up London's night sky 'with the loveliest colour – violet, indigo, blue, green, yellow and red – which eddied and swirled from a chaotic mass into a settled and beautiful colour design...' The scene of the explosion itself, however, was said to resemble the desolation of a battlefield in France.

The impact of such a dramatic event was unmistakable, and only the exigencies of war had allowed the government to override its legal responsibilities for ensuring the safe handling of explosives. Other risks were more diffuse, making them difficult to isolate and deal with. The family of a Manchester toddler who died after eating 'toffy' – chromic acid – from a waste tip had to take what comfort they could from the inquest verdict of misadventure and the recommendation that a police officer should pay a cautionary visit to the dyeworks company concerned. Such events still occur (Box 6.1).

Box 6.1 A brief history of fly-tipping

Extraordinary death of a boy from poison

Yesterday afternoon, Mr William Hardy, deputy coroner, held an inquest on the body of a boy named John Williams aged 12 years, who lived with his parents at 117, Elton Hill Road, Radcliffe. It appeared from the evidence that about twelve o'clock on Saturday the deceased was playing with his elder brother and some other children on some waste lands where rubbish was deposited from Messrs Ashworth and Scholes' Dyeworks, Radcliffe. Soon afterwards the deceased went home crying, and on being asked by his mother what the matter was could only point to his mouth, from which a yellow liquid was flowing. The deceased's brother then told her that another boy had given him some stuff he had found, but he could not tell whether it was toffy or not. An emetic was administered and Dr Holmes, who was sent for, did all he could for the deceased, but the lad died about 4.30 p.m. the same day. Dr Holmes deposed to finding the deceased suffering from the effects of poisoning and in a state of collapse. A substance was shown to him, a portion of which the deceased had eaten, and he said it was chrome. The symptoms which showed themselves before death were quite consistent with chromic acid poisoning. The jury returned a verdict that the deceased died from the effects of chromic acid poisoning by misadventure. The deputy coroner instructed a police officer to see Messrs Ashworth and Scholes, and to ask them to exercise more care in future when depositing rubbish, and to see that no poisonous substance was in it. (*Manchester Evening News*, 2 March 1887)

In January 1972, the Birmingham newspaper, *The Sunday Mercury,* carried a story in which a truck-driver employed by a waste-disposal company reported how some of his workmates were given bonuses of £20 per week if they dumped loads of cyanide, chromic acid, caustic soda, phenol and other toxic substances using documentation which described these loads as being harmless 'suds oil'. The driver's complaint to his employers met with threats of dismissal, the offer of promotion and a £300 incentive to seek employment elsewhere. He declined all these offers and reported his experiences to the local branch of the Conservation Society, who in turn informed the Secretary of State for the Environment. This minister, who a few months earlier had been privately informed of the concern of the Royal Commission on Environmental Pollution over the inadequacy of existing controls over 'fly-tipping', did nothing.

The publication of the story in January 1972 led to considerable media interest in British refuse tips. But during a special Commons debate, a government spokesman, although aware of the imminent publication of the Royal Commission's views, maintained the view that the parliamentary timetable did not allow for legislation to be introduced before 1974.

On February 24, thirty-six one-hundredweight (50 kg) drums were found in Nuneaton (Warwickshire) on derelict ground where children often played. Crystals of 'sodium cyanide' still adhered to the drums from which the warning labels had been removed. In the ensuing public outcry, a Bill was rapidly drafted; the Deposit of Poisonous Waste Act 1972 received the Royal assent on 30 March 1972. (Description of the provenance of the 1972 Act from Ashby, 1978, p.20)

The 1972 Act was eventually superseded by the Control of Pollution (Special Waste) Regulations 1980, which introduced a system of 'consignment notes'. Prepared initially by the waste producer, the consignment note should describe the amount and chemical composition of 'special wastes' and must accompany the load until its disposal at an appropriately licensed site. 'Fly-tipping' would therefore require the loss of a consignment note or by giving false information which, in both cases, constitutes a criminal offence. Among the substances which may constitute 'special waste' are asbestos, chromium compounds and cyanide.

On 9 December 1992, Glasgow Sheriff Court sentenced a director of a waste disposal company to six months in prison for an offence under section 3(3) of the Control of Pollution Act 1974, involving the illicit disposal of 36 000 litres of toluene waste. Although not the first time that a custodial sentence had been imposed (and not suspended) by a British court for an environmental offence – that honour fell to a Liverpudlian caught 'in flagrante' fly-tipping non-toxic rubbish in 1988 – it does perhaps represent the first occasion when the *criminal* nature of the unlawful disposal of chemical waste was publicly recognised. It is undoubtedly a far cry from the cautionary word of a constable.

2.4 The first campaigns for clean air – issues and interests

The transformation of Britain's countryside into what one historian has called 'the drab lifeless grey of industrial wasteland' (Macleod, 1965) brought unprecedented economic progress to Britain, as well as attendant social and environmental ills. Regulation to mitigate these negative features inevitably raised fundamental issues about the nature and purpose of intervention and about its impact on the interests concerned. In very few cases were either the issues or the interests clear cut and it is for this reason that control over pollution was so controversial and difficult to achieve. This proved to be particularly the case in relation to pollution of the atmosphere – a part of the environment which everyone needed but no one could own, not even the state. In this section the interests involved and their attitudes to pollution will be discussed before going on to discuss the outcome in three specific cases – chemical hazards, the control of smoke, and emissions from the alkali industry.

Landowners who made up the agricultural interest that was still a dominant force in Parliament throughout most of the nineteenth century were particularly affected by air pollution. In the fields around places like Widnes and St Helens it was said 'the spring never comes hither ... the foul gases ... have killed every tree and blade of grass for miles around...' Even in light winds the gases from alkali works could carry for distances of up to nine miles, damaging farms 'just as if a fire had gone over them' killing trees and hedges, stunting wheat and putting cattle off their feed. Agricultural rents declined disastrously, and landowners complained that old ancestral seats were being destroyed by pollution from the air. Not surprisingly then, agricultural estate owners were active in seeking measures to protect their property. The House of Lords Select Committee on Injury from Noxious Vapours (1862) had among its members five great landlords and was chaired by Lord Derby, the 'king' of Lancashire. Indeed, the 1863 Alkali Act was for many years known as Lord Derby's Act, reflecting the safeguards it achieved for the land-owning class. Even here, however, the issue was scarcely clear cut since the majority of landowners also had extensive interests in the new industrial order, either directly, as in Lord Derby's case, through the exploitation of coal and other mineral resources from their estates, or indirectly through rising rents from building land in the expanding towns.

The industrialists themselves organised powerful opposition to even the mildest measures to control smoke and other emissions, arguing that the expense and disruption that would be caused would lead to economic stagnation. The Royal Commission on Noxious Vapours (1878) was told that these vapours were the inevitable and largely unalterable price of prosperity, since 'half or two-thirds of your income is derived directly or indirectly from manufacturing industry, and you must take the rough with the smooth.' Faced with the prospect of shutting up trade and the dire warnings of the manufacturers, legislators were cautious and even the great Public Health Act of 1875 required the abatement of smoke from furnaces only 'as far as practicable, having regard to the nature of the manufacture or trade...' Several industries where pollution was worst were exempt, notably coking, brick-making and iron-ore smelting, on the grounds that a proven control technology had yet to be devised.

Where improvements were secured, these proceeded chiefly on the grounds of the more efficient operating of plant. This device protected the position of manufacturers because it was predicated on the view that it was not industry itself, but rather plant operating below maximum efficiency and profitability that was the source of pollution problems. Controlling smoke emissions could be shown to reduce consumption of coal. Hydrochloric acid could be recovered as a marketable product when it had previously escaped from the chimneys of alkali works. These examples show how pollution control measures could succeed when they were allied to the interests of industry or, to use the currently popular phrase, 'preventing pollution pays'.

Local newspapers and essayists who purported to reflect public opinion often echoed this enthusiasm for expanding industrial activity even it meant valleys studded with factories and their chimneys discharging volumes of gas and smoke close to houses. Their presence was welcomed by those living nearby, it was said, because 'It is all good for trade, we want more of it, we find no fault with smoke.' Absence of smoke, by contrast, might well mean 'the quenching of the fire on many a domestic hearth, want of employment to many a willing labourer, and want of bread to many an honest family.' Reliance on the 'general nose of the public' as the primary line of attack against air pollution therefore carried little weight, and neither individual common law prosecutions for nuisance, nor police powers to report individual offenders to local magistrates, were greatly used.

Though there were a number of vociferous local pressure groups such as the Manchester Association for the Prevention of Smoke and the Lancashire and Cheshire Association for Controlling the Escape of Noxious Vapours, public opinion whether organised or not could not make much impact on the situation. Unlike other areas of public health reform, progress on air pollution did not rely on popular passion. This characteristic may be related to the fact that throughout the nineteenth century there was little hard evidence that airborne pollutants did cause ill health. Some actually believed black smoke was an antiseptic. Even reformers like Sir John Simon (Chief Medical Officer of the City of London) and the Rev. John Molesworth, vicar of Rochdale and Chairman of the Manchester Association for the Prevention of Smoke, were forced to argue that ill effects were indirect. Simon spoke of air pollution causing not specific illnesses but 'states of unhealth where a man was constantly below par' and Molesworth argued that airborne dirt made people poorer because of the cost of keeping clean, and by this and other means their tendency to disease was increased.

Given the ambivalence of the growing urban electorate to the presence of industry and its pollutants, it is not surprising that the local authorities of the time felt able to take only limited action against it. Indeed, as MPs pointed out, in the manufacturing districts the manufacturers themselves not infrequently constituted a large proportion of the sanitary authorities.

The multiplicity of local and sanitary authorities affected by any particular pollution incident made problem avoidance more likely than problem solving. Box 6.2 gives an example which is contrasted with the response to Shell's pollution of the River Mersey in 1990. In the Clayton district of Manchester, the Town Council, the Ashton-under-Lyne Sanitary Authority and the Openshaw Local Board of Health all complained about the 'abominable nuisance' caused by discharges from dyeworks to watercourses. No one company, however, could be held to blame in any specific location and the best that could be agreed was to call a meeting of the chemical manufacturers and 'see what could be done'.

Elsewhere, councillors simply appealed to what might be termed resigned pragmatism: if their town was no worse than others why should it try to be better? Any action might simply mean industry removing to a more tolerant area causing unemployment, a weakened local tax base and

Box 6.2 Contrasting approaches to river pollution

The pollution of Manchester rivers

Yesterday, at the meeting of the Ashton-under-Lyne Rural Sanitary Authority, a letter was received from the Town Clerk of Manchester complaining of the offensive gaseous smells experienced in the neighbourhood of Cornbrook. In consequence of the numerous complaints of this nuisance, the Rivers Committee of the City Corporation had caused inquiries to be made, and found that it was caused by chemical refuse discharged into the sewers of Clayton, and flowing thence into the Openshaw sewers, and so into the Cornbrook. He had now to request that immediate action be taken. A letter was also received from the Openshaw Local Board complaining of the 'abominable nuisance' which was caused by Clayton and which they (the Openshaw authority) were called on to abate. The chairman (Mr T. Jackson) said the nuisance was not all caused by Clayton. There were similar works both in Gorton and Ardwick doing just as much to pollute the streams and the atmosphere. However, he would call a meeting of the chemical manufacturers and see what could be done. It was agreed to sanction this course. (*Manchester Evening News*, 11 March 1887)

On 23 February 1990 in Liverpool Crown Court, Mr Justice Mars Jones fined Shell UK £1 million for polluting the River Mersey with 156 tonnes of crude oil which leaked from a corroded section of the pipeline linking Shell's refinery at Stanlow with an oil terminal in Ellesmere Port. Shell made a further £1.4 million contribution to the costs of cleaning up the spill and paid £100 000 for a system to detect such leaks in the future.

increased demands for poor relief. Though many local authorities included air pollution clauses in local bylaws, Nuisance Acts and Improvement Acts, progress was minor and slow. Provisions were weak, fines were low and convictions minimal. Many local acts relied on the concept of 'practicability' (see Section 3.3 below) and the goodwill of manufacturers. Many excluded the very industries most detrimental to air quality – Leeds, for example, excluded dyeworks, ironworks and brickworks and Liverpool excluded all manufacturers who had 'done their best' to prevent the escape of smoke. Despite some claims to have rendered the atmosphere in towns purer, many councils admitted that their record was unimpressive. They could claim neither great rigour nor expertise in dealing with the sources of air pollution, and confidence in them was consequently undermined.

Significant improvements, it seemed, required action by the national government to introduce comprehensive, compulsory legislation backed up by a system of rigorous inspection. Action by Parliament, however, had not only to overcome the opposition of MPs representing manufacturing constituencies, but also had to satisfy the need for legislation that met the criteria of uniformity, legitimacy and practicability while preserving the liberty of the individual.

It is now time to turn to the case of smoke prevention, control of other emissions and reduction of chemical hazards to see how these requirements were met and the various interests were settled. The regulation of the three areas had separate and contrasting histories until the mid-twentieth century, after which they began to coalesce. They thus provide interesting examples of alternative methods of controlling the dangers to health from industry.

2.5 Control of smoke from industrial sources

Until the nineteenth century, individuals suffering from specific nuisances from individual works could seek redress only through common law actions in nuisance. With the expansion of industry powered by furnaces generating steam, the problem became one for towns as a whole and the issue was redefined as one of public rather than private nuisance. The emphasis shifted from prosecution to prevention and the police were given powers to compel owners of steam engines to use appliances that had been invented to enable smoke to be consumed more efficiently within furnaces and not discharged to the atmosphere. Pressure for Parliament to become involved in framing statutes to deal with the problem recurred throughout the nineteenth century, but the decades are strewn with unsuccessful attempts to enact the necessary legislation. When legislation did emerge, the interests of manufacturers were largely protected and the initiative was left to local authorities acting within guidelines set by Parliament and anxious not to disrupt economic activity.

In its regular attempts to frame legislation for the abatement of smoke from industrial premises, Parliament looked for specific criteria. MPs continually stressed the need for technically effective means of consuming smoke; for acceptable means of consuming smoke given the nature of particular industrial processes and especially those where furnaces were frequently quenched and restarted; for a workable definition of 'smoke' itself; and for the determination of whether smoke constituted a health hazard or simply a nuisance. Only the first of these conditions proved easy to achieve and in 1843 the Select Committee on Smoke Prevention accepted that suitable techniques were available that were both easy and cheap to install.

Acceptability was only achieved, however, by confining mandatory powers to new furnaces while making it permissive for existing ones, and by exempting altogether those processes where furnaces were used intermittently, such as iron foundries and pottery kilns. The absence of any definition of smoke ensured that several Bills were rejected on this ground alone, and successful prosecutions were rare even in cases where statute allowed actions to be brought. It is no wonder that local authorities who were responsible for bringing polluters to court relied on the deterrent value of their powers while rarely actually invoking them. From 1853, smoke control Acts contained clauses exempting polluters from prosecution if they could demonstrate that they had used the **best practicable means** (BPM) to abate smoke. This fundamental principle was to play a vital part in the story of pollution control as will be seen in the discussion of the alkali industry. In the case of smoke control, judgement on this issue lay with local authorities who brought prosecutions and also with local magistrates before whom cases were heard. In these circumstances the inclination was to give defendants the benefit of the doubt, but the system did give individual citizens a direct route for complaints through their local councillors. This local accountability for smoke control persisted through the present century, and was extended in the 1956 Clean Air Act to enable local councils to declare smoke-control areas where only smokeless fuels could be used. The same Act, however, took away from councils control of those industrial premises where there were technical difficulties in smoke abatement. Here control passed to the central Inspectorate, the Alkali Inspectorate.

2.6 *The extension of control from 'alkali works' to other 'scheduled processes'*

From the beginning, powers to control emissions from alkali and other chemical works took a different route from that pertaining in smoke control. From the outset in 1863, control was placed, not in the hands of elected local authorities, but in the hands of a central body of professional scientific inspectors (chemists and later chemical engineers) responsible to Parliament through a Minister of the Crown – the **Alkali Inspectorate**. Unlike smoke, accurate chemical measurement of hydrogen chloride gas was possible and such data could, if necessary, be cited as evidence in any prosecution of a polluter. Although initially viewed with suspicion by both industry and Parliament, the Inspectorate gradually established itself and pressed for more effective controls over a wider range of polluting industry.

In 1874 the Alkali Act was amended to include a statutory emission limit on hydrogen chloride emission (in addition to the 95% condensation requirement) and to require the use of BPM to minimise the discharge of other noxious gases (e.g. smoke and sulphur dioxide) from alkali works. In 1881 the jurisdiction of the Alkali Inspectorate was extended to include a further twelve types of chemical works listed in a schedule of the amending statute – hence the term 'scheduled works'. Sulphuric acid works were included and these were also subject to fixed limits on acid emission.

The use of statutory limits in the early Alkali Acts is interesting for it is often stated that these devices are foreign to the British approach. This is not strictly true. The type of plant to which the fixed limits applied gradually became obsolete and control came to be exercised entirely through BPM. Also, the Inspectorate adopted a practice in which each scheduled process was assigned a 'presumptive limit' for its particular noxious gas; if emissions did not exceed that limit, then adherence to BPM could be presumed. But it might be emphasised that these presumptive limits were non-statutory and were intended to serve as general guidelines, each works being subject to control as the Inspector saw fit.

All works operating scheduled processes had to obtain a certificate of registration to be renewed annually. In the case of new works, certification was only granted where the works met the requirements of the Act, but existing works were allowed to continue. In operation, the Alkali Acts relied heavily on the decisions of the Chief Alkali Inspector and his team of local inspectors.

Various piecemeal amendments were consolidated by the 1906 Alkali Etc. Work Regulation Act which remained substantially in force until 1983. It set out specific cases where statutory standards were to be applied to emissions, namely alkali works, hydrochloric acid plant and some sulphuric acid works. It also listed other classes of works which were required to prevent the emissions of 'noxious or offensive gases' and to render those gases that were discharged 'harmless and inoffensive'. In these cases, the measure to be applied in judging the action taken was that it should represent 'the best practicable means'. There is still no statutory definition of BPM, nor is there any record of this important phrase ever having been subject to judicial interpretation – a circumstance which must be attributed at least in part to the Inspectorate's reluctance to prosecute (between 1929 and 1966 only two prosecutions are recorded). The following interpretation by a Chief Alkali Inspector has subsequently become authoritative:

> The expression (best practicable means) takes into account economics in all its financial implications and we interpret this not just in the narrow sense of a works dipping into its own pocket, but including the wider effect on the community. In the long run it is not owners of the works who pay for clean air, but the public, and it is our duty to see that money is wisely spent on the public's behalf. The country's, industry's and work's current financial situations have to be weighed against the benefits for which we strive, and careful thought has to be given to decisions which could seriously impair competitiveness in national and international markets. (MHLG, 1967)

The *modus operandi* adopted by the Inspectorate was that of gradual gentle persuasion, using the 'intellect' and not the 'cane' and rejecting any action that might smack of 'militancy' or 'policeman-like behaviour'. The 1974 study of the Alkali Inspectorate by the pressure group Social Audit was particularly critical of the cosy relationship between local inspectors and industrialists. Box 6.3 describes one long-running conflict between the Inspectorate and a local authority over emissions from a scheduled process.

Box 6.3 The Glossop chimney

The emissions of sulphur dioxide from the molybdenum smelter at Glossop in Derbyshire have been a longstanding source of nuisance and the alleged cause of an excess of respiratory morbidity in the vicinity. In 1987 HMIP issued an improvement notice requiring a 70% reduction in emission by 1992. This reduction has been achieved by reduced output rather than by installing desulphurisation. The 90 m stack serving the plant had received a temporary (10 year) planning consent in 1977. On expiry, High Peak Borough Council took enforcement action requiring its demolition. The company won the subsequent appeal but, in an attempt to give the local authority its own enforceable controls over the plant, the Secretary of State accepted his Inspector's recommendation that the HMIP requirements should be incorporated as conditions of the planning consent. In other words, High Peak could enforce the 70% of reduction irrespective of any subsequent change in HMIP's view as to what should constitute 'best practicable means' or BATNEEC (see Section 3.3) for this particular site. This type of duplication of controls was deprecated in *Planning Policy Guideline 23* published by the Department of the Environment in 1994.

In the background of the photograph in Figure 6.12, the Pennines are clearly visible. Hills surround Glossop on three sides and they serve to negate the extra dispersion achieved by a tall stack.

▲ *Figure 6.12 The Glossop chimney.*

Activity 1

List the most significant sources of air pollution following the industrial revolution of the nineteenth century. What were the major factors in the continuation of pollutants in the air?

3 1900–73: consolidation of the UK regulatory systems

3.1 The urban and industrial revolutions of the twentieth century

In the twentieth century, and especially in the aftermath of the two World Wars, industry and the economy generally underwent far-reaching changes. In terms of both location and function, urban areas and their factories were again transformed. The centripetal forces that had taken population and industry into cities, were, from 1918, replaced by centrifugal ones. As population moved out to the suburbs, land-use patterns were reordered. Locations that had seemed remote enough for noxious industries and other undesirable activities, such as isolation hospitals, filled up with houses, sometimes taking the 'Homes for Heroes' funded by the state and built by local authorities, to the very gates of chemical plants (Figure 6.13). Accidents inevitably occurred.

The chemical industry was itself undergoing considerable change, responding to the demands for new products, many of which were needed to satisfy the suburbanites' desire for comfort and efficiency in their homes – plastics, paints, fertilisers, adhesives, detergents, cosmetics, cars and other consumer durables. New factory sites were needed. Many firms simply expanded around their existing site, no matter how dangerous or unsuitable this might be. The Clayton Aniline Company, for example, was established in 1876 on a one-acre site in east Manchester, but by 1950 the works had expanded to cover 54 acres. Such examples fuelled demands for land-use planning, at least where green field sites were being developed. Industrial estates were created and Trafford Park alongside the Manchester Ship Canal was the world's first. Trafford Park became probably the most heavily industrialised area in Britain and many chemical companies established plant here, especially those concerned with food processing, pharmaceuticals, cosmetics and plastics. Even in this technolocially advanced development, residences for the workers continued to be provided on site. And explosions still occurred in its chemical factories because, as happened with Roberts, Dale & Co. (see Section 2.3), new processes generated unpredicted hazards.

For most of the twentieth century, the growth of large-scale industrial processes outstripped the development of legislation to protect the health of people working in or living near such industries. The output of toxic chemicals and potential pollutants was increasing, but the dangers they

▲ *Figure 6.13 Crumpsall Vale, Manchester 1925 showing hospitals and institutions in the foreground, with houses and chemical works beyond.*

posed were often only attended to when brought to light by a major disaster. It is to the control of such hazards that we now turn.

3.2 *The recognition of the wider hazards of chemical manufacture*

Successive Factories Acts aimed to reduce threats to workers, but in the early nineteenth century they had been chiefly concerned with regulating the hours worked by women and children. From 1844 the Factories Inspectorate was empowered to require the fitting of guard-rails around the flywheels of steam engines. A concern with the risks to the *public* outside the factory gates came much later. Risk to workers and to the public at large from industrial production was in many respects obvious, but responsibility was difficult to determine. By definition, accidents were irregular events resulting often in serious damage. Such cataclysmic incidents could be regarded as 'acts of God' or examples of gross human error. Both the Silvertown explosion of 1917 and the destruction at the Manchester works of Roberts, Dale & Co. were found to have been begun

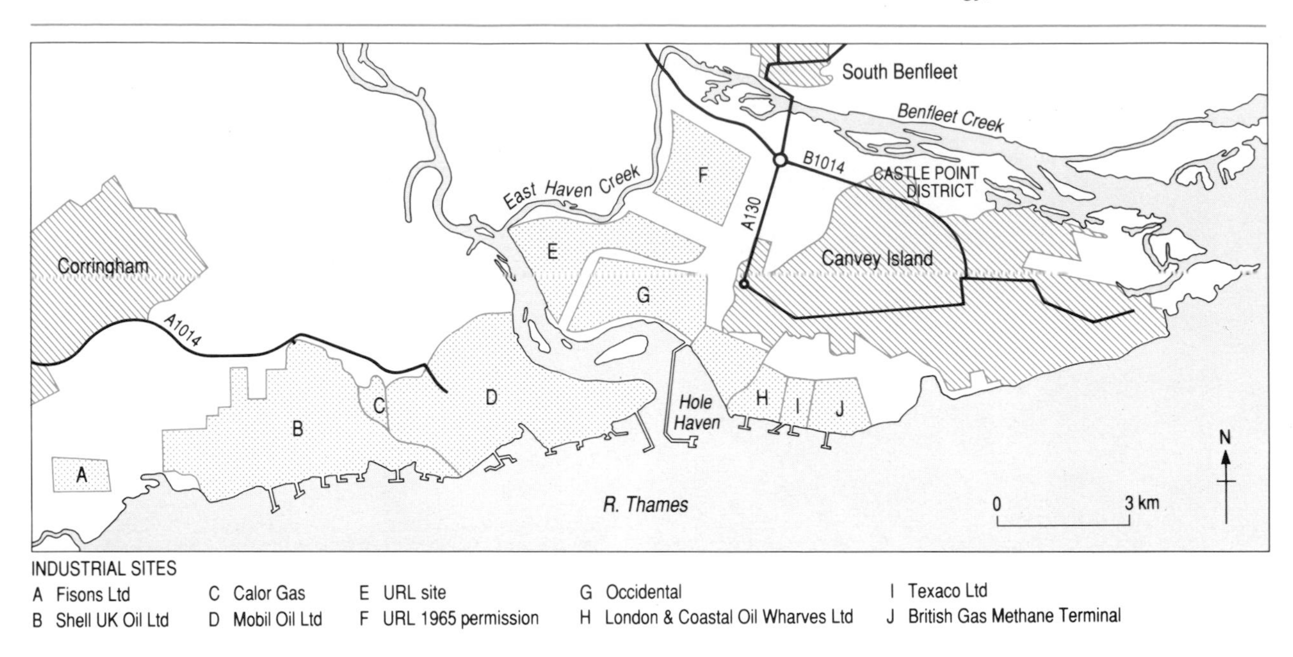

▲ *Figure 6.14 The location of major hazard installations at Canvey Island.*

by a fire caused by a workman's matches. Private insurance was the chief protection of employers and individuals alike in such cases.

When, following the Second World War, the chemical industry increasingly came to rely upon imported oil rather than coal as feedstock, so sites near the oil (and later, North Sea gas) terminals became the favoured location for the new plant. With fewer constraints on the availability of land in these remote areas, massive complexes arose on the Mersey estuary at Stanlow and on the Thames estuary at Canvey Island (Figure 6.14).

In 1967 HM Chief Inspector of Factories applied the term 'major hazard' to that growing number of industrial installations (especially within the petrochemical sector) where accumulations of toxic or explosive material could, under accident conditions, cause widespread damage and injury far beyond the perimeter fence. In 1974, as part of the major reforms recommended by the Robens Committee, the Factories Inspectorate became the principal component of the Health and Safety Executive (HSE), the enforcement arm of the Health and Safety Commission (HSC), which is the agency that has statutory responsibility for the administration of the Health and Safety at Work Act 1974 . This important piece of legislation imposes the duty on employers to do whatever is 'reasonably practicable' (see Section 3.3 below) to protect both workers *and* the public from the hazards posed by industrial accidents.

Concern accelerated in the aftermath of the explosion at the cyclohexane plant at Flixborough on Humberside in 1974 (Box 6.4). Twenty-eight workers were killed and thirty-six injured, but the remoteness of the site meant that, despite widespread damage to property, there was no fatality among off-site personnel. The potential for an even greater disaster was not lost on the Advisory Committee on Major Hazards, set up by the HSC in response to this disaster. The Committee's awareness of the land use and siting implications of hazardous processes and of the consequent need to involve planning authorities is apparent in its three reports (HSC, 1976).

Box 6.4 The Flixborough disaster

Figure 6.15 The Nypro site at Flixborough.

Figure 6.15 shows aerial photographs of the 60-acre Flixborough works of Nypro (UK) Ltd which were devastated by an explosion at about 4.53 p.m. on Saturday 1st June 1974. The cause of the disaster was the ignition and deflagration of a massive vapour cloud formed by the escape of cyclohexane under a pressure of 8.8 kg per cm^2 and a temperature of 155 °C. Twenty-eight of 64 personnel on site (far fewer than would have been present on a weekday) were killed. There were no offsite fatalities, although 53 casualties were recorded and many more were known to have suffered minor injuries. Seventy-two out of 79 houses in Flixborough, about half a mile from the plant, were damaged; and in Scunthorpe, a large town some three miles to the south, 786 houses suffered damage. Were it not for the relative remoteness of the site, a far greater tragedy would have occurred.

The principal effect of what are now the Notification of Installations Handling Hazardous Substances Regulations 1982 (NIHHS) is to require operators of sites storing, manufacturing, processing or using certain specified substances to notify the HSE. The regulations specify threshold or 'notifiable' quantities of each substance (e.g. 500 tonnes ammonium nitrate; 10 tonnes chlorine) and the notification obligation is estimated to apply to some 1600 sites. In theory, these sites would, following notification, receive particular attention from HSE and, where necessary, its specific controls (e.g. improvement and prohibition notices) if informal reminders of employer responsibilities under sections 2 and 3 of the 1974 Act went unheeded. Under the NIHHS Regulations the HSE has a statutory obligation to inform each local planning authority of all 'notifiable installations' in its area. In 1984 the Executive drew up 'consultation zones' (Figure 6.16) for each notifiable installation, and a planning circular published in that year advised planners to seek the advice of the Executive before granting planning consent for development falling within a zone.

Ironically, the full implementation of the NIHHS Regulations was delayed as a result of another disaster. In 1976 a large release of highly toxic dioxin from a chemical plant in Seveso near Milan contaminated a large area and gave rise to a number of cases of chloracne. Pressure from members of the European Community Parliament led to the preparation of

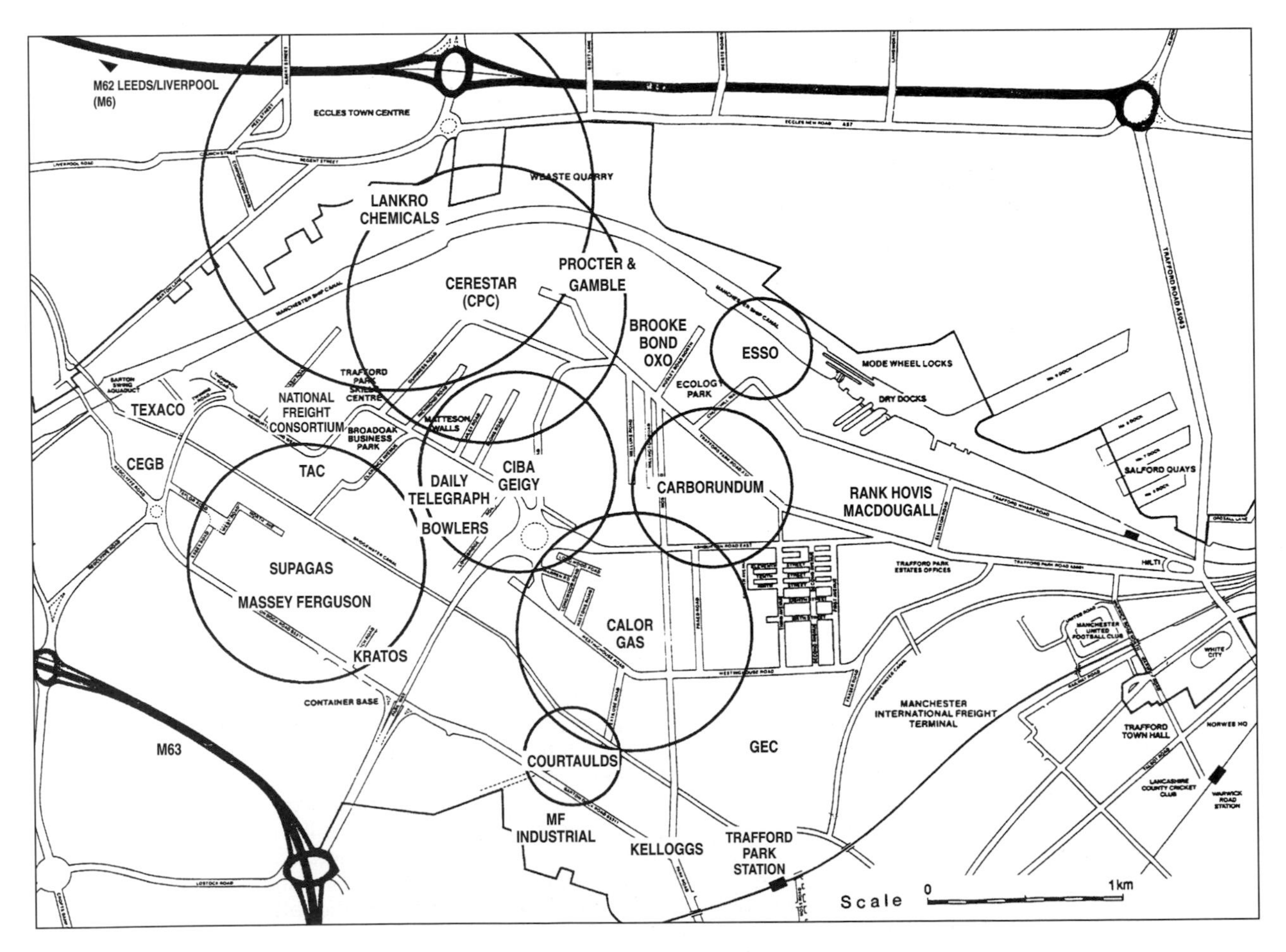

▲ *Figure 6.16 Consultation zones surrounding hazardous installations in Trafford Park.*

the 'Directive on the major accident hazards of certain industrial activities' (82/501/EEC; EEC, 1982). This 'Seveso Directive', as it became known, required controls over some 250 'major hazard' sites in the UK which were more stringent than those specified in the (then draft) NIHHS Regulations. Implementation of the latter was suspended while the Control of Industrial Major Accident Hazards Regulations 1984 (SI 1984 No. 1902) were prepared.

Before embarking upon a detailed discussion of the effect of European institutions on the UK regulatory regimes in the last quarter of the twentieth century, we need to pay further attention to some of the central principles of the traditional domestic systems now being modified or supplanted by continental influences.

Activity 2

What changes occurred in chemical manufacture in the twentieth century and what significance did they have for problems of pollution and hazard?

3.3 *Practicability, BPM and BPEO*

The phrase 'reasonably practicable' has been central to UK factory legislation for over a century. It has been the subject of considerable judicial interpretation, but the ruling of Lord Justice Asquith in *Edwards v. the National Coal Board* is now taken as definitive:

> 'Reasonably practicable' is a narrower term than 'physically possible' and seems to me to imply that a computation must be made by the owner, in which the quantum of risk is placed on one scale, and the sacrifice involved in the measures necessary for averting the risk (whether in money, time or trouble) is placed in the other; and that if it be shown that there is a gross disproportion between them – the risk being insignificant in relation to the sacrifice – the defendants discharge the onus on them. ('Edwards v. National Coal Board', *Law Reports (King's Bench)*, 1949(1), pp. 704–716)

An economist examining this judgement is likely to point to the balance metaphor and see it as an exhortation to apply the techniques of cost/benefit in the determination of an acceptable level of employee safety. A technologist, on the other hand, might counter by arguing that, even if it were possible to compute within a single measure risk, sacrifice, time and trouble, the demonstration of what constituted 'gross disproportion' remained a qualitative judgement and one which justified adherence to traditional 'good engineering practice' and faith in the probity of British Standards to secure an appropriate margin of safety. Figure 6.17 is a graphical representation of the approach adopted by the nuclear safety branch of the HSE to the interpretation of the closely allied term '**as low as reasonably practicable**' (ALARP) (HSE, 1992).

The general duty upon operators of sites that come under the Control of Industrial Major Hazards Regulations to 'operate safely' is not qualified by the 'so far as reasonably practicable' clause (section 3 of the 1974 Act). In practice this means that the 'gross disproportion' between the monetary evaluation of residual risks and further reduction measures must be even grosser when major hazards are being considered.

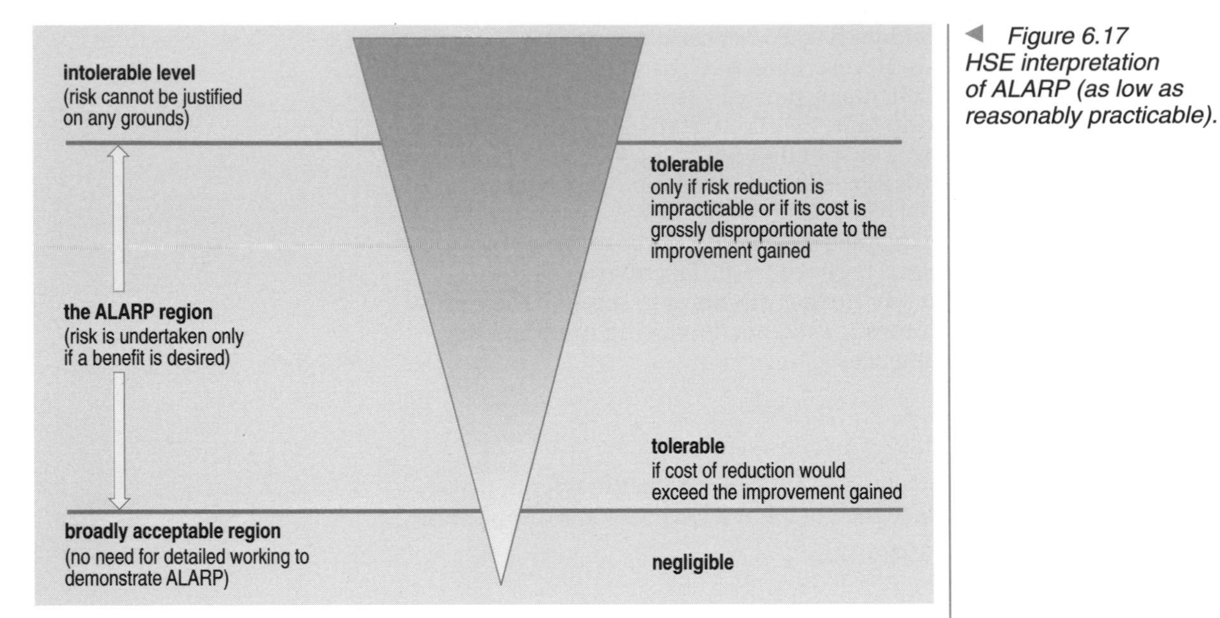

Figure 6.17 HSE interpretation of ALARP (as low as reasonably practicable).

Just as the manufacture of chemicals can never be entirely devoid of risks to employees, neither can it be conducted without a degree of pollution of the environment. Determining an acceptable level of residual pollution (i.e. that which escapes the control measures) again involves economics. But it is doubtful whether the Chief Alkali Inspector was legally entitled to employ the balance metaphor in interpreting BPM as he did in the passage quoted in Section 2.6 above. In a case arising under factory legislation, it was established that the term 'practicable' implied a stricter standard than 'reasonably practicable' and one in which 'questions of cost might be eliminated ... but the measures must be possible in the light of current knowledge and invention.' Thus if this notion of what was practicable were to be translated into the field of air pollution, then it could be argued that the latest proven technology, albeit very expensive, constitutes BPM and should be installed regardless of the consequences for the plant operator. This view has never been supported by the Inspectorate, nor by the Department of the Environment, or its predecessors. It also accounts for the insistence of the UK representatives at European Community discussions, that the phrase '**not entailing excessive cost**' (NEEC) should qualify the requirement that '**best available techniques**' (BAT) should be used to prevent air pollution from industrial plant.

The fifth report of the Royal Commission on Environmental Pollution (RCEP, 1976), represented an influential and authoritative review of the UK system of air pollution regulation which was then in force, i.e. before the full consequences of UK membership of the EC took effect. In Section 4.2 below, we refer to RCEP's cautious endorsement of non-statutory air quality guidelines and its opposition to the use of planning powers which encroached upon the Alkali Inspectorate's jurisdiction. Two of its other recommendations have subsequently proved to be of more lasting relevance.

First, the Royal Commission supported the demand by local authorities for effective powers of 'prior control' over those processes which were not scheduled under the Alkali Act 1906, but which were nevertheless sources of air pollution. When pollution from such sources amounted to nuisance (especially odour from offensive trade premises), local authorities were empowered to prosecute offenders under the Public Health Acts. But retrospective action of this sort was generally recognised to be insufficient. What was required was anticipatory control; that is, such sources should be allowed to commence operation only if the local authority were satisfied that adequate safeguards had been observed to prevent nuisance arising.

Secondly, but more important, the Commission argued that separate legislative controls over pollution of land, the atmosphere and water had arisen through historical accident, had persisted through administrative inertia, and ignored current understanding of the interconnectedness of the environment. As an example of a multi-media pollution problem, consider the following.

The practice of landfilling hazardous wastes poses a threat to underground waters as well as denying that land for other uses; a decision to incinerate such wastes entails extra emission to the atmosphere, and reducing this burden by the use of flue gas cleaning systems (wet scrubbers) in turn entails treatment before any scrubber liquor can be discharged to surface water. As originally conceived, '**best practicable environmental option**' (BPEO) represented an extension of the traditional BPM approach to processes where pollution control involves trade-off between different receiving media. It would be exercised by a small but well-qualified body of scientists and engineers, HM Pollution Inspectorate, with the expertise to identify and the legal powers to insist upon the disposal option which posed the least threat to the environment as a whole while remaining cost-effective. The government took five years to respond to the Commission's proposals and then it saw BPEO as unnecessary and argued there was 'probably not a large number of situations in which major choices exist in practice'. It did, however, agree with the Commission that a comprehensive review of air pollution legislation should be carried out.

Before that review took place, interest in the concept of **integrated pollution control** had increased in Europe. The third Community Action Programme on the Environment in 1983 had as one of its priorities the prevention of pollution, but not by its transfer to another medium. The following year saw the adoption of the Framework Directive on industrial air pollution (EEC, 1984; see Section 4.3 below) and, in Britain, the publication of the tenth report of the Royal Commission on Environmental Pollution, in which this body reiterated its call for BPEO to be incorporated into national legislation (RCEP, 1984). Thus, by 1986, when the review had reached the consultation stage, there were many issues on which local authorities, industrial and environmental organisations and others were able to comment.

Before the results of this consultation procedure were announced, the government declared in 1987 its belated conversion to the integrated approach by amalgamating four pollution control agencies – the Industrial Air Pollution Inspectorate (the successor to the Alkali Inspectorate), the Hazardous Wastes Inspectorate, the Radiochemical Inspectorate and the Water Inspectorate (three specialised agencies within the Department of the Environment) – to form Her Majesty's Inspectorate of Pollution (HMIP). The immediate task of HMIP was to continue the statutory and advisory roles of its constituent inspectorates, but it also had a brief to further the eventual implementation of BPEO. Since the Environment Act of 1995, HMIP has become part of the new Environment Agency.

Activity 3

Summarise the historical development of smoke, chemical pollutants and hazard control in the UK.

4 *The influence of European institutions*

4.1 *Introduction*

Differences between the British and European approaches to pollution control first emerged in connection with various directives concerned with water pollution. At the risk of over-simplifying matters in which national self-interest and administrative convenience were no less influential than scientific or technical considerations, the difference may be summarised as follows. The UK government favoured an approach in which a standard or objective was set for the quality of the pollution-receiving medium, and the maximum permitted emissions from individual sources were then determined accordingly. These objectives are called **environmental quality objectives** (**EQO**). By contrast, the majority of member states advocated a regulatory regime based upon uniform **emission limits** (EL) over individual emissions of a given pollutant to a given medium (i.e. surface water, underground water or the sea).

Clearly the two approaches are not entirely unrelated and represent two means to the same end – that is, ensuring that the concentration of any pollutant does not exceed the level at which damage to human health arises. Why then did they become the subject of such contentious debate?

Britain is an island with fast-flowing tidal rivers which discharge into seas flushed by the North Atlantic Drift. Given its ample rainfall and its numerous upland catchment areas, it is not particularly dependent, especially in the important industrial areas in the North, on riverine or underground sources for drinking water. These natural geographical advantages and the absorptive capacities of tidal estuaries and coastal waters encourage an adherence to the EQO approach. Thus a limit must still be set on any source of discharge such that the particular EQO is not breached, but that limit can take account of the existing quality and the dilution effect of the receiving waters.

This notion of variation in permitted discharges among equivalent industrial sources was viewed with suspicion in continental Europe. Uniform emission limits (EL) are much simpler to administer; they are not dependent upon assumptions and complex mathematical models of pollution dispersion; and they are easier to enforce since measurement of amounts of pollutant at the point of discharge (i.e. at the base of the chimney or the end of a pipe) is far easier than measurement of very low concentration (parts per billion or less) after dilution in the receiving medium. When that medium is a river which crosses national boundaries (the Rhine being the example most commonly cited), the inherent merits of a system based on EL become more apparent: if an EQO is applied throughout the river, a downstream state is clearly disadvantaged as its own discharges have to be reduced to take account of those of its upstream neighbour. If an overall reduction in total load is necessary (e.g. to save the North Sea fisheries from extinction) then this is most equitably achieved by imposing a community-wide tightening of uniform emission limits on all sources of the pollutant in question (whether located on the Humber, the Rhine, the Seine or the Ebro).

It must also be recognised that *uniformity* of control over pollution sources is the approach most consistent with the original aim of the European Economic Community to create a 'common market' by the abolition of barriers to trade. For example, a state whose indifference to pollution safeguards resulted in lower costs of manufactured goods than those of its more environmentally concerned competitors could be said to have an unfair trading advantage. But the 1957 Treaty of Rome, by which the EEC was established, contained no explicit reference to the environment. When the steady stream of environmental directives began in the 1970s, it was necessary to identify the 'free trade' dimension of each proposal. Occasionally, as with directive 86/278/EEC (EEC, 1986), which set a maximum concentration of heavy metals such as lead and mercury in sewage sludge spread on agricultural land, this process requires more imagination than for others (for example, the early directives harmonising requirements over the biodegradability of consumer products like detergents).

It was only with the Single European Act in 1987 that the EEC's environmental legislation, comprising nearly two hundred directives through which the policy objectives of (now) five Community Action Programmes were implemented, was placed on a legitimate and unequivocal basis. The Treaty of Maastricht in 1992 reinforced this basis by ensuring that an unashamedly *green* concept – 'sustainable and non-inflationary growth respecting the environment' – was cited within the very aims of the European Union. In addition, it enshrined within EU environmental law a number of concepts alien to the traditional British approach to pollution control: the 'precautionary principle', i.e. the presupposition in favour of 'high levels of protection' and the 'integration' of an environmental dimension in *all* areas of EU policy.

The Treaty of Maastricht will be remembered for the depth of the UK government's antipathy to the Social Chapter, culminating in the celebrated 'opt-out'. The protocol which excludes the UK from the Treaty's obligations in respect of social policy has implications for subsequent legislation on occupational risk. There is not within continental member states the same tradition of separation of *physical* working conditions (regulated by statutes enforced by the Factories Inspectorate) and other *socio-economic* conditions (determined by free collective bargaining between unions and employers) as has obtained in the UK. In the rest of Europe, hours of work and safety training, together with minimum wage, maternity leave and pension provision, fall within a package of employment rights to be defended by the state.

4.2 Air quality standards

The first limit values for atmospheric pollution (i.e. smoke and sulphur dioxide) adopted by the EC in Directive 80/779 (EEC, 1980), were based on those suggested earlier by the World Health Organisation (WHO, 1972). The principal author of the WHO standards (Table 6.2) was Professor John Lawther, an eminent epidemiologist and an internationally acknowledged authority on the effects of air pollution on human health.

Table 6.2 WHO air quality standards

Substance	Time-weighted average	Averaging time
lead	0.5–1.0 μg m^{-3}	1 year
nitrogen dioxide	400 μg m^{-3}	1 hour
	150 μg m^{-3}	24 hours
ozone	150–200 μg m^{-3}	1 hour
	100–120 μg m^{-3}	8 hours
sulphur dioxide	500 μg m^{-3}	10 minutes
	350 μg m^{-3}	1 hour

A graphical representation of these limit values is given in Figure 6.18. The two axes represent the 98th percentile values of the concentrations of smoke and sulphur dioxide (SO_2). The area bounded by the straight lines indicates compliance with the Air Quality Directive; its shape derives from the fact that, in any location where the smoke level is relatively low (i.e. <150 μg m^{-3}), a less stringent limit value for SO_2 (350 μg m^{-3} rather than 250 μg m^{-3}) applies. A similar concession applies in the limit values for the annual and winter medians.

This linking of the limits for SO_2 and smoke stems from the mistaken but widely held belief in a 'synergistic' effect between these two pollutants. As Professor Lawther explained to the House of Lords Committee: wherever coal or oil is burnt in large quantities, smoke and sulphur dioxide are inevitably emitted simultaneously. Epidemiological studies of the incidence of respiratory diseases (e.g. chronic bronchitis) cannot distinguish between their separate effects. To claim, therefore, that their combined effect exceeds the sum of their individual effects (i.e. synergy) is scientifically meaningless.

At any monitoring station it is necessary to maintain apparatus capable of measuring the average value of the smoke and SO_2 concentrations over successive periods of 24 hours. Having collected these daily mean values over one year, it is possible to calculate the annual mean (of the 365 daily

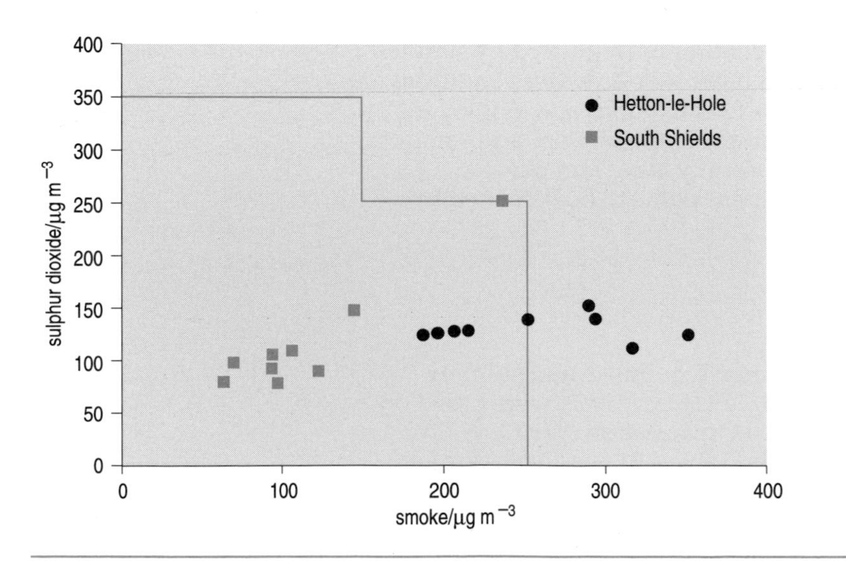

Figure 6.18
98th percentiles – smoke and sulphur dioxide, with examples of specific site data.

averages), the mean for the winter months (1 October to 31 March), and the 98th percentile; hence compliance with the various limit values can be judged. The apparatus required is not particularly expensive or sophisticated, but a network of such monitoring sites, concentrated upon urban and industrial centres, is necessary to ensure that localised violations do not arise.

Such an extensive network already existed in the UK well before the Air Quality Directive. Co-ordinated by the Warren Spring Laboratory of the Department of Industry, it had been set up to monitor the implementation of the Clean Air Acts of 1956 and 1968, especially the domestic smoke control programme. At its height, the network consisted of more than a thousand sites, most of which were serviced by the technical staff of local authority environmental health departments. As the domestic smoke control programmes began to take effect, and as supplies of (very low sulphur) gas from the North Sea came on stream, so the data revealed a steady decline in average concentrations of SO_2 and an even greater fall in smoke. So much so, that by the late 1970s, a far smaller monitoring network was being contemplated. The advent of the Air Quality Directive has necessitated the retention of some 400 hundred stations.

Q The figures below represent the 98th percentiles for the daily concentrations (in units of $\mu g\ m^{-3}$) of smoke and sulphur dioxide recorded at four monitoring stations in one year. At which site(s) has the EC Directive been exceeded, and by which pollutant?

Station	SO_2	Smoke
A	240	240
B	340	140
C	260	160
D	160	260

A C – SO_2 limit exceeded; D – smoke limit exceeded.

The 1980 Directive on air quality can therefore be seen as the result of a common desire among the member states of the EC to reduce the risk to public health posed by the principal pollutants (that is, smoke and sulphur dioxide) resulting from fossil fuel combustion. But this directive and those subsequently concerned with nitrogen dioxide and lead were aimed at ensuring that *concentrations* remained below the levels at which impairment of health was believed to arise. Within Western Europe, there remain only isolated pockets where such conditions prevail; but to the East, in Poland especially, the absence of controls over emissions during the Communist era has meant that environmental conditions in the industrial areas are comparable to those of mid-Victorian Britain.

4.3 *European responses to 'acid rain'*

'Acid rain' is the popular term applied to the deposition of a complex mixture of chemical species resulting from emissions of sulphur dioxide and oxides of nitrogen produced in the combustion of fossil fuels. Acidification of lakes in Scandinavia is not a particularly novel phenomenon nor, for that matter, is the occurrence of localised damage (traditionally attributed to disease or adverse meteorological conditions) to forests. But during the late 1970s, the scale of the harm being suffered by the forests of northern and central Europe became apparent, as did the urgent need for action to avert ecological catastrophe.

Since air pollution is no respecter of state frontiers, intervention would have to be internationally co-ordinated, with the industrial economies of the EC leading by example. The 1980s witnessed a shift in the underlying motivation of the EC air pollution policies away from a desire to safeguard with human health towards a concern for the wider effects on the environment. For example, even though, as we have seen, the EC (save for a few isolated pockets) generally conformed to its health-directed air quality limit value for concentrations of SO_2, total emissions of this pollutant could, perhaps in combination with other chemical species, nevertheless pose a threat to forests, lakes and other parts of the ecosystem of such gravity that substantial cuts in that total would become necessary, irrespective of any immediately apparent reduction in mortality or morbidity from respiratory illness.

As the decade progressed, acid deposition was joined by other global pollution problems, most notably 'greenhouse' warming and depletion of the ozone layer, as regular items on the agenda of the meetings of Environment Ministers. Although a number of international treaties were signed (Table 6.3), consensus was not to be a common characteristic of these meetings.

Table 6.3 International agreements on atmospheric emissions

Treaty	Subject of agreement
Geneva Convention (1979)	transboundary air pollution
EMEP Protocol	monitoring regimes
Helsinki Protocol (1985)	30% reduction in SO_2 emissions
Montreal Protocol (1987)	control of ozone-depleting substances
Sofia Protocol (1988)	control of nitrogen oxides emissions
UNECE Protocol (1991)	30% reduction in emissions of volatile organic compounds

There were many occasions in which the UK representative was to be found in the minority; the refusal to join the **30% Club** (i.e. those nations committed to reducing their SO_2 emissions by 30% between 1980 and 1993) is perhaps the best known. Until the 1988 decision to retrofit three 2000 MW coal-fired power stations with flue gas desulphurisation plant made the UK a *de facto* member, it was argued that adopting 1980 as the benchmark penalised the UK unduly since its emissions had been falling steadily since 1970, and the overall reduction would be more equitably achieved if this were to be taken as the base year. Less well publicised is the British insistence that action to curb acid deposition should not be concentrated on sulphur dioxide, but should give equal weight to reducing the discharge in oxides of nitrogen (produced when nitrogen is oxidised by the high temperatures which arise in power plants and in vehicle engines). This stance stemmed, in part, from the belief, advocated by the then Central Electricity Generating Board, that the die-back of the German forests could not be attributed to acid deposition alone and was better explained by damage from ozone (a form of oxygen which is produced in the lower atmosphere by the photochemical dissociation of nitrogen dioxide).

On the continent, an alternative explanation tended to be preferred. Named the 'Ulrich hypothesis' after its originator, it held that tree damage was attributable to the effect on the roots of toxic aluminium ions made more mobile by increased acidity in the soil. It is now accepted that, irrespective of reduced emissions of SO_2 and NO_x from power plants, protecting lakes and rivers will still require the application of lime in catchment areas where chalk-deficient soils cannot neutralise the sulphur background deposited over the past 150 years.

It should not be assumed that this scientific debate has been resolved or that the differences between the politicians derived solely from conflicting advice from their respective scientific advisers; national self-interest was only thinly disguised. But this should be seen as yet another occasion when political action, this time at the supranational level, cannot await unanimity among scientists.

An awareness of the vast areas of uncertainty in scientific understanding of the complexities of the long-term effects of man-made pollution on the ecosystem prompted the development in West Germany of the '***Vorsorgeprinzip***'. The translation – 'precautionary principle' – used in the English version of the Maastricht Treaty does scant justice to a sophisticated concept which seeks to set pollution control within the wider context of environmental management. Discharges should be reduced as far as technology permits for, even if such control represents an economically inefficient distribution of resources and cannot be justified in terms of immediate health benefits, the prospect of their causing as yet unsuspected and irreversible detriment to the ecosystem demands that we err on the side of caution. *Vorsorge* goes beyond the presumption, enunciated in the EC Second Community Action Programme on the Environment of 1977, in favour of regulating pollution at source rather than remedial measures applied in the environment. The level of regulation is, as we have shown in Section 3.3, usually linked to some notion of economic efficiency, but, in theory at least, *Vorsorge* takes precedence over this constraint.

The Framework Directive (84/360; EEC, 1984) on combating air pollution from industrial plants stemmed predominantly from the determination of the then Federal Republic of Germany to secure decisive intervention with regard to acid deposition. This essentially requires that industrial processes which involve emission of pollution to the atmosphere shall be required to satisfy statutory 'prior authorisation' procedures before being permitted to operate. A series of 'daughter directives' specify detailed requirements of the authorisation of particular classes of processes (Box 6.5 lists the daughter directives agreed by 1995), but the Framework Directive specifies the following general conditions which must apply before an authorisation can be made.

1 All appropriate preventive measures against air pollution must be taken utilising the principle of best available technology not entailing excessive cost (BATNEEC).

2 Any emission must not create 'significant' air pollution in the sense that it results in demonstrable additional damage to human health, building fabric or the viability of ecosystems.

3 Emission limits values must not be exceeded and air quality limit values must be taken into account.

This directive seeks to ensure that harmonised and stringent powers to regulate the major sources (power plants, metals production, mineral processing, chemical manufacture, waste incineration) of air pollution (especially SO_2 and NO_x, but also asbestos) exist throughout the EC. Its impact on UK pollution law needs to be examined.

Box 6.5 Daughter directives under 84/360/EEC

- The prevention and reduction of environmental pollution by asbestos (87/217/EEC)
- The limitation of emissions of certain pollutants into the air from large combustion plants (88/609/EEC)
- The prevention of air pollution from new municipal waste incineration plants (89/369/EEC)
- The reduction of air pollution from existing municipal waste incineration plants (89/429/EEC)

4.4 The Environmental Protection Act 1990

State intervention in the environment can never be understood simply by reference to the wording of legislation. Only after laws have been implemented and enforced for some time can their true effectiveness be gauged. As we have seen, the central concept of the Alkali Acts was never defined in the statutes, but evolved gradually, its interpretation and development falling to successive Chief Inspectors. Whether the 1990 Act will live up to its proponents' claim to be a watershed in UK environmental management remains to be seen. But it is undeniably an ambitious and intriguing mixture of old and new, foreign and domestic influences. Part I might best be summarised as an attempt to develop a distinctly British model of 'integrated pollution control' in a way which allows the obligations of various EC directives to be fulfilled.

European obligations include:

- prior authorisation;
- environmental quality objectives;
- emission limits;
- the BAT element of BATNEEC, i.e. *best available technology* not entailing excessive costs.

Domestic influences include:

- the NEEC element of BATNEEC, i.e. the best available technology *not entailing excessive costs;*
- the use of 'best practicable environmental option' (BPEO) for multi-media pollution problems;
- a 'residual duty' to apply BATNEEC to any aspect of a polluting process not covered by an explicit condition.

Under section 2 of the 1990 Act, the Secretary of State may, by regulations, prescribe substances (Box 6.6) or processes involving those substances to which the substantive provisions shall apply. Regulations under section 3 can be used to establish emission standards, which may vary with the location of sources, for prescribed processes and substances; this section also empowers the Secretary of State to introduce quality standards for substances in different environmental media, i.e. air, water and land.

Box 6.6 Substances prescribed by the Environmental Protection (Prescribed Processes and Substances) Regulations 1991

- oxides of sulphur and other sulphur compounds
- oxides of nitrogen and other nitrogen compounds
- oxides of carbon
- organic compounds and partial oxidation products
- metal, metalloids and their compounds
- asbestos (suspended particulate matter and fibres), glass fibres and mineral fibres
- halogens and their compounds
- phosphorous and its compounds
- particulate matter

Hitherto most effort has been devoted to the authorisation (section 6) of individual prescribed processes and sources of emission. In considering an application, the enforcing authority takes into account the competence and experience of the named applicant as well as the technological and managerial controls associated with the site. In determining the conditions (section 7) of the authorisation, the enforcing authority primary objective is to ensure that BATNEEC is used:

- to prevent or, where prevention is not practicable, to reduce to a minimum, and to render harmless any release of prescribed substances into an environmental medium,
- and to render harmless the release of any other potentially harmful substances.

In addition, the conditions of any authorisations must be framed so as to ensure compliance with any central policies relating to specific processes or substances and specified by the Secretary of State; for example:

- any obligations, under EC or other international law, relating to environmental protection;
- any emission limits;
- any quality standards or objectives;
- any quota system limiting national or industry sector totals of any substance.

Notwithstanding the non-statutory 'presumptive limits' used by HMIP's predecessors as indicators of compliance with 'best practicable means', the new regime is clearly far more comprehensive and prescriptive than that which it replaced. As before, the impossibility of total prevention of emission is implicitly recognised in the 'render harmless' obligation. This will continue to be met, where emissions to the atmosphere are concerned, primarily by ensuring dispersion via stacks of sufficient height. However, environmental pressure groups are likely to argue that reliance upon dilution runs counter to the 'precautionary principle' which, following Maastricht, is now embedded within the Treaty of Rome.

Despite the amount of attention devoted to the concept, BPEO is represented in the 1990 Act as just another objective underlying conditions of authorisations for processes designated for HMIP control and which are likely to involve releases to more than one medium. BPEO is not defined in the statute, but the Royal Commission, having conceived the idea in its fifth report, devoted its twelfth report to an elaboration of the concept.

> The BPEO procedure establishes, for a given set of objectives, the option that provides the most benefit or least damage to the environment as a whole, at acceptable cost, in the long term as well as the short term. (RCEP, 1988)

The concept should not be seen to be simply concerned with the choice of medium to receive the pollutant at the 'end of the pipe'; it could be used to illuminate policy choices, e.g. reprocessing versus dry storage of irradiated nuclear fuel, or investing in desulphurisation of coal-fired power plant versus gas-fired stations or even reduced energy consumption as a means of meeting national targets on sulphur emissions. The derivation of a 'BPEO Index' (Box 6.7) represents HMIP's approach to the implementation of this multi-faceted concept.

Box 6.7 Derivation of the BPEO Index

- The operator selects a range of processes or abatement techniques which are 'consistent with' the standards set out in HMIP guidance notes.
- For each process option identified, all releases to the environment are quantified.
- Using dispersion modelling techniques where appropriate, 'predicted environmental concentrations (PECs)' are calculated for each release in each medium.
- The tolerability of each in the long term will be judged by reference to the relevant standard 'environmental quality standard (EQS)'.
- For that great majority of pollutants with no EQS, HMIP defines a 'regulatory assessment level (RAL)'.
- Any PEC exceeding the appropriate EQS or RAL would be deemed 'intolerable' and the process option would be rejected.
- Any PEC below a specified action level – typically 10% of the EQS or RAL – would be deemed negligible and justifying only minimal additional expenditure.
- In the other cases, HMIP would focus on pollutants with the greatest impact, existing background levels being taken into account.
- For all significant releases, a 'tolerability quotient', i.e. the PEC divided by the EQS/RAL, will be calculated; the sum of the tolerability quotients for each medium is the BPEO Index for that medium (see Figure 6.19).

Retention of the term 'practicable' in BPEO implies a continued need to seek a balance between control and damage costs. By way of an illustration of HMIP's early approaches to the problem, Figure 6.20 relates to the derivation of BPEO for reducing SO_2 and NO_x from a hypothetical 2000 MW coal-fired power station. It shows four options.

Option A No flue gas desulphurisation (FGD) or NO_x abatement

Option B Spray-dry FGD + low NO_x burners

Option C Spray-dry FGD + selective non-catalytic reduction (SNCR)

Option D Spray-dry FGD + selective catalytic reduction (SCR)

Of the four options, D (spray-dry FGD + SCR) is the least environmentally damaging – ammonia is introduced into the gas stream, which in the presence of a catalyst reduces the NO_x to N_2. However, HMIP would accept option B as BATNEEC for this process since this option is located at a point on the graph where a further annualised cost of £25 million yields a relatively small decrease in pollution potential as measured by the BPEO Index.

Industrialists were not slow to point out the importance placed upon existing background concentrations in the calculation of tolerability quotients. Given its limited brief under existing legislation, HMIP is obliged to exclude from the calculation of the BPEO Index the cost associated with any solid waste generated. This is a not insignificant consideration for large combustion plant where FGD plant can consume large amounts of

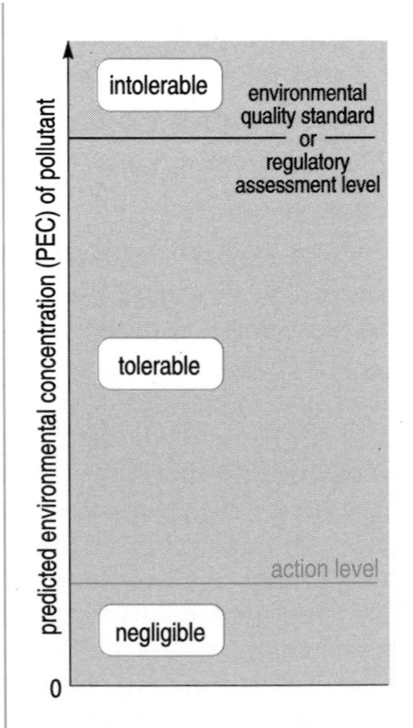

▲ *Figure 6.19 Tolerability, BPEO.*

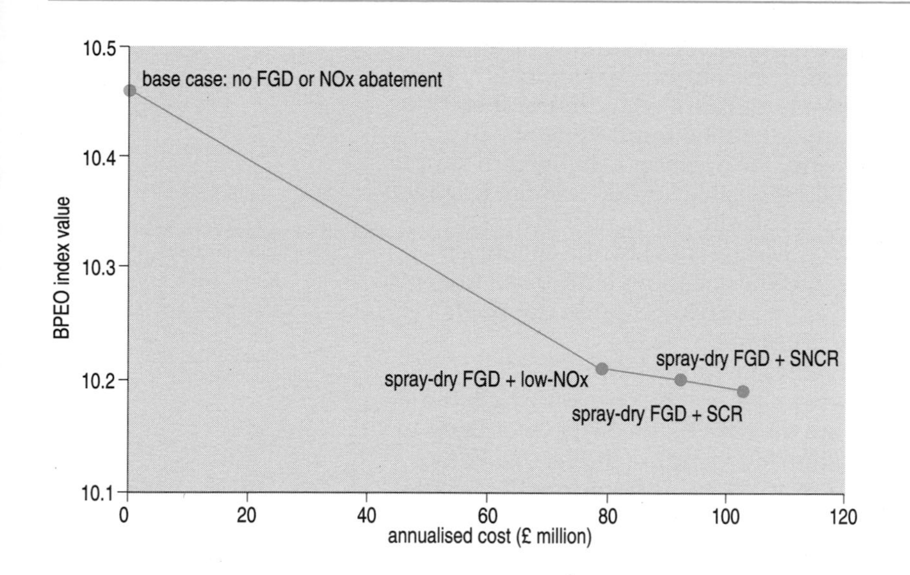

Figure 6.20
BPEO Index: an example.

limestone and generate quantities of gypsum (calcium sulphate) far greater than can be utilised in the manufacture of plasterboard.

Despite these major methodological issues, attempting to add a measure of quantitative argument to the authorisation procedure is generally seen as an advance on relying upon technical experience and judgement which characterised the former regime. Operators of prescribed processes might question whether the enormous amount of time required to complete an IPC application adequately and the huge fees payable (the average application fee was £11 250 per process in 1993/94) can be justified. However, the statutory obligation (section 20 of the 1990 Act) on the enforcing authorities to maintain public registers on applications and subsequent authorisations ensures that the earlier criticisms of decision-making behind closed doors and a too cosy relationship between regulator and industry are no longer sustainable in respect to pollution control.

Activity 4

Discuss the impact of membership of the EU on the British approach to regulating industrial sources of pollution and hazard.

4.5 Controls over major sources of chemical hazard

Reference has already been made to the Notification of Installations Handling Hazardous Substances Regulations 1982 in Section 3.2. The obligation to notify the Health and Safety Executive of the use of certain substances predated the Seveso Directive (EEC, 1982) although it forms part of the latter's UK implementation. But it was with the advent of the Control of Industrial Major Accident Hazard Regulations 1984 (CIMAH) that compliance with the aims of the Seveso Directive was really achieved.

The CIMAH Regulations seek to regulate those 'top-tier' sites which pose the greatest threat to the off-site population. There are some 2050 large inventory top-tier sites and over a thousand with a small inventory. The primary requirement of CIMAH is the preparation of a survey, often termed the 'safety case', of risk-minimising measures. The survey is validated by the Health and Safety Executive (HSE), who can use their powers under the 1974 Act to ensure any identified improvements. In addition to the submission of a safety case, the operator is obliged to supply the relevant county council (or the body responsible for the local fire brigade) with information to enable it to draw up an off-site emergency plan. This requirement for the release of information was resented by the chemicals industry which, as with information on routine emissions authorised under pollution legislation, views public disclosure as giving commercial rivals access to data which can enable them to learn trade secrets.

Given their greater potential for a release with catastrophic consequences, the large inventory top-tier sites are the ones on which the HSE concentrates its attention. These are the sites that have been obliged to release information to the local population surrounding the plants. Before CIMAH, third parties enjoyed no right of access to such information as the nature of the inventory, the risks posed (e.g. explosion, toxic cloud) and emergency plans (see, for example, Figure 6.21). However, the amount of information released has proved, in many cases, to have been the minimum required to satisfy the regulations.

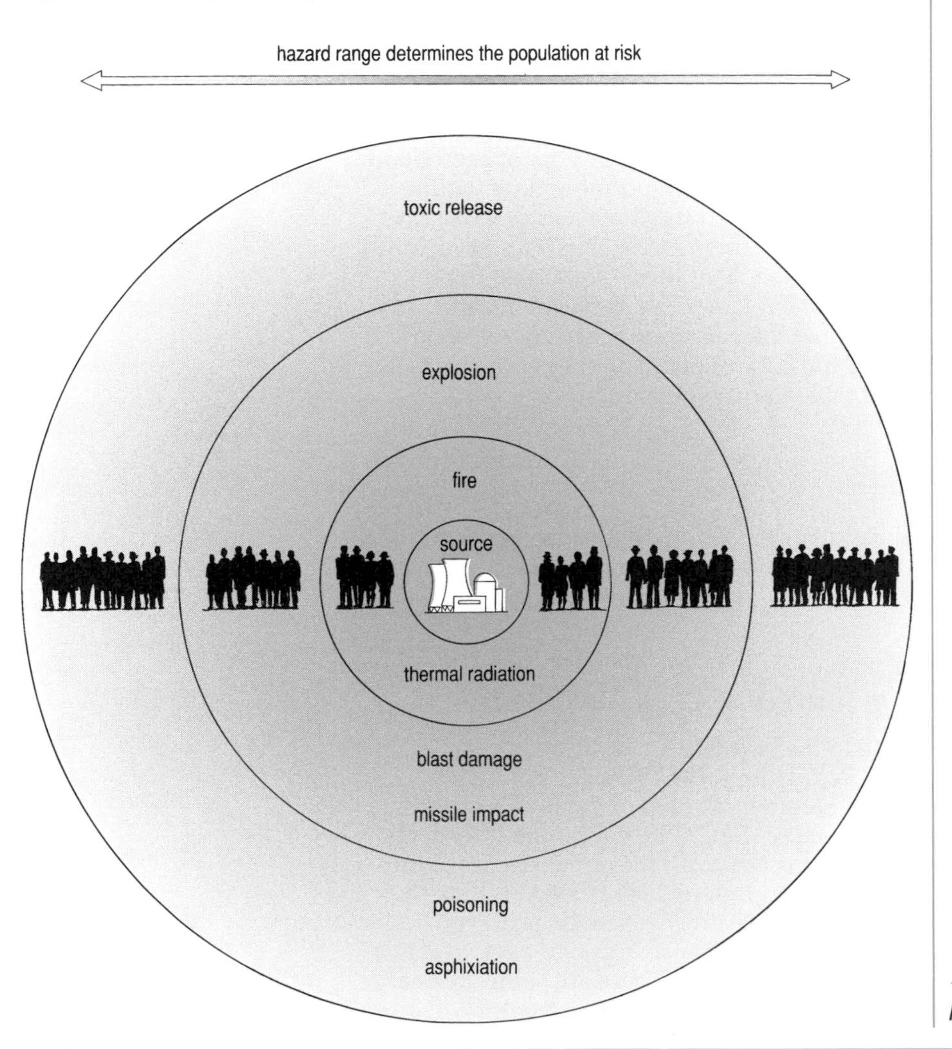

Figure 6.21
Main types of hazard event.

Data on the chemical inventories, the processes and the control mechanisms enable estimates of the risk (probability of incident and its consequences) to be calculated, a process termed **risk analysis**. Box 6.8 gives, as an example, the quantification of the release of chlorine from a storage depot. These estimates are central to any determination of planning consent (under the Town and Country Planning Act 1990) for a new hazardous plant or for housing or other vulnerable development in the vicinity of an existing hazardous installation. The control of information has been a critical factor in the political conflict surrounding major hazard sites. In many cases local residents' groups have been denied access to relevant data and later had their fears dismissed by industry on the grounds of their incomplete database. Where data have been more generally available, industry has used its greater resource of technical expertise to undermine the credibility of 'pressure group' analyses. Open information flows are more likely when the planning process involves a public inquiry, where central government's obligation to defend its final decision encourages it to require industry to support its case with empirical evidence, irrespective of the value of such data to commercial competitors.

The multi-disciplinary techniques of **risk assessment** (Figure 6.22) had been exposed to public scrutiny during the hundred day 'Windscale' inquiry into planning consent for the thermal oxide nuclear fuel reprocessing plant at Sellafield in 1977. The Canvey Island public inquiries and the subsequent HSE studies marked the first time that its application to non-nuclear risks was subject to public examination and debate in Britain. Serious attention by social scientists into the factors which determine why certain risks are accepted while others generate resentment and public outcry can be said to date from the seminal study entitled 'Social benefit versus technological risk' by Starr (1969) into the diffusion of nuclear technology. Since then **risk acceptability** has become one of the most challenging issues for contemporary social science.

In the 1970s, the nuclear industry in many industrialised countries became increasingly frustrated by the public's refusal to accept the overall risk posed by nuclear power (see Chapter 7). Calculations showed that the risk of death to an individual member of the public from incidents at UK nuclear installations to be 10^{-6} per year – a figure which is not only negligible in comparison with probability of death from natural causes and from non-technological hazards such as drowning (Figure 6.24), but, more to the point, is arguably outweighed by the attendant benefits of nuclear power.

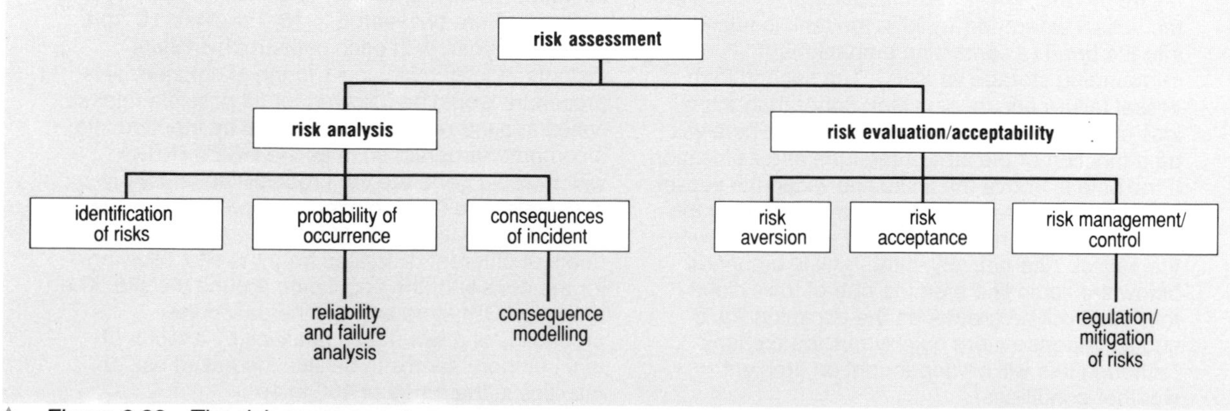

Figure 6.22 The risk assessment process.

At first, attempts were made to distinguish between 'objective risk', the output of formal, probabilistic risk assessment, and what was labelled 'perceived risk' or that intuitive apprehension sensed by lay persons either unfamiliar with, or indifferent to, modern science and technology. But then sociologists of science pointed out that formal risk analysis of complex technologies involved prior assumptions which were not capable of empirical validation and which must be thought of as an inevitable part of science and departures from proper procedures of good science. Empirical research revealed that social attitudes to risk, far from being simply irrational, were determined by the extent to which decision-making procedures (i.e. planning and related approvals) and the overall

Box 6.8 Risk analysis in practice

The nature of the problem

A local planning authority have been asked to grant planning permission for the construction of a DIY store at the edge of an urban area. While such a proposal may seem innocuous in itself, the proposed site is less than 400 m from a large industrial complex which includes a major oil refinery, a bottled gas plant and a water treatment works where chlorine is stored in 30-tonne pressurised tanks. The latter installation causes most concern about fatalities within the proposed store.

Characteristics of the plant

Chlorine is brought into the plant aboard 20-tonne rail wagons and is stored in three pressurised tanks (Figure 6.23a). In addition there is a network of pipes which feed into and out from the tank. It is possible, therefore, to make some assessment of the likely failure modes for the system and this done in the simplified fault tree shown in Figure 6.23(b).

The 'top event' is a release of chlorine into the environment. This can occur as a result of a number of failure possibilities which can act in unison or independently of each other. If we take the vessels first then it is possible to see two main failure modes. The first is a catastrophic failure of the vessel where the liquid in the tank is released into the bund (a containing embankment surrounding storage vessels). The second form of vessel failure occurs as a leak rather than a total loss of containment. The rate of release here will be a function of the size of the hole and its location. If the hole is above the liquid line within the vessel then the rate of release will be determined by the flow of gas and the rate of liquid evaporation within the vessel. Alternatively, if the tank is breached below the liquid line then the rate of toxic cloud formation will be greater as the escaping liquid should vaporise more quickly outside the tank (although this will be dependent on ambient weather conditions).

Another failure mode could be through a pipeline breach which again may be catastrophic (guillotine) or gradual (split and gasket failure). For all cases the ability of plant operators to cut off the flow of chlorine to the point of release will be of importance.

The next two failure possibilities can be considered together as they are concerned with the import and export of chlorine. In both cases, perhaps the greatest risk of release arises as a result of a failure in the transfer hose on the coupling mechanism. There is also the risk of vessel impact (due to road or rail tankers being brought close to the tank) and supply vessel failure (which can be considered in the same way as the pressurised tank itself). Finally, it is possible to consider any other likely initiating events such as pump or vaporiser failure.

Risk analysis

Once the fault tree has been drawn up, it is possible to attach probabilities of failure to each event. The data for this part of the risk analysis are usually drawn from reliability data banks which record details of different types of failure of pump, flange vessel, etc. and their frequency. For example, for a catastrophic failure we may find that the rate could be somewhere between 5×10^{-6} and 4×10^{-6} per vessel per year, with each catastrophic failure instantaneously releasing 1 tonne of chlorine. This procedure would be followed for all possible release cases and the resultant data would be inputted into a computer program such as the HSE's RISKAT, which would generate the probabilities and likely consequences for a chlorine release from the site. Further calculations would indicate the probability of death at different distances from the site, allowing for the density of the population around the site. The proposed DIY store actually increased the probability of a few dozen fatalities by a factor of ten, but there seems to be little increased risk of fatalities in the range of 400 to 10^4.

distribution of hazards were felt to be equitable.

How the state regulates those sources of hazard for which 'hard' scientific understanding is far from complete is now seen as the key issue. The regulatory process generally adheres to the 'techno-rational' approach in which formal estimates of probability and consequence are central to the decision-making process. In 1994 the planning consent for a hazardous waste incinerator, which had attracted considerable local opposition, became the subject of ruling by the Court of Appeal. One crucial passage (dismissing the planning authority's legal challenge to the Secretary of State's decision to approve the development) gets to the nub of the problem on which this chapter has been focused:

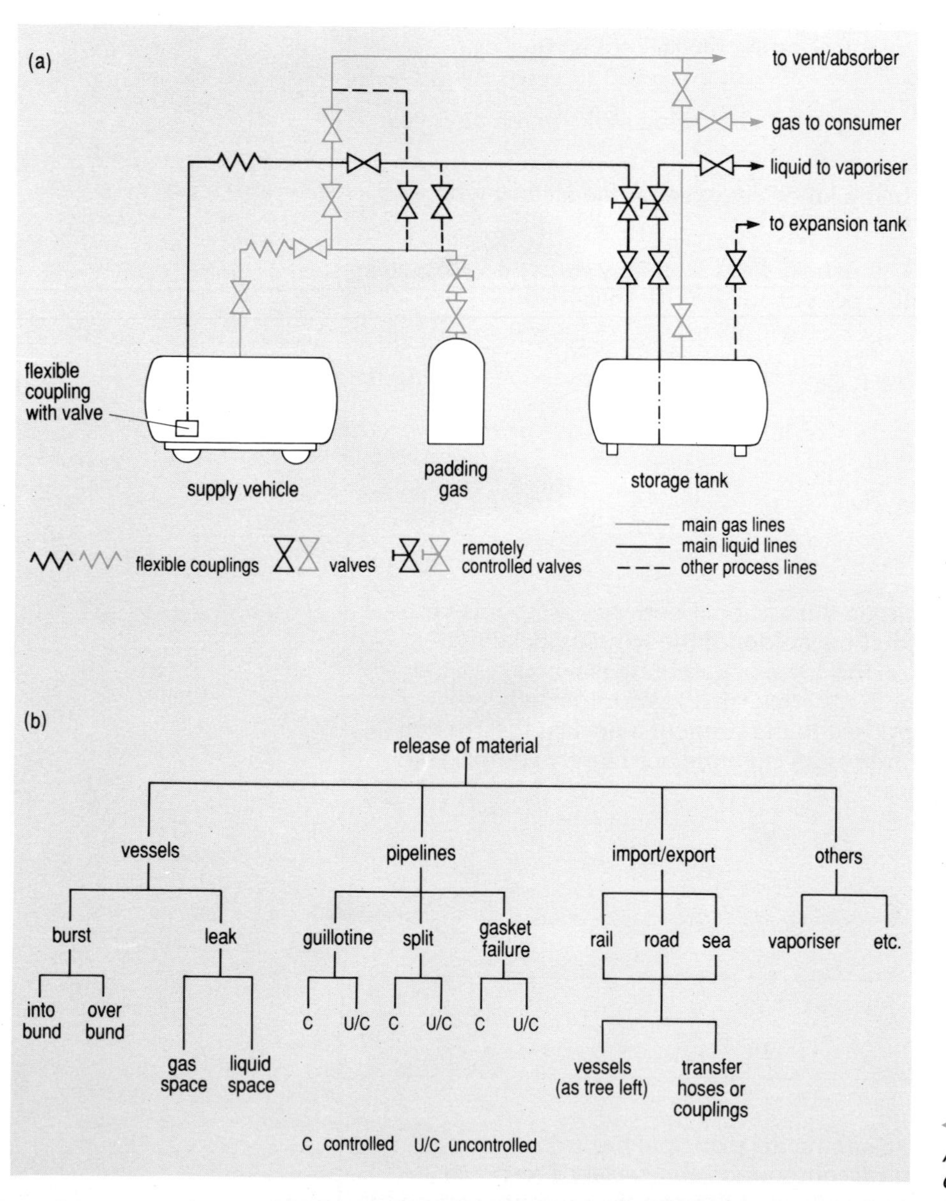

Figure 6.23(a) A chlorine installation (not showing pipe sizes, instrumentation, etc.). Liquid chlorine is pushed from the supply tank by pressure from the padding gas supply through a series of flexible couplings, pipes and valves to one of three storage tanks. Gas given off by the liquid chlorine in the storage tanks can then be piped to the consumer. Tanks and flexible couplings, pipes and valves (joined by gaskets) will each have some propensity for possible failure.

Figure 6.23(b) A simplified fault tree for a chlorine installation.

> Public concern is, of course, and must be recognised by the Secretary of State to be, a material consideration for him to take into account. But if in the end that public concern is not justified, it cannot be conclusive. If it were, no industrial development – indeed very little development of any kind – would ever be permitted. (Lord Justice Glidewell in *Gateshead MBC v. Secretary of State for the Environment*, Court of Appeal, 12 May 1994)

But of course, the questions still remain: justified by whom and against what criteria?

Q The probability of any one entry winning the National Lottery jackpot (i.e. randomly selecting six numbers from forty-nine) is almost one in 14 million or alternatively 7 chances in 100 million or 7×10^{-8}). Using Figure 6.24, calculate how many separate lottery entries I must buy in order that the probability of my winning the jackpot is roughly comparable to:

A the probability of my dying from a lightning strike in the next year (about 6×10^{-7} per year);

B the probability of my being killed next year at the factory where I work (about 4×10^{-5} per year);

C the probability of my dying next week from any cause, if I am a 50 year old man (about 1×10^{-2} per year or 2×10^{-4} per week).

A **A**: about 9.

B: about 600.

C: about 3000.

Activity 5

Since the probability of winning the National Lottery's jackpot is far lower than the probability that the holder of the winning ticket will die before enjoying the prize, this form of gambling is wholly irrational. The **NIMBY** ('not in my back yard') syndrome, whereby residents are implacably opposed to the siting of a nuclear installation or chemical works in their area, is an example of a closely comparable form of irrationality. Discuss.

5 *Conclusion*

This chapter has attempted to examine the principal hazards associated with the manufacture of chemical compounds and society's response to those hazards. The dangers of uncontrolled fire and the nuisance caused by

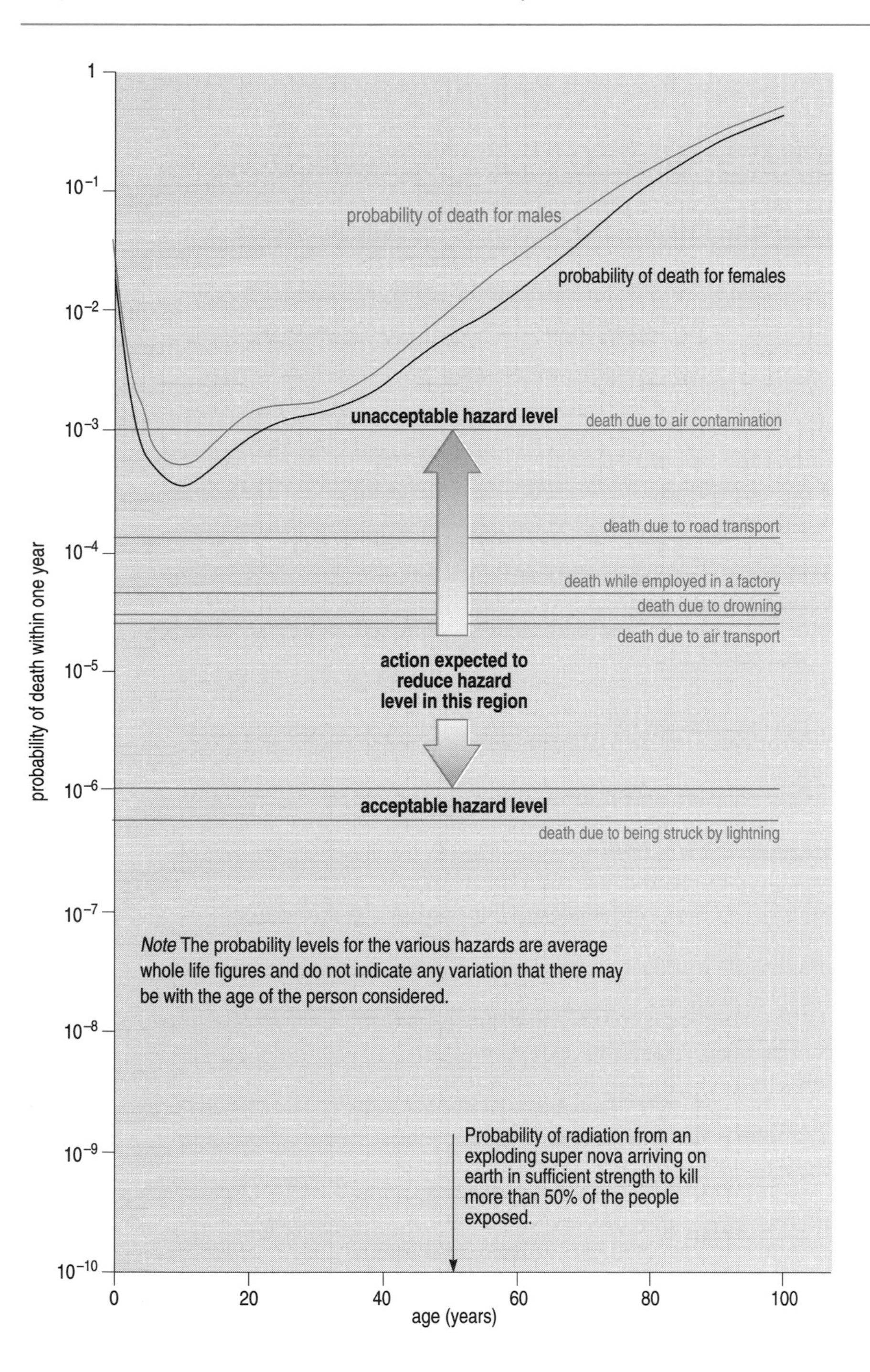

Figure 6.24 Non-technological hazards.

smoke have been apparent since man's first attempts to smelt those minerals which produced bronze. Awareness of other hazards, such as the toxic nature of dioxins and benzene, is a far more recent phenomenon. Some consequences of global pollution, such as ozone depletion, arise only at the levels of industrialisation experienced in the developed world today. In addition, understanding of such problems requires a level of scientific knowledge and techniques which has only recently been available.

As the variety and scale of hazards have expanded with the inexorable pace of industrialisation, so have society's attempts, sometimes spurred by major calamities, to address their consequences. The torts of trespass and nuisance have been used since before the reign of Henry II to give redress to owners of land whose enjoyment of which has been compromised by smoke, fumes, odour, fire and explosions arising from a neighbour's property. The rational allocation of land and the separation of incompatible land uses had been used to mitigate the consequences of industrial hazards for centuries before the present one, when these practices formed the basis of what became a succession of town and country planning legislation measures.

However, the processes of mediation between environmental protection (and thus the needs of the public) and the interests of the chemicals industry have not always guaranteed equitable outcomes in policy terms. In the notable example of the use of risk analysis in decision-making, the disproportionate power of the chemicals industry to control the flow of information and to deploy technical expertise to the advantage of its own case is evident enough.

But when the consequences of industrial emissions are truly global, the location of their sources are of secondary importance. Only the reduction of emissions by all states, in accordance with international agreements backed by the moral authority of international law, can alleviate the problem. Natural dispersion of chemical species in rivers and the atmosphere and the resulting 'transfrontier pollution' serve to strengthen further the role of supranational bodies, such as the European Union, in addressing contemporary environmental problems.

The historical focus adopted in the chapter was intended to dispel the compelling but erroneous view that there was once some 'golden age' to which we can return once our politicians have established the 'right' regulatory regime and our scientists have perfected the ultimately 'clean technology'. Each stage of industrialisation, each advance in chemical manufacture, has brought its attendant problems, but these have been seen as clearly outweighed by the unimaginable increase in human welfare which freedom from famine and disease afford.

It is only with the revolution in environmental consciousness in the past few decades that this 'balance' has been called into question. Both environmental pressure groups and other less formal local alliances have shown an increasing willingness at public inquiries to subject plans for new chemicals plant to the close critical analysis of what seems to them to be a Faustian bargain. This is not to imply that the chemicals industry must always be perceived as being pushed reluctantly into ever tighter legal constraints on their activities. More recently, some of the most astute companies have attempted to keep a step ahead of such controls, realising that generally acknowledged higher standards of environmental practice can increase their processing efficiency or their marketing capability.

For example, take the case of the use of catalysts – important agents which have long been used to speed up chemical reactions. Research has now produced a new range of catalysts for certain processes which are not only recoverable and can be reused (unlike conventional catalysts), but because they are more efficient, can be employed in smaller quantities, so reducing waste at source. Thus environmental bonuses have been neatly tied into reductions in operating costs.

At the same time, some chemicals manufactures are building formal alliances between their suppliers, customers, hauliers, distributors and others involved in the supply chain, in order to meet increasing demands for safer and more environmentally acceptable products. Here 'green' credentials can offer a very positive marketing advantage.

However, in spite of such individual examples of improved environmental practices and the claims by such bodies as the Chemical Industries Association that 'a voluntary approach is the most efficient and cost effective approach' to environmental constraints and the achievement of sustainable life-styles, other less euphoric voices can be heard. Indeed, there is a growing recognition, particularly among the scientific community, that the scale of ecological problems is such that any discussion of the costs and benefits (practicability) of pollution and its control can only serve to delay the implementation of *effective* remedial measures. What is required, it is argued, goes far beyond the traditional system of consents, inspections and technical advice exercised over specific industries such as that of chemicals and within the confines of individual states or the EU. It has little to do with the self-imposed disciplines of individual companies or industrial organisations. The need is for environmental and resource constraints to be incorporated into all policies, but especially those relating to industry, energy and transport, preferably inside an agreed world framework. It also requires politicians and public alike to cease to delude themselves that irreversible ecological damage can be averted without infringing individual liberty.

References

EUROPEAN ECONOMIC COMMUNITY (1980) 'Directive on air quality limit values and guide values for sulphur dioxide and suspended particulates', EEC Directive 80/779 (Air Quality Directive), *Official Journal of the European Community*, L229 30.8.90.

EUROPEAN ECONOMIC COMMUNITY (1982) 'Directive on the major accident hazards of certain industrial activities', EEC Directive 82/501 (Seveso Directive), *Official Journal of the European Community*, L230 5.8.92.

EUROPEAN ECONOMIC COMMUNITY (1984) 'Directive on combating of air pollution from industrial plants', EEC Directive 84/360 (Framework Directive), *Official Journal of the European Community*, L188 16.7.84.

EUROPEAN ECONOMIC COMMUNITY (1986) 'Directive on the protection of the environment and in particular of the soil, when sewage sludge is used in agriculture', EEC Directive 86/278, *Official Journal of the European Community*, L181 4.7.86.

HEALTH AND SAFETY COMMISSION (1976) *Advisory Committee on Major Hazards, First Report*, HMSO, London.

HEALTH AND SAFETY EXECUTIVE (1992) *The Tolerability of Risk from Nuclear Power Stations*, HMSO, London.

MINISTRY OF HOUSING AND LOCAL GOVERNMENT (MHLG) (1967) *103rd Annual Report on Alkali, etc., Works 1966*, HMSO, London.

ORGANISATION FOR ECONOMIC CO-OPERATION AND DEVELOPMENT (OECD) (1976) *Economic Measurement of Environmental Damage*, OECD, Paris.

ROYAL COMMISSION ON ENVIRONMENTAL POLLUTION (1976) *Fifth Report: Air Pollution Control: An Integrated Approach*, Cmnd 6371, HMSO, London.

ROYAL COMMISSION ON ENVIRONMENTAL POLLUTION (1984) *Tenth Report: Tackling Pollution – Experience and Prospects*, Cmnd 9149, HMSO, London.

ROYAL COMMISSION ON ENVIRONMENTAL POLLUTION (1988) *Twelfth Report: Best Practicable Environmental Option*, Cmnd 310, HMSO, London.

STARR, C. (1969) 'Social benefit versus technological risk', *Science*, **165**, pp. 1232–8.

WORLD HEALTH ORGANISATION (1972) *Air Quality Criteria and Guides for Urban Air Pollutants: Technical Report Series No. 506*, WHO, Geneva.

Further reading

ASHBY, E. (1978) *Reconciling Man with the Environment*, Oxford University Press, London.

ASHBY, E. & ANDERSON, M. (1981) *The Politics of Clean Air*, Oxford University Press.

BRIMBLECOMBE, P. (1987) *The Big Smoke: a history of air pollution in London since mediaeval times*, Methuen, London.

DRAKE, C. D. & WRIGHT, F. B. (1983) *Law of Health and Safety at Work: the new approach*, Sweet and Maxwell, London.

FRANKEL, M. (1974) *The Alkali Inspectorate: the control of industrial air pollution*, Social Audit, London.

HAIGH, N. (1987) *EEC Environmental Policy and Britain*, 2nd edition, Longman, London.

KARGON, R. H. (1977) *Science in Victorian Manchester: enterprise and expertise*, Manchester University Press, Manchester.

MILLER, C. E. & WOOD, C. M. (1983) *Planning and Pollution*, Oxford University Press, Oxford.

ORGANISATION FOR ECONOMIC CO-OPERATION AND DEVELOPMENT (OECD) (1976) *Economic Measurement of Environmental Damage*, OECD, Paris.

ROBENS, Lord (1972) *Report of the Committee on Health and Safety at Work* (Cmnd. 5034), HMSO, London.

WOHL, A. (1983) *Endangered Lives: public health in Victorian Britain*, Dent, London.

Answers to Activities

Activity 1

The major air pollutant for 150 years up to 1960 was the smoke from domestic fires. In the nineteenth century the steam engine added its own prodigious quantities of gas and ash. Industrial processes, especially those associated with the growing chemical industry, produced severe pollution problems both from production itself and from the disposal of waste. The severity of these problems was intensified by the density of towns, and the continued location of industrial plant within urban areas and closely intermingled with houses, commercial premises, and other, often potentially dangerous, factories.

The continuance of these pollution problems was in many respects due to Britain's sophistication as an industrial nation and to her continued development of new products and materials based on her chemical industry. It is impressive that the growing urban population as well as the industrial sector could still be supplied with fuel. In many cases, alternative less polluting methods of production were simply not available. In others, however, the technological argument carries little weight – indeed the technology to reduce emissions of hydrogen chloride gas from soda and potash works had been available for more than a quarter of a century before the Alkali Act of 1863 introduced its controls.

Government intervention was limited by the prevailing policy of *laissez-faire*. Confidence in market mechanisms to operate in the public interest and concern about the detrimental effect of state regulation discouraged government action. Social welfare might in particular circumstances be regarded as a special case, but interference with the industrial sector posed fundamental problems. Even when the Alkali Inspectorate was set up, it was obliged to consider economic as well as technical feasibility. Local government was more interventionist than central government, but councillors were also sensitive to the 'town's interest' especially because councillors were often also industrialists. Apart from any direct interest, there was also concern about the effects of pollution control on local employment, which, if it fell, would lower rateable values at the same time as raising the need for welfare.

Local electorates were sometimes moved to form pressure groups to agitate for control of smoke and noxious emissions, but environmental concerns were not politically significant issues. Pollution was too all-pervasive and other problems were too acute for the former to gain political significance. Serious though the problems were, their place on the political agenda was low. Individuals who wished to take action had to rely on the Alkali Inspectorate, the nuisance action in the magistrates' court, or a private complaint to the police.

Major sources of air pollution in the nineteenth century were:

1 domestic chimneys;

2 smoke-fired steam engines;

3 alkali works;

4 other industries and especially brick-making, pottery, coking and smelting.

All these potential sources were intensified by the concentration of population and industry into towns of unprecedented size.

Pollution persisted because of the following major factors.

1 It was difficult to establish origins of pollutants because of concentration of industry in congested areas.

2 Pollutants in air disperse over a wide area and form compounds with other pollutants making identification difficult.

3 There was a lack of knowledge both about pollutants themselves and their environmental and health effects.

4 Design expertise to reduce the pollution potential of industrial processes was slow to develop and to be adopted. Cost–benefit of pollution control was at best rudimentary.

5 Alternative materials to reduce air pollution (e.g. to replace coal as a domestic fuel) were not available.

6 Positive attitudes to pollution and risk often prevailed – at worst they were accepted as inseparable from life in an industrial society.

7 Local government in industrial towns emerged only from 1835. Business people were regarded as the most suitable to serve as councillors; pollution control was therefore often weak, relying on the co-operation of local manufacturers.

8 National government followed the general principle of *laissez-faire*. Although this ideology was modified to permit intervention in some areas of welfare, industry was regarded as an area where government should not intervene.

Activity 2

The twentieth century saw the development of the chemicals industry away from its initial reliance on local salt and lime-producing alkalis for the growing soap, glass and textile industries. The post second world war boom in petrochemicals gave the industry its present emphasis on organic chemicals based largely on oil-derived feedstocks. These changes altered the nature of the problems posed to the environment – away from the more visible kinds of air pollution towards water pollution, and away from regular discharges to the risk of a major hazard through explosions such as that at Flixborough on Humberside.

The changes in industrial processes were also accompanied by locational and structural changes in the chemicals industry. A large number of small specialist firms were amalgamated into a few, very large integrated companies. Sites expanded many times to accommodate new plant. While some still occupy large sites surrounded by residential areas, others occupy relatively remote sites on large estuaries open to tankers. These locational changes make the calculation of the risk posed by a major hazardous event even more difficult. Environmental controls may be applied more easily to an industry made up of large corporations, especially as they can be more concerned with their safety record and public image. Their major contribution to the national economy, however, means that regulation to protect the environment must also recognise the economic aspects of the public interest.

Activity 4

On entry to the European Community in 1973, the process was very much two-way. Compliance with certain directives setting standards for the quality of surface water, groundwater and drinking water was alien to the British tradition of pragmatism. But with solid waste, Britain was able to assume the lead in shaping the directives so as to accord with existing domestic arrangements.

But gradually the environmental aims became inculcated within an antipathy shown by Thatcher governments towards the wider economic and political objectives of the EC. A demand that the EC should confine itself to the creation of a 'common market', as described in the 1957 Treaty of Rome, resulted in a greater scrutiny of those environmental directives which could not be easily interpreted as promoting 'free trade'. Air quality objectives were introduced into British law with very few politicians aware of the significance of this innovation. Given public apprehension of major hazards, like the dioxin release at Seveso in Italy, directives aimed at reducing their risk were accepted with little opposition from the Council of Ministers. But when the continental member states demanded massive reduction in emissions of atmospheric pollutants from industrial sources, irrespective of the ability of the individual plant operators to pay for the necessary technology, this represented too drastic a change from the traditional UK practice. In the 1980s Britain acquired the title of 'the dirty man of Europe'.

With the Single European Act in 1987, the EC no longer had to conceal its environmental objectives under a 'free trade' veneer. In 1992 the Treaty of Maastricht went even further; it ensured that the quality of the environment and the pursuit of sustainability became part of the very aims of the European Union. Once again this radical change went largely undebated by the Westminster Parliament. This was partly because most attention was focused on the 'Social Chapter', which although primarily concerned with the social and economic welfare of workers, also sought to safeguard the quality of the working environment. It must also be remembered that Maastricht also introduced the principle of subsidiarity – action is to be taken at the EC level only where that is more effective than national action. To those worried about how environmental policy has become a supra-national concern, this principle represents a powerful brake. But as we have seen, often it was the *transboundary* movement of certain forms of pollution which led the EC originally to engage in these matters. An EC-wide attack on acid rain, for instance, or a campaign to clean up the North Sea cannot be reconciled with subsidiarity. This principle is unlikely to lead to a reversal of the major innovations such as integrated pollution control, but it could be invoked to reduce the pace of environmental protection in the future.

Activity 5

The rationality of reactions to extreme events must not be viewed solely in terms of their (very low) probability. It is necessary also to pay attention to the consequences, both real and perceived, for the risk-taker and for the wider society. Consider first the case of the National Lottery.

The mathematics of probability enable us to have an entirely objective measure of the chance of any one entry winning that week's jackpot. Death certificates, recording cause and age at death, have been a statutory requirement for over 150 years. It is possible to make meaningful statements about death rates for men or women of a given age (that is, such statements could in theory be verified by reference to the Registrar General's data). A fifty-year-old man or woman has a chance of dying in any given week that far exceeds the chance of their (solitary) lottery entry winning the jackpot. But even if our hypothetical gambler is an actuary and fully aware of these and other probabilities, his or her behaviour cannot be dismissed as irrational. The satisfactions gained – quite apart from the prospect of enormous wealth – from the weekly purchase of a ticket are many and varied:

- the crescendo of excitement as the Saturday evening draw approaches;
- the knowledge that, even if they do not live to enjoy any prize, their children will be greatly enriched;
- the knowledge that a fraction (albeit small) of the take money goes to 'worthy causes'.

By the same token, residents' responses to the prospect of a nuclear or chemical plant in their locality cannot be understood simply in terms of the very low probability of their being killed following an accident at that plant. The plant may well be an important source of employment and its products may be vital to the national economy. But 'risk aversive' residents are hardly being irrational by calling for the site to be built elsewhere: they can then enjoy those benefits without incurring the risks. Among the many attributes of catastrophes at chemical plant – other than risk of individual death – which add to their unacceptability are:

- multiple fatalities (i.e. one incident involving 100 deaths is viewed with far greater horror than 100 separate incidents each involving a single fatality);
- the prospect of long-term damage to subsequent generations by mutagenic substances escaping into the environment;
- a sense that the benefits are widely distributed whilst the risks are spatially concentrated.

Successive occurrences at chemical works (Flixborough, Seveso, Bhopal) tend to undermine public confidence in the basis of 'scientific' risk analysis. This tendency has been reinforced by the succession of transport disasters (King's Cross Fire, Clapham rail disaster, the sinking of the ferries *Herald of Free Enterprise* and *Estonia*). Popular belief that 'accidents will happen' is very resistant to probabilistic reasoning.

Chapter 7 The politics of radioactive waste disposal

1 Introduction

As you read this chapter, look out for answers to the following key questions.

- What are the distinctive features of radioactive waste management?
- What are the sources and characteristics of radioactive waste?
- Why has radioactive waste shifted from being a non-problem, to a technical problem, to a political problem?
- What are the political consequences and possibilities of radioactive waste management policies?
- To what extent have the political responses to radioactive waste posed an environmentalist challenge to conventional politics?

You have already encountered the nuclear industry at a number of points in this series. Right at the beginning, in Chapter 1 of *Sarre & Reddish, 1996*, the controversy surrounding the radioactive discharges from Sellafield in Cumbria was introduced, and in Chapter 2 in this book the problems of identifying and measuring risks from that plant were discussed. In Chapter 1 of this book the process of energy conversion from nuclear fission was described and nuclear energy was placed in the context of the world's energy balance. Some of the problems with nuclear energy, such as its high cost, limited reserves of uranium, accumulating wastes and the danger of proliferation of plutonium, were considered in Chapter 2.

In terms of risk and potential environmental impacts the nuclear industry has much in common with other large-scale technological systems such as the chemical industry considered in the previous chapter. As we saw, such industries are capable of devastating destruction and widespread impact. The various stages of the nuclear fuel cycle* present different degrees of risk to the environment. In this chapter we shall focus on the 'back-end' of the nuclear cycle: the management of nuclear wastes. In many ways this is the nuclear industry's most intractable problem. Nuclear waste also has some distinctive characteristics which make it as much a political as a technical problem, as we shall see. Our aims will be:

1 To indicate the distinctive features of radioactive waste management compared with the management of other hazardous and toxic materials. The comparison can be partly understood by identifying the environmental impact of nuclear waste. This is a question that can be addressed by scientific inquiry but is also a matter of perception.

2 To identify the sources and characteristics of radioactive waste and to examine in what respects radioactive waste constitutes an environmental problem. A distinctive feature of the nuclear industry and, increasingly, of nuclear waste, has been its political salience. It provides a good example of the ways in which environmental problems become politicised and a cause of conflict.

3 To show how radioactive waste has shifted from being a non-problem to a technical problem to a political problem. To do this we shall need to look back at the history of the nuclear industry from its military origins in the United States to the development of electricity generation and reprocessing. Contemporary political conflict over nuclear waste management will be examined to understand the local, national and international dimensions of the political problem of nuclear waste. The focus will be on the UK but reference will be made to other countries.

* We use the phrase 'nuclear fuel cycle' in this book without endorsing the nuclear industry view that nuclear materials are best recycled within a closed circuit (after mining). Anti-nuclear groups tend to describe the same process as a 'chain'.

4 To evaluate the political consequences and possibilities of radioactive waste management policies. From the discussion of conflict in the UK some interesting conclusions can be drawn. It will be clear that political conflict has circumscribed the technical and locational options for radioactive waste management. The nuclear industry has been retreating in the face of political opposition. This opposition has derived its power from an ability to mobilise effective coalitions notwithstanding differing 'political cultures'. Awareness of the potential hazards of nuclear waste has been translated into action that has influenced policy-making.

5 To examine the extent to which the politics of nuclear waste illustrates a developing environmental challenge to conventional political assumptions and arrangements. Radioactive waste is a major environmental problem, presently confined to the 'nuclear nations' but extending down the generations. It indicates the constraints and possibilities of political action both locally and internationally. The political problems surrounding nuclear waste suggest that solutions to this and other global environmental problems will be hard to achieve.

Before beginning our analysis, we should emphasise that because our focus is on the politics of nuclear waste, we only present enough technical information to show what the debates are about. We do not intend to try to resolve the debates or to argue for any particular technical interpretation or solution of the problem of radioactive waste disposal.

2 *What are radioactive wastes?*

2.1 *The nuclear fuel cycle*

In order to understand why radioactive wastes are such a distinctive and possibly intractable political problem we must first define the origins and categories of these wastes. Only by knowing something about the materials we are dealing with, and their environmental implications, can we begin to explain the political significance achieved by nuclear waste. Nuclear wastes are the unwanted by-products of the processes of the 'nuclear industry'. The nuclear industry in this sense includes the civil production of nuclear energy and its attendant facilities, the military production and use of nuclear materials, and the creation of radioactive isotopes for use in industry and medicine. The wastes may be routinely released into the environment, or they may be 'managed', i.e. stored or buried in order to minimise their release into the environment. Radioactive materials are also released as a result of accidents such as Chernobyl, and such releases may, broadly speaking, be regarded as unplanned wastes. Finally, there is the deliberate release of materials from nuclear explosions either through testing or war.

Radioactive wastes are created at each stage of the nuclear fuel cycle (Figure 7.1). You will also find that Figure 4.11 in Chapter 4 can be used to indicate the environmental pathways available at various stages in the

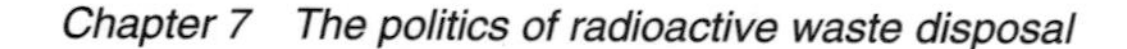

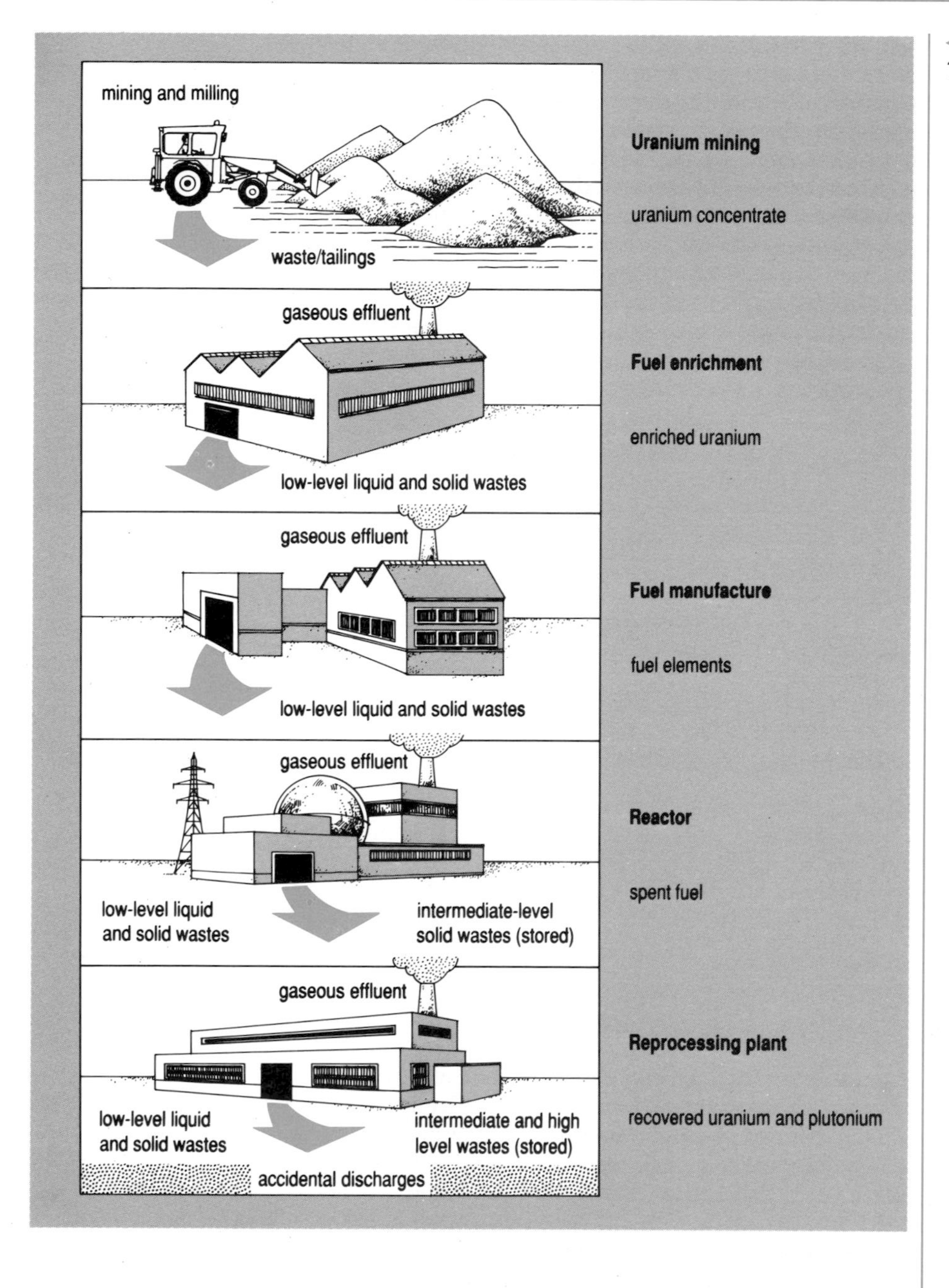

◄ Figure 7.1
The nuclear fuel cycle.

cycle. At the first stage, uranium mining and milling, the ore is extracted, leaving tailing piles with long-lived radionuclides that can contaminate wide areas as in, for example, Australia, Namibia, New Mexico or Saskatchewan. Fuel enrichment (not needed for all fuel types) and fabrication, by which the uranium concentrate is converted into fuel elements, create relatively little waste. The main waste-creating processes are *electricity generation* and *reprocessing*. Most countries with nuclear electricity generation have an 'open' cycle without civil reprocessing facilities, hence the fuel rods when removed must be stored as spent fuel. Since they are highly radioactive they are usually stored on site along with other highly contaminated materials such as ion exchange resins (used for purifying the cooling water), sludges and fuel cladding materials. In those countries with reprocessing, notably France (at Cap de la Hague, near Cherbourg, and Marcoule on the lower Rhone) and the UK (mainly at Sellafield), spent fuel is removed from the power stations and transferred to the reprocessing facilities. The process has been described in Chapter 2, Section 4.4. Reprocessing concentrates the highly radioactive wastes but it also creates large volumes of mildly contaminated trash which must be managed. Apart from radioactive wastes created during the civil nuclear fuel cycle there are those wastes which are produced by military nuclear operations and bomb tests, and wastes from industrial and medical processes. Finally, once nuclear facilities are closed down, the reactors and buildings remain radioactive and are a source of decommissioning wastes. The first of the UK's Magnox power stations at Berkeley, Gloucestershire, was permanently shut down in 1989, leaving a problem of waste management. Subsequently, Hunterston 'A' in Scotland is being decommissioned and the Trawsfynydd station in North Wales has also been permanently closed down as a result of safety considerations. Recognition that the costs of decommissioning would not be acceptable to the private sector was a factor in the decision taken in 1989 not to privatise the nuclear power stations along with the rest of the electricity industry. It was also a major factor in the proposal to retain the Magnox stations in the public sector as the rest of the industry was prepared for privatisation.

2.2 *Classification of wastes*

The distinctive problem with these wastes and the reason they are an environmental hazard is, of course, that they are radioactive. The process of radioactive decay was explained earlier in Chapter 1, Section 4.5. For the purposes of our present discussion there are two essential points about radioactivity that you need to grasp. The first concerns the measurement of radioactive decay.

Q Can you recollect how radioactive decay is measured?

A It is measured in terms of activity and of 'half-life', the time it takes for half the total activity to undergo a nuclear transformation. Activity is measured in becquerels, the number of nuclear transformations per second. Half-lives vary very widely among radioactive elements. As we saw in Chapter 1 of the first book in this series (*Sarre & Reddish, 1996*), some have a very transitory half-life (e.g. iodine-131 has eight days), while others have half-lives extending over unimaginable gulfs of time (e.g. plutonium-239 has 24 000 years; uranium-238 has 4 510 000 000 years). It is therefore important to know the half-lives of materials being dealt with, particularly of those nuclides with high activity levels.

The second point, also identified in Chapter 1 of *Sarre & Reddish, 1996*, is the method of classifying ionising radiation.

Q What are the three types of ionising radiation?

A They are alpha, beta and gamma radiation. Alpha radiation is the least penetrating but most densely ionising and, if ingested, can cause the most cellular damage. Beta, and particularly gamma radiation, can penetrate more readily causing damage to external body surfaces.

At one time the classification of radioactive wastes in the UK was based on the *method of disposal.* Thus, low-level wastes were those that could be safely disposed of by existing routes. This suggested that the definition of wastes could alter over time as new disposal routes were introduced. To avoid such confusion and to achieve consistency, wastes were classified according to their *activity* and the half-lives of radionuclides. In the United States classification emphasises the *sources* of the waste, the most dangerous being irradiated spent fuel. There is no internationally agreed system of classification, which inevitably leads to difficulties when wastes are transported across frontiers without a standardised description or definition. It is interesting to note that despite contentious moves by the European Commission to harmonise other matters within the European Union, notwithstanding a 1994 Commission White Paper on radioactive waste, no directives on this traded commodity have been issued. In Europe and Canada the following three broad categories are recognised:

1 **High-level wastes (HLW).** These are heat-generating wastes in the form of spent fuel or, in those countries with reprocessing, in the form of liquid wastes. Though volumes of such wastes are small they contain around 95% of the total radioactivity and are therefore very dangerous. HLW are technically the major waste management problem and are politically the most contentious, especially in those countries, like the USA, where they are stored in the form of spent fuel at reactor sites.

2 **Intermediate-level wastes (ILW).** These wastes arise from processes closely related to energy production and reprocessing and include fuel cladding, control rods, sludges, resins and filters. They are subdivided into long-lived ILW (mainly alpha emitters with more than 30 year half-lives) and short-lived ILW (predominantly beta and gamma emitters).

3 **Low-level wastes (LLW)** are those wastes which are lightly contaminated, including clothing, paper, plastics, building debris and other materials. Eventually the buildings outside the reactor itself will become LLW. They are high-volume wastes especially where there is labour-intensive reprocessing. Box 7.1 adds some technical information.

The volume of radioactive wastes arising is difficult to estimate and depends on estimates of nuclear electricity generation and decommissioning as well as the capability and capacity to incinerate or compact the more voluminous but least radioactive LLW. One estimate suggests a UK total of about 80 000 cubic metres of ILW by the turn of the century and about a quarter of a million cubic metres by 2030 (NIREX, 1983, 1989). It is estimated that around 20 000 cubic metres of ILW arises from foreign reprocessing contracts and that most of this is covered by return-to-sender clauses which provide a contractual commitment to return all wastes arising from reprocessing to the country of origin. It has also been

Box 7.1

You may recollect that radioactivity is measured in terms of the becquerel (Bq) which has replaced the curie (Ci). In the UK LLW are now defined as wastes not exceeding 4 GBq/tonne of alpha or 12 GBq/tonne of beta or gamma which is about 3×10^8 Ci/tonne. In the USA all waste not classified as spent fuel, liquid HLW or transuranic (in simple terms, long-lived but non heat-producing) waste is LLW, though this is placed in different subclasses (a, b, c) according to activity.

agreed in principle that HLW having an equivalent amount of radioactivity may be returned in substitution for the ILW or LLW. If enacted, this would mean lower volumes of waste being returned (albeit of equivalent radioactivity) while ILW and LLW from foreign countries would be managed in the UK in perpetuity. Substitution would only be possible by agreement between contracting parties and only when a permanent repository is available for the disposal of ILW in the UK.

Volumes of LLW are estimated to be around a half million cubic metres by the end of the century and 1.5 million cubic metres by 2030. These volumes will be increased by about a fifth to take into account defence wastes, which are small amounts equivalent to about 2 million tonnes of packaged material by 2030. The total amount of radioactive wastes arising every year is just over a million tonnes, comprising 1.0 million tonnes of LLW, 0.16 of ILW, and 0.004 of HLW, in contrast to an annual production of between 5 and 10 million tonnes of industrial wastes containing hazardous materials and 29 million tonnes of domestic wastes that is disposed of every year (House of Commons Environment Committee, 1986, p. xix). The volume of HLW in the UK is much smaller, amounting to around 1200 cubic metres which, according to the nuclear industry, is a volume sufficient to occupy a pair of semi-detached houses. For a variety of reasons, including deliberate obfuscation, it is difficult to establish really reliable figures for future radioactive waste arisings.

2.3 *The management of radioactive wastes*

If volume was the main concern, the problem of radioactive wastes would be relatively easy to solve, compared with household refuse for instance. It is, of course, the activity of the wastes that requires them to be isolated from the accessible environment. HLW is heat generating with long-lived radionuclides so that it requires cooling and subsequent safe management for a period far longer in the future than we believe human civilisation has existed in the past. Deep geological disposal in a repository with multiple barriers to prevent escape of the radionuclides in the first instance and reliance on the geological barrier in the long term, is regarded by the nuclear industry as the most acceptable method of management for HLW and for the longer lived ILW. Opponents of the nuclear industry argue that it has not been established as an acceptable solution and they advocate surface storage until a safe method of management can be agreed. Both France and the UK have developed pilot vitrification plants to encapsulate HLW in glass within metallic containment for storage or long-term disposal.

Activity 1

Identify some of the arguments for and against deep geological disposal of HLW.

This debate is at the heart of the political problem of radioactive waste. In nearly every country with HLW efforts have been made to identify suitable locations for deep disposal; in every case such efforts have met with opposition. There is, as yet, no deep repository opened to receive HLW.

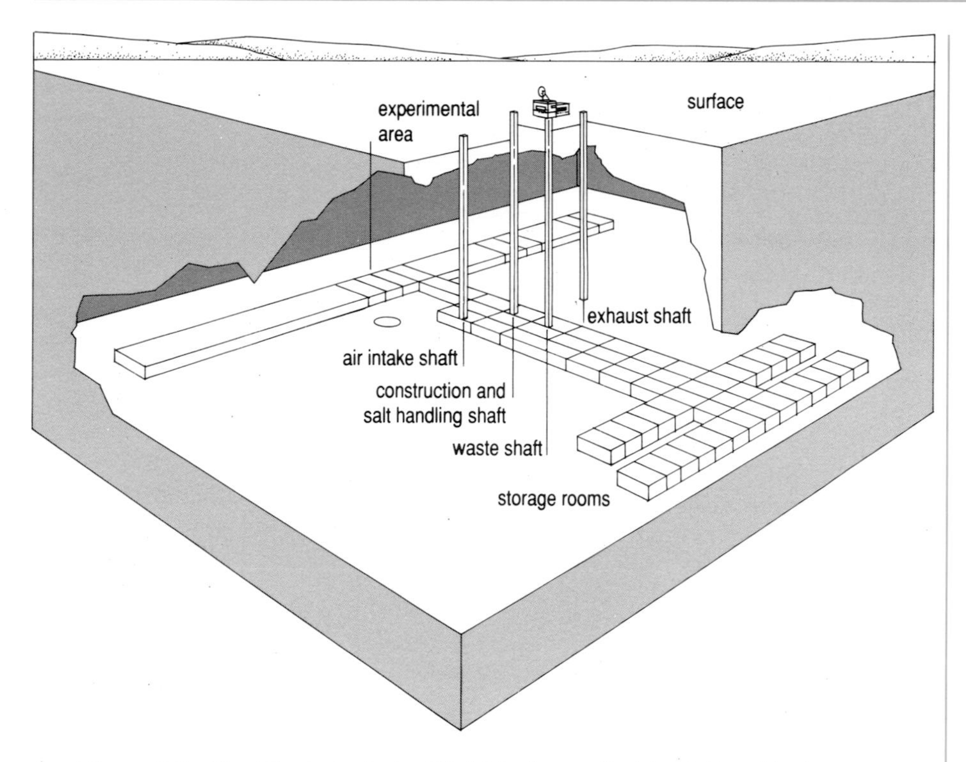

Figure 7.2 The Waste Isolation Pilot Plant near Carlsbad, New Mexico.

By 1990, a purpose-built deep repository in the desert of south-east New Mexico, USA, designed to receive transuranic wastes from the military sites, had been constructed but not opened as a result of technical problems (Figure 7.2). It is planned to open the facility by the turn of the century. Plans for siting deep repositories in the USA, UK, Germany and elsewhere have been consistently delayed both for technical reasons and as a result of political opposition.

Rather more progress has been made in the management of LLW and short-lived ILW. In the UK, shallow disposal of LLW has been undertaken at Drigg, near Sellafield, since 1959. The need for a further shallow repository was a source of political conflict as we shall see in Section 4. In France LLW was managed in a surface permanent storage facility near the reprocessing works at Cap de la Hague and is currently managed at a new site near Troyes to the east of Paris. In the United States three shallow burial facilities are in operation in South Carolina, Nevada and Washington state and three have been closed as a result of technical problems involving leakage from the sites. The threat of closure of the remaining three sites has precipitated the formation of groupings of states (called regional compacts), and the difficult and controversial task of identifying new sites for the disposal of LLW. In Sweden a purpose-built repository for LLW and short-lived ILW has been fully in operation since 1992 about a kilometre under the Baltic Sea to the north of Stockholm. This 'Rolls-Royce' solution to the problem has been achieved in a country which had decided to phase out its nuclear energy programme by the year 2010, although pressures have been brought to drop, or put back, this self-imposed deadline, following a review by a Special Parliamentary Commission.

2.4 *The problem of radioactive wastes*

In almost every country with a substantial nuclear industry the technical problems of managing nuclear wastes have been compounded by the increasing political difficulty in securing particular solutions. The nuclear industry provokes a pervasive, growing and consistent fear different from the anxiety caused by any other industry. It is not so much the possibility of harm (which may be recognised as remote) but the probability that the consequences will be catastrophic that arouses such fears. Local people are understandably worried about the potential dislocation that may be caused to their community if large volumes of radioactive rubbish is deposited near to where they live.

Activity 2

Do you remember the pie diagram in Chapter 1 of Book One of this series (*Sarre & Reddish, 1996*, Figure 1.13) showing the sources of radiation exposure of the UK population? Before referring back to it can you think roughly how much of the exposure, in percentage terms, came from natural sources? How much came from radioactive waste?

Now check your answers with the diagram. Why should we be cautious in interpreting these statistics?

Over four-fifths of the average radiation dose received by people in the UK comes from natural sources and about a tenth from medical sources (*Sarre & Reddish, 1996*, Chapter 1, Figure 1.13). Only 1.3% arises from the 'nuclear industry' and defence, of which it is calculated about 0.1% is from nuclear waste discharged into the environment. On this basis it might be supposed that nuclear waste can be dismissed as an environmental problem of any significance. It is the possible consequences of accidental leakage of wastes into the environment, from the repository or during transportation, that causes anxiety. People are anxious about nuclear activity and do not discriminate between the risks or consequences whether from a reactor melt-down or leakage from a low-level waste repository. All radioactivity is perceived to be dangerous regardless of source or intensity.

Q Why do you think people are so concerned about radioactivity produced as a result of the nuclear fuel cycle? Jot down your answer and compare it with the one given below.

A The risk from radioactivity produced by the nuclear industry is an *avoidable* risk; there are alternatives to nuclear energy that are less potentially harmful to human health. Although the production of electricity from fossil fuels contributes to the enhanced greenhouse effect, as we saw in Chapter 2, the claims that nuclear facilities are safe are belied by the evidence of leakages and accidents. The concentrations of radionuclides are such that, if an accident occurs, the consequences are catastrophic and *irreversible* extending over long time-scales because radioactivity cannot be halted. The characteristic of low probability/high consequence risk with irreversible and catastrophic consequences is one that the nuclear industry shares with other high-technology hazardous industries such as chemicals, as discussed in the previous chapter. But, in certain respects, the nuclear industry induces unique anxieties.

Q Can you think of some of these?

A Radioactive materials are immune to outside influences; they are indestructible and only become harmless through decay, which may take thousands of years. Radiation is dangerous by proximity as well as by ingestion or inhalation. It cannot be seen, touched or smelled – this very invisibility arouses fear. The nuclear industry is perceptibly linked to the destructive capability of the atomic bomb. Furthermore, though concentrations and activity may vary, radioactivity is perceived to be dangerous *whatever its source.* The possible links between leukaemia clusters and nuclear activities, discussed in *Sarre & Reddish, 1996,* Chapter 1 and in Chapter 2 of this book, illustrate the increasing concern about the health effects of long-term exposure to low levels of radioactivity.

Ultimately all nuclear activity becomes radioactive waste. Military wastes and those from the civil nuclear fuel cycle as well as manufacturing and medical radioactive wastes must be acceptably controlled either through containment or, for very low activity wastes, through dilution and dispersion in the environment. This latter strategy has its own pitfalls. In the USA, attempts to resolve the disposal problem of low activity, short-lived radioactive wastes by deeming them 'Below Regulatory Concern' have fallen foul of the charge that this change of definition is a sleight of hand to allow such wastes to be co-disposed with household refuse in local landfill sites. As the nuclear industry has developed and public anxiety has grown, so it has become increasingly difficult to achieve acceptable solutions to the management of nuclear wastes. The emergence of radioactive waste as a political issue is the subject of the next section.

2.5 *Summary*

In this section we have covered the first two aims of this chapter. In certain respects nuclear wastes are distinctively different from other hazardous and toxic materials. In particular, the fear of radioactivity from whatever source has made the management of radioactive wastes increasingly controversial and a major cause of political conflict. Radioactivity arouses greater fear than any other form of socially induced risk, and is a unique problem.

Nuclear wastes are the residual materials from the production of nuclear electricity, reprocessing, atom bomb manufacture and certain industrial and medical processes. They constitute an environmental problem in that they must be managed to ensure that radioactive materials do not enter accessible environmental pathways and present a health risk. Radioactive wastes are classified according to source and levels of activity. These wastes are treated differently from other hazardous wastes. High-level wastes, in the form of spent fuel or liquid wastes from reprocessing, though very small in volume, are heat-generating and highly dangerous and deep geological disposal is the proposed method of management. By contrast low-level wastes have large volumes but short half-lives with low radioactivity and are diluted or dispersed or buried in shallow repositories. Whatever the statistical risks, public concern is now so developed that the issue of nuclear waste facilities has become a source of political conflict.

3 The development of nuclear waste from a non-problem, to a technical problem, to a political problem

3.1 Early years: a non-problem

The origins of the problem of radioactive waste are found in the United States where the major military nuclear programme created the first large quantities of nuclear wastes. For centuries indigenous North American (and Australian) peoples used uranium as a colouring agent for glassware, or as an ingredient in paint for warfare or ceremonies. Since the late nineteenth century uranium and other radioactive elements had been used in Europe and North America for scientific experimentation, for medical purposes and for cosmetics. Many small companies dabbled with commercial uses of radioactive materials. As in all businesses, some went bankrupt. But what was left behind in this case was radioactive contamination and a legacy of radioactive residues which, along with those arising from the other uses, particularly laboratory research, created the first nuclear waste 'burden'. At the time these wastes were not regarded as a difficulty – and this perception became part of a later problem.

The problem revealed itself in differing ways over time. First to emerge were the radiation-induced illnesses of the workers and scientists who worked largely unprotected, many contracting various cancers directly attributable to exposure to radioactivity. Although some scientists, such as Marie Curie, an early pioneer in radioactive research, did die of cancer, this did not inhibit her colleagues and peers in continuing this experimentation, such was the fascination of the research. As research continued, radioactive waste built up.

By the mid-1930s, some of the best brains in chemistry and physics were devoted to nuclear science, in Europe, North America and elsewhere. As a result of a coincidence of history, many of these scientists came together in the late 1930s and the early 1940s in the United States and Canada. This collection of scientists made it possible for the United States to develop the atomic bomb, in secret, by 1945 and thereby created the institutional base for the post-war development of atomic energy for military and civil uses.

The American atomic bomb project ('The Manhattan Project') required the setting up of a range of large scale research and development establishments, such as at Los Alamos (New Mexico), Chalk River (Ontario, Canada) and Hanford (Washington state). For reasons of secrecy, primarily, but also safety, these major facilities were located in isolated areas. Following the end of the second world war, the United States also developed nuclear warhead testing sites in the Marshallese and Aleutian Islands in the Pacific Ocean and later in the Nevada desert, north of Las Vegas. Along with a proliferation of other smaller sites across the continental United States, this atomic bomb programme bequeathed an inestimable quantity of radioactive waste and on-site contamination. The problem was being created incrementally and went largely unrecognised and unremarked for some five decades. As a former manager in the US Atomic Energy Commission (USAEC; the body given responsibility for the management of the project) put it: 'Chemists and engineers were not interested in dealing with waste. It was not glamorous; there were no careers;

it was messy, nobody got 'Brownie points' for caring about nuclear waste. The Atomic Energy Commission neglected the problem' (Wilson, 1979).

The extent of the problem of radioactive wastes arising from the Manhattan Project has been the subject of several studies. One of these, published in 1980, indicated that at least 76 separate sites were formerly used in the project and 17 remain operational today. Many sites have high levels of radioactive contamination in the soil or in groundwater, leaving a substantial economic and environmental burden for the United States. In 1995, the US Department of Energy, the successor to the USAEC, published two detailed reports which revealed the extraordinary extent of the clean-up requirements at its inherited military sites (US Department of Energy, 1995). It was estimated that the cost of site clean-up and restoration would be (in 1995 rates) more than $500bn over 75 years to complete a task without technical precedents. Over 400 000 cubic metres (about 106 000 000 gallons) of HLW, 107 000 cubic metres of transuranic waste and some 780 000 cubic metres of LLW or mixed radioactive chemical waste will need to be disposed of. Additionally, remediation methods will need to be found to decontaminate soil and ground water offsite (Rothstein, 1995).

Hanford, a vast nuclear reservation in the sagebrush desert along the Columbia River in the north-western United States, provides an example of the creation of radioactive wastes in these early years. Hanford was the major centre of the plutonium manufacturing process. During the war there was serious consideration of the use of radiological weapons, based on nuclear waste derived from plutonium manufacture, in order to disperse fission products over enemy territory. More disturbing was the revelation that massive quantities of radioactive iodine-131 had been routinely released from exhaust stacks over the Hanford area. In one 1949 experiment, known as 'Green Run', iodine-131 was deliberately released over Hanford, and the resulting 200 by 40 mile plume was secretly measured. This experiment, unknown to the local population, resulted in radioactive contamination which exceeded the US Atomic Energy Commission's own 'tolerance levels' by 11 000 times over Hanford (and those tolerance levels were considerably higher than today's far more conservative permissible-exposure standards).

This casual approach to the risks was also present in the early systems for managing the radioactive wastes produced in Hanford's growing number of nuclear plants. One method was the pumping of liquid radioactive wastes directly into two deep wells that were drilled into the aquifer that passes underneath the site. Other wastes were deposited in earthen ditches, trenches, cribs, ponds, swamps, underground drains, as well as double-shelled steel tanks. Mainly arising from the underground disposal of liquid radioactive wastes, both the Columbia River (and areas downstream through which it flowed towards the Pacific Ocean) and the aquifer were contaminated. Hanford has continued to provide a serious problem in the management of radioactive wastes. For example, there was a very serious leak of over 115 000 gallons of liquid radioactive waste from storage tanks in 1973. An official report stated in 1986 that there were over 200 areas within the Hanford reservation alone that were contaminated with radioactive leaks or spillages. Hanford remains the storage site for large volumes of all levels of radioactive wastes from the military programme, and has been earmarked as an appropriate site for the bulk decommissioning of wastes from dismantled reactors, both those already on site and others around the USA.

Hanford is just one example, albeit a significant one, of how wastes were managed in the early years. Elsewhere in the United States and in the

Soviet lake is most n-polluted spot

NEW YORK, Aug. 17 (AP)

Karachay lake, a 40-hectare body of water in the Soviet Union, could be "the most polluted spot on the planet", the result of years of radioactive waste dumping, a published report said.

American experts who recently visited a Soviet nuclear weapons production complex found extremely high levels of radioactive contamination and a history of very high doses to production workers, the New York Times reported yesterday.

The Soviets have confirmed that in 1951, atomic bomb makers began pumping radioactive wastes into Karachay lake – about 1,450 km east of Moscow – a practice that continued until the accumulation reached 120 million curies.

That amount is about two and a half times greater than the release from the Chernobyl nuclear accident in 1986.

"This has got to be the most polluted spot on the planet," said Mr. Thomas Cochran, a physicist and senior staff scientist at the National Resources Defence Council, a private environmental group in Washington DC.

The dose rate on the lake shore near the outlet pipe is 600 roengen per hour, according to a new report by Mr Cochran and Mr Robert Standish Norris, also at the Council.

At that rate one hour's exposure would kill a person within weeks, the Times reported.

The Times said Soviet documents described a "sanitary alienation zone" in contaminated areas in which people are forbidden to live or travel.

The size of the zone was not known.

Figure 7.3 Extract from The Hindustan Times, New Delhi, 18 August 1990.

USSR, France and Britain nuclear defence resulted in the commissioning of plutonium-producing reactors and associated plant and the creation of nuclear wastes. The management of these wastes was a non-problem politically for several decades (Figure 7.3).

Activity 3

Can you suggest reasons why radioactive waste was a non-problem in these early years? Then read the suggested answer at the end of the chapter before continuing.

In these conditions it is possible to give unfettered primacy to the 'positive product' of the production process. The pollution and wastes that result are ignored. In a liberal democratic society such as the United States, such projects as the secret Manhattan Project and subsequent military nuclear programmes illustrate these characteristics. They are also found in centrally controlled economies. Information on the widespread industrial contamination, including toxic and radioactive wastes, was known before the political revolutions in eastern Europe in 1989 but the enormous extent of ecological damage has subsequently been widely publicised.

3.2 *A matter for the experts*

The Soviet Union, United Kingdom and a little later France and China followed the United States in the development of nuclear weapons programmes. By the 1950s these and many other countries (except China, whose 'civil' nuclear programme was started in earnest only in the 1980s) had started civil nuclear energy programmes, heralded as the development of 'Atoms for Peace'. Electricity produced from nuclear fission was portrayed as cheap ('too cheap to meter' was a popular slogan at the time), clean and safe. While there was little concern about environmental impacts there was anxiety over the potential dangers of the spread, or proliferation, of military nuclear capability arising from the production of plutonium in civil nuclear programmes. This led to the establishment of international bodies to oversee the spread of civil nuclear technology. After the United Nations Atomic Energy Commission, which had a brief life in the late 1940s, another institution, the International Atomic Energy Agency (IAEA) was established under the wing of the United Nations, with strong support from the United States, in 1957. Regional nuclear bodies, such as the EEC's Euratom and what became the Organisation for Economic Co-operation and Development's (OECD) Nuclear Energy Agency, were established around the same time. Today all three bodies are concerned with the oversight of radioactive waste management strategies, policies and technologies. In the late 1950s radioactive waste was barely mentioned in their 'mission goals'.

During the 1950s, while international and inter-state atomic co-operation developed, exemplified by the Atoms for Peace conferences sponsored by the UN in Geneva, the main focus was upon reactors and fuel development and the diversified uses of atomic energy, not upon the inevitable waste arising. Public concern focused on the possibility of reactor accidents and was fanned by the accidents that actually occurred at Chalk River, Ontario, Canada in 1952, Windscale in 1957, Three Mile Island in Pennsylvania in 1979 and at Chernobyl in 1986. Such accidents can, and in the case of Chernobyl did, cause widespread contamination. They are dramatic and have contributed to a growing unease about the safety and environmental impact of nuclear energy, as well as providing a damaging blow to the economic case for nuclear reactors. The casualties arising from nuclear accidents are difficult to estimate. In the case of Chernobyl there were over 31 immediate deaths and around 130 000 people were evacuated from a heavily contaminated exclusion zone. The eventual death toll is unpredictable with estimates varying from 180 to 280 000 depending on the assumptions made. It is very difficult to compare the scale of such a disaster with other major industrial accidents. The leakage of methyl isocyanate from the Union Carbide pesticide plant at Bhopal, India, in 1984 caused 2500 deaths and over 250 000 suffering injuries of varying severity.

Nuclear waste accumulates gradually and does not generally give rise to spectacular incidents. There is the exceptional case of the explosion at badly maintained waste trenches at the secret Soviet military nuclear establishment at Kyshtym in 1957. Public concern about nuclear waste emerged slowly and developed at local and national level, only later becoming an international political problem. We shall use the example of the United Kingdom to illustrate this process of politicisation, though similar trends may be observed in other western democracies such as the United States, Germany, Sweden and, though to a lesser degree, France.

In the UK low activity liquid and gaseous wastes from atomic facilities were routinely discharged into the air, sea and local rivers. In the mid-1950s the UK Atomic Energy Authority (UKAEA) considered using abandoned mine shafts in the Forest of Dean, about 130 km from its main research centre at Harwell (Oxfordshire), for deep disposal of solid wastes. The plans were opposed by a local group known as the 'Free Miners' who claimed ancient rights to the Forest. This episode was the first example of local reaction in response to a proposal for nuclear waste disposal. This 'not in my back yard' (NIMBY) local protest against a proposal to use a new ('greenfield') site has been a key political component of the opposition to on-land disposal of nuclear waste, as we shall see.

At that time, no civil nuclear programme had yet been developed, so the volumes of waste were small, and the option of sea disposal, with no local opposition to confront, proved attractive. Consequently, the UKAEA withdrew the forest disposal proposal and returned in earnest to sea disposal in the north-east Atlantic Ocean, which had begun in 1949. After a brief and inconsequential public flourish, radioactive waste became once

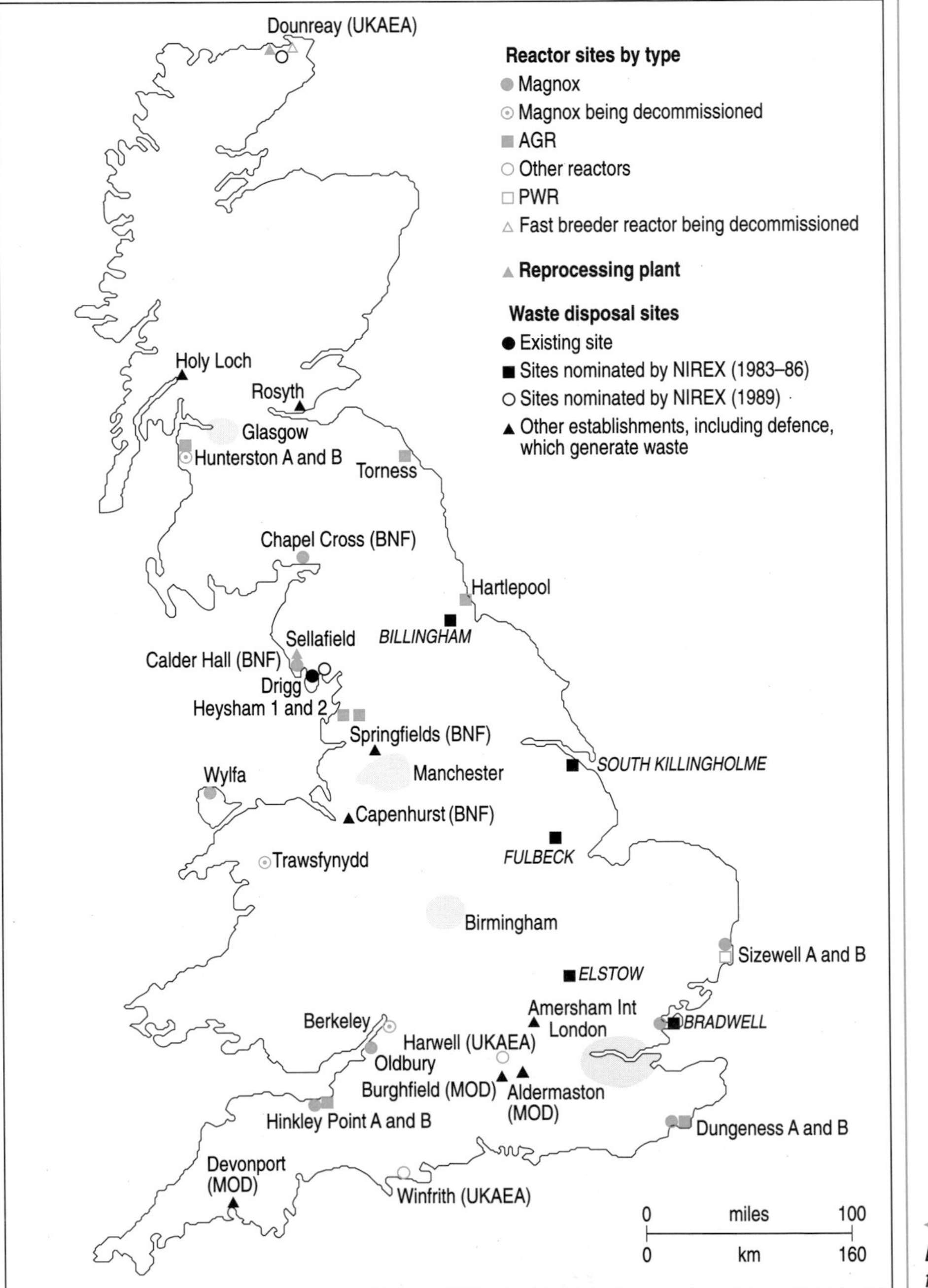

Figure 7.4 Location of nuclear facilities in the UK.

again a technical matter firmly in the hands of experts. For example, the 1955 White Paper, *A Programme of Nuclear Power* (HMSO, 1955) which launched the British civil nuclear programme, dismissed the problem of waste, stating 'the disposal of radioactive waste products would not present a major difficulty.'

Throughout the 1960s and early 1970s the civil and military nuclear programmes in the UK expanded. The Magnox reactor programme of nine power stations was completed and the advanced gas-cooled reactor (AGR) programme, fraught with technical difficulties and delays, was being brought into operation (for locations of nuclear sites in the UK see Figure 7.4). The military designated programme was also expanding, as were operations at the Sellafield complex, which includes the reprocessing plant formerly known as Windscale and the Calder Hall Magnox power station. There was no local or national protest against the radioactive waste being concurrently created. The matter was addressed only peripherally in Parliament. While there was popular protest against nuclear weapons (the Campaign for Nuclear Disarmament's Aldermaston marches best exemplifying this) and against the Vietnam War in the late 1960s, the civil nuclear cycle was not a focus of popular opposition at this time.

Political concern for nuclear waste first became fully established during the mid-1970s. This was a decade when 'environmentalism' grew as a social and political phenomenon (*Sarre & Reddish, 1996*, Chapter 3). It was also a period of energy crisis brought about by the national coal miners' strike in 1972 and the Middle East conflict leading to the rise in oil prices in 1973. Energy shortage brought nuclear energy and attendant reprocessing into greater prominence as a secure source of electricity. But the dangers were gradually becoming articulated, as is illustrated by the extracts from a letter to *The Guardian* signed by 43 senior scientists, environmentalists and former government policy advisers (Figure 7.5).

Q Why was radioactive waste left mainly to the experts during the period up to the mid-1970s?

A Attention became focused on the benefits of nuclear energy as the civil nuclear programme developed. Nuclear waste was not regarded as a serious problem, even by most experts. Wastes could be either routinely discharged into the environment, stored on site or disposed of at sea or on land at the low-level site at Drigg near Sellafield. By the mid-1970s some experts were beginning to express concern about radioactive waste, as the *Guardian* letter shows. The outstanding additional question remains: Why did radioactive waste from military programmes still remain in the shadows when civil wastes began to be assessed negatively?

3.3 *The politicisation of radioactive wastes*

The first authoritative and accessible account of the problems posed by radioactive waste in Britain is contained in the Sixth Report of the Royal Commission on Environmental Pollution (RCEP) on *Nuclear Power and the Environment* (HMSO, 1976). Out of 50 recommendations in the Report, 14 were devoted to nuclear waste. Two new organisations were recommended. One, a national waste disposal company, eventually became, in 1982, the Nuclear Industry Radioactive Waste Executive (NIREX), a body formed by partners in the nuclear industry to manage ILW and LLW. The other, an advisory body on waste management, became the

Nuclear dustbins for the centuries

Sir, – The debate, last year, about the choice of reactor system in the development of a nuclear power programme obscured the problems posed by the production of dangerous long-lived radioactive wastes which have to be kept away from the living environment for hundreds, and some for many thousands of years.

We are concerned that the production of electricity by any nuclear fission power plant is inescapably coupled with the production of hazardous radioactive materials and that all known methods of storing the waste require regular supervision. The problem of disposing of them safely has not been solved in spite of some 30 years of nuclear technology. The conversion of liquid wastes into stable solids would merely make storage easier; it will not remove their dangerous properties.

Each reactor continually produces waste for disposal and only stops when it is closed down. Apart from these wastes the commitment to fission involves the containment of vast amounts of radioactive materials in reactors and in processing plants and in transit between plant and storage points at all times.

We noted the Government's decision not to order the American designed light water reactors but instead to adopt the British steam generating heavy water reactor (SGHWR) for the next programme of nuclear power stations and that this programme is a reduction in size on the original proposals.

We were pleased when it was announced in April last year that the Royal Commission on Environmental Pollution was to conduct an inquiry into the safety of the storage and disposal of all radioactive waste from nuclear power stations. The Government did not, however, wait for any report of these inquiries before making its decision to proceed with the nuclear power programme.

We do not think that any society, or its government, can guarantee safety to future generations from these deadly materials for one hundred years, let alone the many thousands that are involved. We cannot guarantee them against foreseeable dangers – we know that the best intentioned men are fallible and can make mistakes; we know that men with bad intentions indulge in acts of sabotage, theft and terrorism and we also know that plutonium is the raw material of atom bombs.

Our nuclear power programme represents a Faustian bargain in which we are jeopardising the safety of future generations and their environment for our own short-term energy benefits and the comforts that go with them. We consider that it is immoral and unwise to pursue a technology which will leave such a dangerous legacy to posterity.

We are, therefore, opposed to the building of any more nuclear power stations at least while there is no safe and proven method of disposing of the long-lived radioactive wastes.

We are convinced that there are other ways of tackling our energy problems. We should be much more economical in the use of the energy already available to us. We should make use of all heat produced and not waste it as we currently do from most of our buildings, power stations and municipal incinerators. Priority should be given to the development of techniques that utilise non-polluting energy sources such as the sun, wind, waves and organic wastes. These are the areas in which we should concentrate our major research and development efforts.

Yours faithfully,

D. Bryce Smith (Professor of Organic Chemistry, University of Reading); **Sally Cane** (Youth Organiser, Population Countdown); **Irene Coates** (Chairman, The Conservation Society); **Gerry Davis** (author and TV script writer, co-author of Doomwatch); **P. D. Dunn** (Professor of Engineering Science, University of Reading); (Dr) **P. N. Edmunds** (Consultant Bacteriologist, Fife Health Board); **John Foster** (Department of Political Economy, Glasgow University); **Peter Hain** (Former Chairman, Young Liberals); (Prof) **John Hawthorn** (Department of Food Science and Nutrition, University of Strathclyde); **J. Owen Jones** (Director, Commonwealth Bureau of Agricultural Economics); **T. W. B. Kibble** (Professor of Theoretical Physics, Imperial College of Science and Technology); **F. Lafitte** (Professor of Policy and Social Administration, University of Birmingham); **Gerald Leach** (Former science correspondent, the Observer; Fellow of the International Institute for Environment and Development); **Laurie Lee**; (Dr) **John Loraine** (Medical Research Council, External Scientific Staff, Dept. of Community Medicine, University of Edinburgh); (Dr) **Richard Marshall** (Department of Physics, University of Keele); (Rt. Rev) **Hugh Montefiore** (Bishop of Kingston); (Prof.) **J. K. Page** (Building Science, Sheffield University); (Dr) **Kit Pedlar** (Opthalmic surgeon; co-author of Doomwatch); (Dr) **Robert Perrin** (Lecturer in Soil Science, University of Cambridge); **Lady Robson** (Chairman, Liberal Party Environment Panel); **G. F. C. Rogers** (Professor of Engineering Thermodynamics, University of Bristol); **Richard Sandbrook** (Director, Friends of the Earth); (Dr) **E. F. Schumacher** (Founder-Chairman Intermediate Technology Development Group; President, the Soil Association); **Cedric A. B. Smith** (Weldon Professor of Biometry, Department of Human Genetics and Biometry, University College, London); (Sir) **Kelvin Spencer** (Former Chief Scientist, Ministry of Power); **M. W. Thring** (Professor of Mechanical Engineering, Queen Mary College, London); **Graham Tope**, and 13 others.

London.

▲ *Figure 7.5*
A letter to The Guardian, *7 January 1974.*

Radioactive Waste Management Advisory Committee in 1978. The most oft-cited recommendation of the RCEP Report was:

> . . . There should be no commitment to a large programme of nuclear fission power until it has been demonstrated beyond reasonable doubt that a method exists to ensure the safe containment of long lived, highly radioactive waste for the indefinite future. (para. 27)

The nuclear opposition took this to support their case that nuclear power should henceforth be halted. The nuclear industry, concentrating on the words, 'demonstrated beyond reasonable doubt', took it as an endorsement of their intention to continue exploration of practical geological disposal sites for the accumulating nuclear wastes. The RCEP lent support to the latter interpretation by calling for a national land disposal facility for the more highly active wastes but was scathing about waste management policy:

> The picture that emerges from our review of radioactive waste management is in many ways a disquieting one, indicating insufficient appreciation of long term requirements either by government departments or by other organisations concerned. In view of the long lead times that will almost certainly be involved in the development of appropriate disposal facilities, we are convinced that a much more urgent approach is needed, and that responsibilities for devising policy and for executing it need to be more clearly assigned. (para. 427)

The magnitude of the problem was first established by the public inquiry into a planned new reprocessing plant (the Thermal Oxide Reprocessing Plant, THORP), modification of the existing reprocessing plant, and construction of a vitrification research plant at Windscale in 1977 (Box 7.2). The role of Sellafield (i.e. Windscale plus Calder Hall) is discussed in *Sarre & Reddish, 1996*, Chapter 1, and the problem of discharges is considered in Chapter 2 of this book. The THORP plant was designed to process spent fuel from the AGR reactor programme and also (crucially for its economics) spent fuel imported from other countries, raising new questions of international co-operation, trade and cost–benefit assessments. The public inquiry into the Windscale proposals, which lasted 100 days, was the first of four major inquiries into nuclear projects. The other three were the inquiry into the first pressurised water reactor (PWR) proposal at Sizewell, which lasted a record 333 days during 1983–85, the Hinkley Point C inquiry into

Box 7.2

Reprocessing separates out uranium and plutonium from spent fuel. The liquid high-level wastes that remain are stored in stainless steel tanks. Vitrification is a process whereby the liquid wastes are mixed with glass-making materials and converted into solid glass-like form. These blocks will be stored and cooled and it is intended that they will eventually be emplaced in a deep repository. The Sellafield vitrification plant proposal was similar to the plant operating in France.

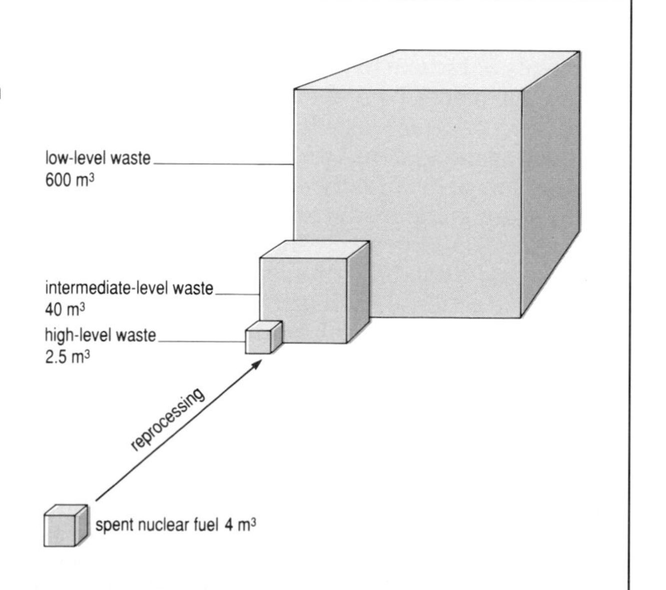

The figure shows volume of waste from spent nuclear fuel before and after reprocessing. Volumes are per year per 1000 MW_e pressurised water reactor.

the second PWR, lasting 182 days in 1988–89, and the inquiry into the proposed rock characterisation facility (RCF) at the Sellafield site of the NIREX repository in 1995–96 (for locations see Figure 7.4).

The Windscale inquiry has been described in, for example, Pearce *et al.*, 1979; Williams, 1980. If the proposals were to be approved, objectors stressed that they would lead to a manifold *increase* in the volumes of ILW and LLW. There would also be a sevenfold *decrease* in the volume, though not the radioactivity, of HLW. The liquid HLW would be vitrified into blocks that could eventually be buried in a deep repository. In theory, the separated plutonium and uranium recovered from reprocessing would be recycled in new fuel. In practice, both products have proved an additional storage problem. Plutonium is a particularly expensive 'reserve' fuel following the world-wide demise of the fast breeder reactor which, as originally envisaged, could have consumed significant quantities.

The scale of the potential trade and the problems of dealing with imported wastes were controversial and attracted Parliamentary questioning and some public protest. But the Windscale inquiry was essentially a debate among experts about technical matters which had increasingly prominent political overtones.

Windscale had first achieved notoriety in 1957 when a fire in the plutonium production plant released radioactivity into the atmosphere. Studies released by official sources during the 1980s indicated perhaps between 33 and 200 people may have suffered cancers from this radiation release. In 1973, 35 workers were contaminated by an accident causing a reprocessing plant to be shut down permanently. Concern about reprocessing foreign waste began to interest the Press, provoking such front-page headlines as PLAN TO MAKE BRITAIN THE WORLD'S FIRST NUCLEAR DUSTBIN in the *Daily Mirror* in October 1975. Later, in 1983, the television documentary, *Windscale – the Nuclear Laundry*, drew attention to the possibility of childhood leukaemia associated with the plant.

The public and political debate surrounding Windscale, beginning in the later 1970s, has had a formative influence on the politicisation of radioactive waste. It provided an opportunity for technical debate of the issues and created a small group of nuclear experts not beholden to the nuclear industry or government who could provide scientific support to opponents of radioactive waste plans in future years. It provided a forum to educate the media who, therefore, could call on journalists with a credible expertise in nuclear waste to conduct investigative research. It also showed that nuclear issues could provide a compelling story line. The planning proposals were eventually approved, but the Windscale inquiry established reprocessing in the media and wider public consciousness as an international enterprise creating substantial volumes of low-activity wastes and also underlined the problem of managing HLW. Disposal of HLW was the first of a series of radioactive waste issues which developed during the 1980s in which the nuclear industry and government were forced to retreat in the face of opposition to their plans.

3.4 *HLW – a victory for local opposition*

The Royal Commission on Environmental Pollution's proposal that a comprehensive search should be undertaken to assess the suitability of various geological strata for the prospective deep disposal of vitrified HLW was followed up by the government in the years 1976–81. The UKAEA commissioned the Institute of Geological Sciences to undertake a borehole drilling programme at various sites in crystalline rock formations in Scotland, Cornwall, mid Wales and Northumberland. A range of other sites in different formations in England (Leicestershire, Nottinghamshire and Avon) were also considered but dropped from the programme after objections were raised. Opposition also developed against the programme in the chosen sites.

The conflict over the HLW drilling proposals first established four of the features which subsequently characterised the politics of nuclear waste.

1 The use of specialist counter-expertise to oppose nuclear facilities. The availability of technical arguments to support their case meant that the opposition could confront the nuclear industry on more equal terms, at least in regard to knowledge, if not resources, at inquiries as well as by conducting a campaign through the media.

2 The development of local groups able to mobilise the local community and public opinion against an external threat. Efforts were made to get the support of local councils and the conventional armoury of protests (marches, petitions, newsletters) was used in campaigns. These local pressure groups were inspired by a NIMBY defence of amenity and environment and, in the case of Ayrshire and mid Wales, they had distinct nationalist overtones, too.

3 The ability of these local groups to gain the support and assistance of other opposition groups in different areas of the country to advise on expertise, strategies and tactics of opposition and fund-raising. This empowerment of one local group by others transcended purely local NIMBY concerns and became a major feature of later conflicts over nuclear waste disposal sites.

4 The difficulty of establishing nuclear facilities in 'greenfield' locations. In every case where drilling was proposed in areas without existing nuclear facilities, the plans were vigorously opposed. Only in Caithness, an area heavily dependent on the Dounreay fast breeder reactor plant, was permission to drill granted. As we shall see, after another decade of futile effort to achieve greenfield locations for nuclear waste facilities, the government had to fall back on its 'nuclear oases' where, at least, some support can be secured from communities familiar with and dependent upon the nuclear industry.

The government abandoned its HLW borehole drilling programme in 1981. The conflict had been essentially local although some liaison between opposition groups did take place. The next conflict over radioactive waste brought in a national and international political dimension.

3.5 *The end of sea dumping*

For many years Britain had organised the dumping of nuclear wastes (four-fifths British, the rest from Belgium, Switzerland and Holland) in the ocean, first off Alderney and latterly in the north-east Atlantic. Disposal by the United States in the Atlantic and Pacific Oceans had ceased in 1970. In 1972 the Convention on the Prevention of Marine Pollution by Dumping of Wastes and Other Materials (The London Dumping Convention or LDC) was adopted and was ratified by the UK in 1975. The LDC sets the norms, guidelines and recommendations for the disposal of wastes, including radioactive wastes, at sea. It expressly forbids ocean bed 'emplacement' of

HLW. The campaign to oppose the dumping of LLW and ILW at sea was initially led from 1978 by Greenpeace, at that time a newly established environmental pressure group which, at first, attempted to blockade ships bringing spent nuclear fuel to Sellafield from Japan. Its campaign was based on a desire to disrupt a trade of which Greenpeace disapproved rather than one it had proof was environmentally damaging.

Greenpeace secured dramatic impact by attempting to stop the dumping in the Atlantic using inflatable dinghies and publicising film of the crew of one dinghy being thrown into the water by a waste barrel dropped from the disposal vessel. There followed a series of 'media stunts' designed to ensure the issue maintained its prominence. Having achieved publicity, Greenpeace then turned its attention to influencing the LDC from its position as an observer using various reports to undermine the credibility of sea dumping. By 1983 the combination of lobbying and spectacular protest was having an effect, and the LDC agreed by 19 votes to 6 with 6 abstentions (the UK opposing) to a two-year moratorium on sea dumping. The UK did not have to accede to the decision but its position was further compromised when the major transport unions, lobbied by Greenpeace and led by the National Union of Seamen, agreed to ban dumping in the Atlantic. Protests against the UK's intended dumping occurred in Spain, Holland, Denmark and New Zealand. In August 1983, the government conceded defeat and abandoned the annual dumping. In 1988 the government formally announced it would not dump drummed waste at sea while keeping an option open on the possible future sea disposal of irradiated waste from closed-down reactors and submarines.

In 1995, the issue of the sea-disposal of nuclear waste again became a politicised issue with the rediscovery that such wastes had been dumped near to Alderney in the Channel Islands. Among environmental groups there was also concern about the shipment of radioactive wastes and plutonium from the French plant at La Hague on shipping lanes close to the Channel Islands.

Q Why did radioactive waste become a political issue during the 1970s?

A An obvious reason is that the problem was getting physically larger as the nuclear programmes expanded and attention began to turn to the problems at the back-end of the nuclear fuel cycle as Britain began to trade in spent nuclear fuel and waste. At the same time scientific experts and government (for example, the RCEP) were beginning to consider the problem and to formulate policies. The emerging environmentalism created a climate of public concern and debate and the media focused on the problem. Public confidence in the nuclear industry began to weaken and consequently undermined the public acceptability of the proposals for the management of nuclear wastes.

Above all, it was the opportunity presented by specific proposals that established nuclear waste as an issue of political conflict. The Windscale inquiry focused scientific and technical attention on nuclear waste whereas the HLW drilling proposals transformed the radioactive waste into a subject of popular local political protest. The long-established sea dumping of nuclear waste became a suitable target for a campaign led by environmentalists able to exploit the opportunities of media publicity and lobbying.

3.6 Summary

The third aim of this chapter is to show how radioactive waste has shifted from being a non-problem to a technical problem to a political problem. During the early period when nuclear activity was essentially military, radioactive waste was, politically at least, a non-problem. Shrouded in secrecy, the accidental, experimental and routine releases of radioactive wastes were not revealed to the public until much later, if at all. As the civil nuclear programme developed, governments and international agencies formulated guidelines and policies for the management of wastes but these were predominantly a technical matter arousing little political or public interest. This was largely true of the Windscale inquiry, the first occasion in which nuclear waste was subject to open public scrutiny and debate.

Arising from the technical debates of the RCEP and the Windscale inquiry, the HLW borehole drilling programme first precipitated local opposition to nuclear waste. Mutual aid between communities was a factor in increasing support and preventing the drilling programme. Use of publicity and lobbying techniques was a key factor in defeating the sea dumping programme. By 1983 radioactive waste was a subject of open public debate and conflict. The protagonists had become evenly matched and the context for the conflict over on-land shallow disposal of LLW and ILW had been set.

4 Radioactive waste – the political consequences

4.1 A problem of time and place

With the abandonment of the HLW drilling programme and of sea dumping of ILW and LLW, the options for radioactive waste management had been considerably narrowed. Accordingly, the government turned its attention to on-land disposal of ILW and HLW. In October 1983, less than two months after the ending of sea dumping, the Secretary of State for the Environment announced that 'effective disposal, in ways which have been shown to be safe, is well within the scope of modern technology' (*Hansard*, 25 October 1983, col. 156). NIREX simultaneously identified two sites which it considered 'most worthy of detailed investigation' (NIREX, 1983, p. 1). One was a disused anhydrite mine owned by Imperial Chemical Industries at Billingham on Tees-side for the disposal of long-lived ILW. The other site was a wartime munitions store at Elstow near Bedford owned by the Central Electricity Generating Board (CEGB) where a shallow repository would be constructed for the management of short-lived ILW and to provide capacity for LLW in addition to Drigg (for locations see Figure 7.4). This proposal inaugurated a political conflict lasting nearly four years, 1983–87. This conflict will be used to evaluate the political consequences and possibilities of radioactive waste management, the fourth aim of this chapter.

The fundamental cause of the political conflict is the anxiety about the effects on the local environment and health extending down the generations. Radioactive waste raises political problems of equity between places and between generations.

Q Can you indicate the nature of these problems of equity? Consider, first, equity between places and then equity over time between generations. In each case it may help to think in terms of costs and benefits.

A The problem of equity between places may be called the problem of *spatial equity.* A radioactive waste facility is a LULU (locally unwanted land use) except possibly in those places where such facilities already exist and are accepted by the local population. The costs of radioactive waste disposal (blighting effects, health risks, and amenity concerns) are imposed on specific localities. The benefits of nuclear electricity are spread widely over the community in general. Nuclear waste brings few local benefits apart from a small number of jobs and, in some cases, compensation in the form of community facilities, infrastructure and tax relief. Certainly compensation is regarded as a necessary component in any radioactive waste management proposal. But, the perceived costs outweigh the benefits, a point that is politically confirmed by the opposition to nuclear waste facilities at greenfield sites, illustrated in the case study in the next subsections. The problem of spatial equity is not confined within national frontiers; an example is the long-standing concern expressed by the Irish Republic over the potential threat posed by Sellafield.

Equity between generations is called *intergenerational equity.* The benefits of nuclear technology accrue to the present generation and, through technology transfer, arguably to immediate future generations. There can be little certainty that the technology will continue to be of benefit in the future while the cost of waste management will inevitably fall on generations far into the future. It is easier to estimate short-term benefits than to calculate long-term costs. The needs of the future tend to be discounted or ignored unless they have some palpable impact on the present.

The nuclear industry, governments and environmentalists agree on the need to manage radioactive wastes safely for local communities and future generations. Whereas the nuclear interests argue that safety is assured by the disposal methods proposed, their opponents claim that waste management proposals lack the technical verification necessary for public acceptability. The problem of uncertainty and credibility was at the heart of the conflict over radioactive waste in the UK during the mid-1980s.

4.2 A premature proposal

Billingham and Elstow were announced without prior consultation with the communities or any public discussion of the disposal methods and strategy which had led to the particular sites being selected. The two local communities mobilised public opinion and marshalled expertise in order to challenge the proposals on grounds of technology, strategy and acceptability.

The technology of deep burial at Billingham and shallow disposal at Elstow was justified by reference to scientific consensus and experience elsewhere. NIREX argued that 'there is a wide consensus that the use of existing technology can lead to the disposal of radioactive wastes in a very safe manner' (NIREX, 1983, p. 3). However, there was no technology comparable to that proposed in operation anywhere else. In the United States burial of wastes in salt vaults had been debated and a facility was under construction, but not in operation, in New Mexico. As for shallow burial, there was no experience of an engineered repository in a clay formation as proposed at Elstow. The French site was essentially an engineered surface facility and in Sweden an engineered repository was being constructed in crystalline rocks under the Baltic. In the United States where three sites were in operation and three had been closed, shallow burial was undertaken without the protective barriers proposed by NIREX. Drigg was similarly unprotected, though problems with its operation were used by opponents to undermine the credibility of the quite different engineered repository proposed for Elstow.

The selection of specific techniques and sites for disposal in advance of publicly evaluating other options indicated an absence of an overall strategy for radioactive waste management. NIREX and the government were exposed to the criticism that they had acted prematurely on the basis of a pragmatic rather than rational assessment of the possibilities. The fact that the Department of the Environment did not produce its policy statement, *Radioactive Waste Management: the National Strategy*, until the Sizewell B inquiry during 1984, seemed to confirm the absence of a coherent approach to the problem.

The absence of a publicly debated analysis including such factors as risk assessment, volumes and activity of waste arising and management options, enabled opponents to claim that the proposals lacked acceptability. They argued that any public inquiry into specific site proposals must include, and preferably be preceded by, an analysis of the generic issues and a comparative assessment of potential sites. The premature announcement of the two sites enabled the local communities to appeal to general principles rather than simply exhibit a NIMBY reaction. This gave widespread credibility to their campaigns against the proposals (Figure 7.6).

4.3 The power of local action

Opposition to the proposals in principle was backed by campaigns against the specific choice of Billingham and Elstow. In each community public meetings were held shortly after the announcement, and pressure groups, each with the acronym BAND (Billingham/Bedfordshire Against Nuclear Dumping) were set up. At Billingham the repository already existed, connected by road and rail within a conurbation of half a million people. The proposal to dispose of long-lived alpha-emitting wastes there provoked great anxiety. Billingham's BAND organised a petition of 85 000 signatures, lobbied Parliament, and held vigils, public meetings and demonstrations (Figure 7.7). Imperial Chemical Industries (ICI) who had at first co-operated with NIREX were sensitive to the pressures of local opinion and announced in March 1984: 'Having carefully considered the implications that the proposal would have for its business, ICI has concluded that it would not be in the company's best interests and is therefore opposed to it.' This was to prove decisive since NIREX was thereby denied access to the mine and the government would be sensitive to the power of local opinion.

The conditions in Bedfordshire were different. The Elstow site was in a rural area, though close to a town of 50 000 people, and its owners, the CEGB, with no significant investment in the area, a partner in NIREX, and a major creator of radioactive wastes, were unlikely to capitulate to local opposition. Moreover, Elstow was destined to take the less dangerous short-lived ILW and also LLW for which a disposal route at Drigg already

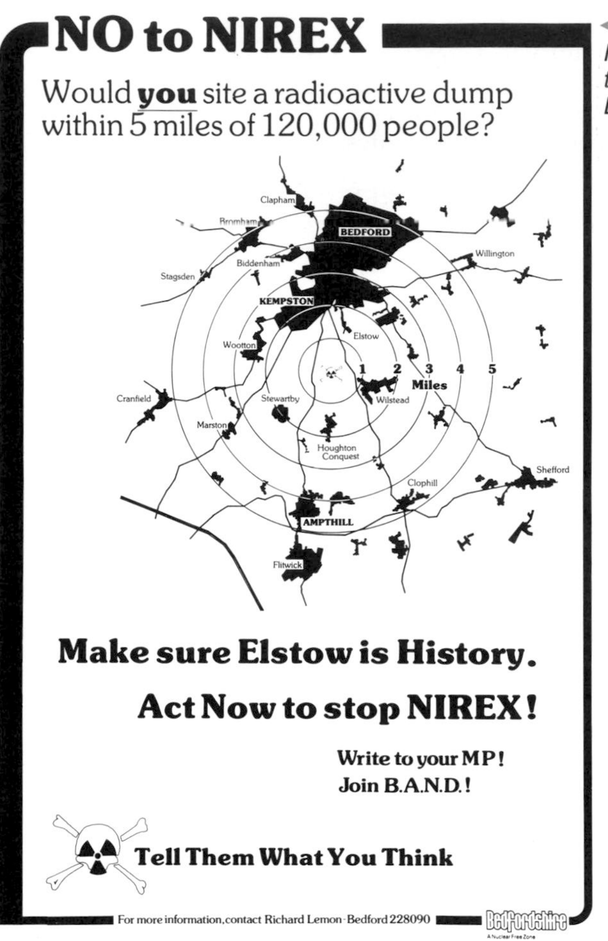

Figure 7.6 Publicity material used in the campaign at Elstow, Bedfordshire.

existed. The campaign against the Elstow proposal, after an initial flurry, lacked the national political impact generated at Billingham. But the Bedfordshire County Council were employing technical consultants and developing technical counter-expertise to be used in the event of a public inquiry. In particular they challenged the selection of Elstow on grounds of hydrogeology, land use and location, arguing that a comparative assessment of sites should be undertaken.

Billingham's campaign was unequivocally successful. In January 1985 the Secretary of State for the Environment announced that NIREX had agreed 'not to proceed further with that site'. Bedfordshire's more muted campaign had secured an important concession that NIREX should 'select and announce two further sites for investigation in addition to Elstow'.

Q Reflecting on what you have just read, what do you think were the main political reasons for the government's abandonment of the Billingham proposal and its agreement to a comparative assessment of sites for shallow burial?

A Whatever the technical arguments, the proposals had been presented without prior consultation and in the absence of a publicly debated national strategy. Two specific sites had been selected, one for each

▲ *Figure 7.7 Billingham Against Nuclear Dumping take a petition to Downing Street.*

category of wastes, without any reasoned justification of the site selection process. This provided the local communities with an opportunity to oppose the sites on grounds of general principles and the lack of a comparative assessment. They were able to generate technical expertise, mobilise local opinion and argue on technical and ethical grounds rather than self-interest. The rationality and political impact of their arguments persuaded the government to retreat. By abandoning Billingham, only shallow disposal on land was left as an option under active consideration.

4.4 Collaboration and coalition

It was over a year before three (rather than two) further sites for investigation were announced in February 1986. They were a site owned by the former Central Electricity Generating Board (CEGB) in an industrial and port area at South Killingholme on the Humber estuary; a former airfield owned by the Ministry of Defence at Fulbeck in the farmlands of south Lincolnshire; and a site near one of the early Magnox power stations at Bradwell on the Blackwater estuary in Essex (for locations see Figure 7.4). Local action groups sprang up in each area using the familiar acronym (HAND, LAND, NAND, EAND; Humberside/Lincolnshire/Nottinghamshire/Essex Against Nuclear Dumping) and emulating the tactics – press publicity, petitions, and fund-raising events – already employed in Billingham and Elstow (Figure 7.8).

▲ *Figure 7.8 Lincolnshire Against Nuclear Dumping protesters at Fulbeck.*

The opposition to individual sites was consolidated by alliances between the four areas which transcended purely local interests. Three of the county councils (Bedfordshire, Humberside and Lincolnshire) formed a County Councils Coalition, which pooled technical information, employed Parliamentary lobbyists and agreed tactics and a longer-term strategy to defeat the proposals through the public inquiry process. For local political reasons, the fourth county council, Essex, did not join the coalition. The four local action groups, together with Billingham's BAND, established a national organisation, Britain Opposed to Nuclear Dumping (BOND), to co-ordinate publicity and political action. Both the Coalition and BOND emphasised the case against the concept of shallow disposal as proposed by NIREX, leaving the local groups to identify the case against the specific sites. The political context for success for these alliances proved propitious.

In March 1986 the House of Commons Environment Committee published a Report on Radioactive Waste which confirmed the Coalition's view that the UK was 'still only feeling its way towards a coherent policy' and that the NIREX proposals revealed a 'premature and uncoordinated approach' (HMSO, 1986, paras 3 and 58). The Report also criticised the method of classifying wastes, the lack of adequate research and development into disposal options and the failure to overcome public anxiety by greater openness and discussion. Not surprisingly the Coalition was gratified 'to find such an overwhelming weight of opinion coming round to our way of thinking'.

On 26 April, 1986 the world was transfixed by the disaster at Chernobyl. The immediate consequences were over 30 deaths and the evacuation of the surrounding area, while the radiation cloud contaminated a vast area to the north and west affecting livestock and food supplies. The longer term effects are still being calculated but it is already clear that premature deaths and genetic effects will persist far into the future. Public opinion on nuclear energy, already sceptical as a result of the leakages and fears of leukaemia at Sellafield, was further hardened by Chernobyl. In the wake of the disaster and, in the light of the House of Commons Report, the government withdrew its proposals to bury short-lived ILW in shallow repositories.

The government intended to pursue its remaining option, the shallow burial of LLW, by means of a Special Development Order (SDO). An SDO is a device whereby Parliament, not the local authority, gives permission for a development, in this case exploratory drilling at each of the four sites. This decision was regarded by the local communities as a means of avoiding an early debate on the general issues, enabling exploration to identify the best site and to hold a subsequent public inquiry which focused mainly on the comparative advantages of that site.

With the SDO in operation, contractors were appointed to undertake exploratory drilling at each site. When they arrived at South Killingholme, Fulbeck and Elstow in August 1986, and at Bradwell in September, they were confronted by well-organised blockades which prevented entry to the sites. Eventually the contractors secured entry to the sites but not before the blockades had generated sufficient publicity to ensure that nuclear waste became a national political issue.

Members of the County Councils Coalition toured nuclear waste facilities in Sweden, Germany and France and pronounced these countries 'far more advanced than the UK in the development of policies and practices for the disposal of radioactive waste' (County Councils Coalition, 1987, p. 1). They recommended that co-disposal of both ILW and LLW in a deep repository was a cost-effective and safe means of managing nuclear wastes and would command public acceptability.

It was by now clear that shallow burial lacked political credibility. Opposition had been generated crossing political, social and geographical boundaries. Although each site was in a safe Conservative-held constituency, the government was undoubtedly sensitive to the hostility the proposals had generated. Accordingly, on 1 May, 1987, the eve of the announcement of a general election, the then Secretary of State for the Environment, the late Nicholas Ridley, withdrew the proposals for shallow burial. In an exchange of letters with the Chairman of NIREX he accepted the case for co-disposal of ILW and LLW in a deep repository as 'the responsible course of action'.

Activity 4

Can you suggest why the comparative assessment of four sites weakened rather than strengthened the government's position?

In terms of future policy two points had been established. One was that disposal in a deep repository would be the *technical* solution to the problem. The other was that disposal in a greenfield location was unlikely to be a feasible *political* option.

4.5 *Retreat to the nuclear oases*

With the surrender of the shallow burial option, the government and NIREX had to begin all over again in the search for a technically suitable and politically feasible solution. It was evident that certain political lessons had been absorbed when NIREX published a discussion document, *The Way Forward*, towards the end of 1987. It aimed 'to promote public understanding of the issues' (p. 4) by open discussion on site selection, environmental protection and local participation in the development. Although the technical options were limited to deep disposal of ILW and LLW, three concepts (under land, off shore and under sea) were under consideration. Potentially suitable sites were likely to be found in areas of hard rock in Scotland, islands mainly off the Scottish coast and sedimentary rocks stretching down the eastern side of England and down the north-west from the Solway Firth to the Wirral. There was, therefore, a large territory in which to find a politically suitable site.

The consultation revealed, predictably, no unanimity of view. There was general support for the concept of deep disposal in principle but opposition to specific sites. There was a sense of political inevitability in the eventual choice, in March 1989, of Sellafield and Dounreay as sites for further investigation. In *The Way Forward*, Sellafield, as the main source of ILW, was considered 'worth careful scrutiny'. During the consultation the Highland Regional Council had recognised the potential of Dounreay, and Caithness District Council had indicated that NIREX should be invited to test for a repository. Copeland District Council, which includes Sellafield, welcomed the new approach to disposal, and British Nuclear Fuels (BNF) announced plans to investigate the possibility of an offshore sub-seabed repository. In making the selection of these two remote nuclear oases Nicholas Ridley commented, 'it would be best to explore first those sites where there is some measure of local support for civil nuclear activities' (*Hansard*, 21 March, 1989, cols 505–6) (Figure 7.9).

In both areas there was nevertheless opposition to the proposals. In Cumbria this was part of a general opposition to Sellafield and the expansion of the THORP plant through imports of spent fuel which would increase the ILW and LLW arisings. There was also support for THORP, especially among the workforce at BNF, and this was crucial in the final decision to go ahead with the plant in 1995.

In Caithness, the local group Caithness Against Nuclear Dumping (CAND) applied pressure on the local councils to oppose test drilling but was careful to confine its opposition to NIREX rather than the Dounreay fast breeder reactor plant where jobs were being lost as the project was run down. CAND held a referendum of the local population, and in a 58% turn-out there was a three to one majority against a nuclear waste repository in the area. The level of support was much the same five years later in December 1995, when another local Caithness referendum, in a 52% turnout of a 21 000 electorate, resulted in 65% voting against plans to reprocess further foreign fuel at Dounreay.

Dounreay and Sellafield, nevertheless, offered the most promising political prospects for a repository, and permission to drill was granted by the appropriate Secretary of State. Provided one or other site was found to be technically satisfactory, it appeared that a political solution had been found.

Thus in September 1995, the first stage of public examination of NIREX's technical proposals for a deep repository at Sellafield began. A public inquiry was held into a proposal to build a rock characterisation facility (RCF), i.e. a subterranean rock laboratory costing around £200m. However, local and international protesters (including Irish MPs and the Irish Government) in their evidence widened the proceedings to examine the second stage, the repository itself, the go-ahead for which was dependent on a positive outcome on the RCF application. The earliest any

Man who hopes to see N-dump under his garden

❑ JOHN VASS

Today, we present one side to the nuclear waste disposal sites controversy and tomorrow the other.

First, the case as put by a leading supporter of a national waste repository at Dounreay – a Highland Regional councillor who expects it to go, not in, but extending under his back garden, for he lives within a mile of the Dounreay reactor.

He is consultant engineer Roy Gilbert, who worked at the Dounreay plant for many years. He claims to have been 30 years in the nuclear industry. He says the prospect of sleeping above a nuclear dump holds no fears for him.

He believes the nuclear power industry has clearly demonstrated there is no supportable objections on safety grounds and that the proposed low and intermediate radioactive waste disposal facility is innocuous in general industrial terms – and quite insignificant, radiologically.

He said: "I have difficulty seeing the hazards suggested by the opponents of the Nirex project. It is an inert material we are talking about, and the low-level waste which is discarded – overalls, gloves and suchlike – is really quite safe to handle. And when the intermediate waste is mixed in cement slurry as it will be, that too will be perfectly safe."

He added: "However, I can understand people being frightened about it because of the very able public-relations campaign run by the groups opposing Nirex.

"People are swayed by all the emotive stuff they put out, but it's codswallop, mainly. Also, they are bombarded with scare stories in the media.

"If people want to oppose the Nirex project on political grounds, that is fine by me but what I cannot accept are those who say their opposition is really for the sake of their children and their children's children."

> **Quote** "From my point of view, I do believe that, if you look carefully at the 'expert opinion' on offer, you will find that a vast majority of that which is anti-nuclear is anything but expert, but is riding a hobby-horse of some political shade."
>
> *Councillor Roy Gilbert*

He went on: "Some 10 to 15 years ago, the protest organisation 'Friends of the Earth' and their militant-offspring 'Greenpeace' embarked on a major long-term campaign to besmirch, denigrate and discredit the nuclear power industry. They have gathered other smaller, less professional but no-less vocal groups around them. This would have been of little consequence but for the enthusiastic attention they received from the mass communications media."

Mr Gilbert went on: "However, it is not good enough to ignore facts in preference to a perception which has been deliberately and pejoratively generated.

"For a small budget, the professional protesters – those who make a comfortable living campaigning against most commercial, industrial or developmental ventures, have been enormously successful in gaining acceptance of their erroneous image on nuclear waste disposal."

But Mr Gilbert accepts that Nirex have to mount a major public relations campaign and put across the proposals in acceptable form to the public to counter the anti-Nirex propaganda.

"The essence of the problems surrounding the Nirex proposals is, quite simply, to be found in public perception," he states.

With Government funding being withdrawn from the prototype fast reactor programme in 1994, and its associated fuel re-processing plants in 1997, the Atomic Energy Authority are desperate for ways to raise money to sustain these.

> **Quote** "I cannot see that a nuclear-waste repository would have the slightest effect on farming and fishing."
>
> *Councillor Roy Gilbert*

Although the AEA feed the National Grid with vast quantities of electrical power, they receive just over a penny a unit – although it costs many times that to produce – so there is no chance of that becoming a money-spinner. Many people in the industry, therefore, see the prospect of a nuclear waste repository as a major source of revenue, as well as a provider of several hundred jobs.

Mr Gilbert says he accepts this is an obvious and strong influence on the thinking of many people in Caithness.

Nirex have not yet started drilling exploratory boreholes at Dounreay, although preparatory work is in hand, said Nirex spokesman Mr Harry Hudson yesterday.

"Drilling is not expected to start for another month," he added. Nirex had intended a series of public meetings to explain their general plans. However, these have been called off because of the sudden illness of the company's managing director.

But in a recent briefing, the company pointed out that a further programme of seismic surveying is planned, together with detailed transport and environmental studies to create a three-dimensional "picture" of the site at Dounreay. They point out these investigations will take them into next year and it could be 1992 before a final decision is taken on the suitability of Caithness as a site.

Similar investigations are under way at Sellafield in Cumbria.

Nirex expect their proposals will go to a Public Inquiry in 1994 and, assuming permission is granted, would expect the repository to be open for business in 2005.

▲ *Figure 7.9 An example of pro nuclear waste feelings from* Aberdeen Press and Journal, *31 July 1990.*

waste could be placed in such a repository – estimated cost £1.8bn – would be around 2012 (Royal Society, 1994; NIREX, 1995).

Q In what respects can Sellafield and Dounreay be said to represent a political solution to the problem of radioactive waste? In considering your answer consider why these sites are likely to be publicly acceptable.

A Three broad reasons can be identified for the emergence of these two sites. (1) The strategy for radioactive waste management had lacked coherence and credibility. Disposal methods had been identified and sites had been chosen before options had been debated. The defeat of the shallow burial proposals left deep disposal as an option which attracted considerable support. Moreover, the consultation provided by *The Way Forward* gave an opportunity for debate about strategy. (2) The conflict over Billingham and also over shallow disposal had demonstrated that it would be politically extremely difficult to secure a greenfield location for the disposal of nuclear waste. (3) Accordingly, it would be prudent to confine the search to places where some solid support for the nuclear industry could be guaranteed. Sellafield and Dounreay fully satisfied this criterion.

Q What factors might weaken the opposition in these areas?

A Again, there seem to be three factors to consider. (1) The local communities are divided between supporters (usually the workforce and dependents) and opponents of the nuclear industry. (2) A consensus favouring Dounreay or Sellafield might be the natural consequence of the relief felt in all other communities at not being selected. If it is accepted that a solution is necessary then the majority will be grateful that it is found somewhere else. (3) The very success of the campaigns against Billingham and shallow disposal make it more difficult to resist future proposals. Any government would proceed more cautiously, seeking public approval for its actions. Provided there are no technical obstacles and the consensus favouring deep disposal is maintained, Sellafield or Dounreay will remain the most likely candidates for a radioactive waste repository.

4.6 *Summary*

A case study has been used to evaluate the political consequences and possibilities of radioactive waste management policies, the fourth aim of this chapter. After the abandonment of HLW drilling and sea disposal, the government and NIREX turned to on-land disposal of ILW and LLW. Two sites, at Billingham for long-lived ILW, and Elstow for short-lived ILW and LLW, were announced without public consultation before a national waste management policy had been debated. This provided the local communities with an opportunity to oppose the strategy in principle as well as in terms of specific site selection criteria. The Billingham proposal was dropped and a comparative evaluation of four potential shallow burial sites, including Elstow, was agreed. By cooperating in a campaign against the concept of shallow burial the four communities were able to apply combined political pressure culminating in blockades which, eventually, secured the withdrawal of the plans. These events left co-disposal of ILW and LLW in a deep repository as the technical option most likely to achieve public acceptability. The political possibilities suggested the nuclear oases at Sellafield or Dounreay as the likely candidates.

5 The political challenge

5.1 Parallels and contrasts

There are over 400 reactors in the world producing nuclear electricity, with over two-thirds of them in North America and western Europe. The major producers in these areas are the United States (109 reactors), France (56), Germany (34), UK (22), Canada (18), and Sweden (12). The only other substantial producers are Russia (29) and Japan (49). The number of reactors does not necessarily reflect the installed capacity (as of 1995). For example, Germany produces more electricity from fewer reactors than the UK since its average reactor size is larger. In addition to civil nuclear energy, there are seven nations (USA, Russia, UK, France, China, India and Israel) with developed nuclear armaments and others with the technical basis for developing them. Nuclear wastes from electricity generation, military programmes and reprocessing are, therefore, a problem confined mainly to the developed countries. But the fallout from bombs, the dangers of proliferation and the possibility of reactor accidents give the nuclear industry, and the residues it creates, a global environmental dimension.

The political response to the environmental problems created by the nuclear industry provides both parallels and contrasts with other global environmental problems. Our fifth aim is to examine the extent to which the politics of nuclear waste illustrates a developing environmental challenge to conventional political assumptions and arrangements. This needs to be considered in two stages. First we must establish how far the political features of nuclear waste in the UK, discussed in the previous section, are paralleled elsewhere. Second, we need to consider whether the politics of nuclear waste is unique or has features in common with other environmental political issues. We can only give a brief outline of these issues here but the book by Blowers *et al.*, referred to in the 'Further reading' section, examines some of them in more depth.

5.2 Experience in other countries

The political priorities for nuclear waste depend on the technical nature of the problem, the political institutions and the political geography of the country concerned. The contrasts and similarities to the UK's experience can be illustrated by brief reference to four major producers, United States, Sweden, France and Germany.

United States

The politics of nuclear waste in the United States has four key elements. First is the continental scale of the country which provokes conflict between the east, with the vast majority of the reactors, and the west, which harbours most of the remote locations for disposal. Second, the absence of civil (as opposed to military) reprocessing means that the most urgent political priority is to find an acceptable location for the deep disposal of highly radioactive spent fuel. Third, the separation of the military from the commercial waste streams has meant that political opposition to nuclear waste projects has focused mainly on the civil programme, which is most open to constitutional challenge. And fourth, political challenge to policies is encouraged by the participative nature of the political institutions of the country.

Search for a suitable site for the disposal of spent fuel has been going on since the 1960s but the Nuclear Waste Policy Act (NWPA) of 1982 set out the strategy for developing a deep geological repository for HLW. Nine sites in the west which were under review at the time were reduced, first to a short-list of five and then to three remote sites: at Hanford in Washington state, and sites in western Texas and Nevada. A necessary compromise had been that a site in the east of the country would be identified before construction could begin at the chosen western site. The announcement of 12 sites in seven eastern states in 1986 provoked intense opposition, forcing the federal government to withdraw the proposals. This, inevitably, intensified opposition from the three western states. State-wide opposition was sufficient to overwhelm the local support for the project in Hanford. In Texas there was tough opposition from the local farming community. In Nevada there was scarcely any local population but the state opposed the plan. The possibility that the three states could mobilise support from corridor states through which the waste would be transported was averted by an Amendment to the NWPA which limited investigation to the Nevada site, thus limiting the opposition and providing a political solution if the site proved technically suitable. However, vigilant opposition to the use of a site under Yucca mountain adjoining the military nuclear test site has led to extensive and expensive delays in attempts by the contractors to the Department of Energy to carry out full site characterisation prior to starting construction of the laboratory.

Technical problems rather than political opposition were responsible for the delays in commissioning the Waste Isolation Pilot Project (WIPP) near Carlsbad in the desert of New Mexico (see Figure 7.2). This new project was welcomed in an area where the potash industry was failing and in a state where Los Alamos, the Sandia Laboratories and other nuclear activities were a powerful interest. Elsewhere nuclear waste projects have been resisted by state governments, although there may be local support from nuclear dependent communities. There has also been the development of a new form of nuclear dependent community, exemplified by the battle between two leaders of the Apache people at the Mescalero Reservation in New Mexico. The government-appointed nuclear waste 'negotiator' persuaded tribal leader Silas Cochise, great-grandson of warrior chief Cochise, to accept a government-funded project to create an interim store for spent fuel on the Apaches' 'Sacred Mountain'. A sum of $250m was offered to the Apaches as a form of compensation for taking the waste. Joseph Geronimo, grandson of the legendary Leader Chief, opposed taking 'White Man's Gold'. But by a vote of 593 to 372 the tribal elders accepted, thus creating a possible voluntary nuclear oasis.

Low-level nuclear waste has also created political conflict at local and state level. In South Carolina, Nevada and Washington state, growing opposition to the three operating shallow burial facilities led to the Low Level Radioactive Waste Policy Act of 1980. Under the Act the three sites would eventually be closed and new greenfield sites would need to replace them. The new sites would receive the LLW arising within each regional compact set up for the purpose of managing LLW. It has been difficult to establish the host state for a site within each compact, and political feasibility will determine the selection of individual sites. The most likely places will be remote locations, areas in economic decline and areas with present or past experience of the nuclear industry.

Sweden

Nuclear power stations at four locations on the coast supply about half Sweden's electricity. The strength of environmental concern about nuclear power was demonstrated in the referendum of 1980 which effectively committed the government to phase out the nuclear programme by around the year 2010, the end of the life of the latest reactors. The total future burden of radioactive waste can, therefore, be estimated and plans laid for its management. Four basic principles are applied to nuclear waste management in Sweden: very high safety requirements; the use of best available technology; the highest degree of national independence; and no burdens on future generations (except some site monitoring).

The waste management policy includes a central store (CLAB) for operating on the coast near the power station at Oskerhamn. A hard rock laboratory is under construction nearby which will test disposal solutions. A deep disposal site for HLW will eventually be selected from among various candidates. ILW and LLW are disposed of in the purpose-built sub-seabed repository next to the Forsmark power station about 160 km north of Stockholm. Sweden exemplifies an approach that applies 'to the entombment of radioactive waste the sort of devotion that the ancient Egyptians gave to the burial of their Pharaohs' (*The Times*, 20 May 1986). The 'Swedish solution' has attracted wide acclaim in other countries but is still treated with scepticism by opponents of the nuclear industry in Sweden. Despite the efforts to manage wastes according to rigorous standards, opposition from national and local environmental groups has focused especially on the test drilling programme for a HLW repository.

France

France is also heavily dependent on nuclear energy (around 70% of the country's electricity) but, unlike Sweden, the state maintains an aggressively pro-nuclear stance. As the world's second largest producer of 'nuclear' electricity and a military nuclear power, France has the full range of nuclear cycle facilities located on coasts and rivers throughout the country. Four potential sites in different rock formations have been identified for a HLW repository. By 2006, the French should be in a position to choose between two potential repository sites where rock laboratories will have been developed. LLW and short-lived ILW are already disposed of at a permanent surface disposal facility at Cap de la Hague, and another permanent facility is now operating at Soulaines in the Champagne region east of Paris.

There have been protests against the HLW plans, and the Soulaines site was challenged in the Supreme Court. But France has not experienced the continuing and coherent opposition to the nuclear industry that has been typical in other western countries. There is a strong industry–state relationship through which the nuclear industry is controlled and subsidised. Policy is developed through a highly centralised system which minimises political debate over the implementation of nuclear projects. Opposition is further limited by the favourable publicity given to nuclear energy and the investment and inducements that are provided to local areas which host nuclear plants. Where necessary the state has not been averse to the use of force to suppress dissent, as in the case of a mass demonstration against the Superphenix fast breeder reactor in 1977 when a demonstrator was killed. Opposition to nuclear energy and radioactive waste projects has been slowly emerging but is likely to remain passive in the face of the country's heavy dependence on the nuclear industry for electricity and defence. However, some concern has emerged over possible transport risks involved in shipping HLW back to Japan, or other reprocessing wastes back to Germany, linked to political campaigns in the recipient countries.

Germany

By contrast western Germany has experienced political convulsions over nuclear energy and waste. It gets around a third of its electricity from nuclear stations but there are no reprocessing facilities. Protest against aspects of the civil nuclear programme is often linked to opposition to the deployment of nuclear weapons and more general environmental concerns. During the 1970s mass protests were held against the construction of new power stations, and during the 1980s consistent protests were a major factor in the cancellation of the Wackersdorf reprocessing project in Bavaria. During these decades there were successful protests against the neutron bomb and Cruise and Pershing missile deployment.

A proposal for a combined reprocessing and nuclear waste facility at Gorleben in Lower Saxony close to the border between the former East and West Germany provoked a series of demonstrations in a conservative rural area during the late 1970s and 1980s. The reprocessing proposal was withdrawn and proposals for permanent nuclear waste management have been deferred. Evidence that LLW waste consignments exported to Belgium for treatment contained highly radioactive materials revealed a major scandal involving illegal payments and corruption and created additional fears about nuclear proliferation. Consequently, the German government has become highly sensitive to nuclear issues. Protest groups have demonstrated an ability to frustrate the plans and programmes of the nuclear industry. These protests culminated in a number of large German power utilities cancelling future options on reprocessing contracts abroad. Additionally, there have been attempts formally to incorporate environmental NGOs and the German Green Party into the formulation of future nuclear policy via an innovative round-table forum called the 'energei Konzensus', which involved senior power utility representatives and the main political parties, Socialist Democratic Party, Free Democratic Party and Christian Democratic Union. Inevitably this attempt to establish a common position involved political compromise, one deal involving going ahead with the interim fuel store at Gorleben.

5.3 Summary: political outcomes of radioactive waste

These brief sketches are confined to major western producers of nuclear waste. It is possible to make some general observations about political outcomes but more evidence from a wider range of countries would be needed before generalisations could be made with reasonable assurance. With this reservation in mind we can derive some general conclusions about the politics of radioactive waste.

In nearly every country nuclear waste management policies and proposals have generated opposition once they become known in detail or location. There are contrasts between countries in the character of the opposition. In some cases such as Germany, nuclear waste is one aspect of opposition to the nuclear industry and nuclear weapons. In several countries there are both national environmental groups and local action groups opposed to proposals. In the United States opposition is mainly focused at state and local level and in France the opposition is, as yet, relatively undeveloped. The target of the opposition will depend on the political priorities in individual countries. In countries committed to reprocessing, like Britain and France, HLW is confined to one or two sites and need not be disposed of for several decades. The large volumes of ILW and LLW that are accumulating in these countries are the main waste problem, leading to a search in the UK for appropriate disposal sites. In the United States, Sweden and other countries spent fuel is accumulating at power stations, posing problems of storage space and cost, dangers to the workforce and the risk of proliferation. In these countries the development of a HLW repository has become the political priority for nuclear waste management.

Where it has developed, the opposition to nuclear waste proposals has derived its strength from its broad appeal in two ways. First, it has been able to combine the defence of specific local interests from external threat with an appeal to more general concerns about the long-term threat to the environment. This focus on issues that affect society as a whole has enabled the opposition to generate widespread interest in the problem among the media and the public. Second, the ability of local groups to form coalitions, emphasising the general principles rather than the detailed local aspects of proposals, has endowed the opposition with conspicuous political power, as demonstrated in the UK, for example, during the 1980s. This power has been used to frustrate policies, defer plans and halt the development of specific sites in a number of countries.

A form of political action has emerged outside the realm of party politics that combines the local with the national and international, the short with the long term, and self-interest with the public interest. Centrally determined policies have been undermined by local activists, fighting along a broad front using counter-expertise, publicity and lobbying. Governments have been willing to concede since thereby they may be relieved of politically unpalatable decisions.

The success of the opposition has narrowed the options for radioactive waste management. It has made political calculations pre-eminent in the search for solutions. Governments have been willing to see controversial projects dropped or delayed to avoid political sanctions. The pragmatic retreat to nuclear oases offers a way of minimising the political problem though it may not offer the best technical solution. Deferral leaves decisions to be taken by future generations. On the other hand, decisions once taken remove options for the future. Solutions require heroic assumptions about safety and security of repositories that have no empirical basis. These considerations make continuing conflict over nuclear waste more likely than consensus.

6 The future

6.1 Prospects for future nuclear waste disposal

Political consensus must rest on an agreement on the appropriate technical means of radioactive waste management. The arguments for and against deep disposal were presented earlier in Activity 1. Some environmentalists argue that disposal of the waste should not be undertaken until the technical problems have been resolved. They advocate continuing on-site storage and monitoring of wastes as the responsible methods of ensuring the safety of the present and future generations. From the anti-nuclear perspective of these groups, on-site storage ensures the visibility of the problem of radioactive waste and emphasises the lack of an acceptable solution. This would act as a brake on the further expansion of the industry.

The interim storage of HLW at a central location is an option being developed in some countries. Sweden already has a facility for the storage of spent fuel and the United States has undertaken an extensive search for possible sites. In the UK and France HLW from reprocessing is stored and vitrified. In all these countries storage is a preliminary stage for cooling prior to final disposal of the wastes in a deep geological repository.

Deep geological disposal of the most dangerous wastes is now favoured by a broad consensus of government, scientific and public opinion. Deep disposal is justified on the grounds that it concentrates the waste at a few (usually remote) locations where the waste can be monitored and (at least initially) possibly retrieved, although this option involves considerable transportation. By contrast, storage requires permanent routine monitoring at many sites, exposes workers and the local community to the risk of radioactivity and increases the dangers of nuclear proliferation.

Disposal requires the selection of sites that are acceptable on both technical and political grounds. Political acceptability depends partly on the technical criteria adopted and these are likely to become more rigorous under the impact of sceptical public and scientific opinion. The accepted technique is to emplace waste that has been carefully packaged in containers in a repository sealed by multiple barriers in a suitably stable hydrogeological medium.

Technically suitable sites will be increasingly hard to find and, in some countries, the technical standards may be impossible to meet anywhere. We saw in Section 3 how the problem of nuclear waste shifted from being a technical to a political problem. In Section 4 we discussed how the political options were narrowed to a search for nuclear oases where disposal could be achieved either because of the presence of the nuclear industry or the need for economic development. While nuclear oases may offer a pragmatic solution they may not always be available, or, if available, may prove to be technically unsuitable. Nuclear waste does not create many jobs and it brings long-term blight and risk so that it may be unwelcome even in nuclear oases. This is the case in Caithness where community support for the UKAEA's Dounreay plant contrasts with the overwhelming opposition to NIREX.

Any political solution to the problem of radioactive waste will need to consider the principles of spatial and intergenerational equity that we discussed in Section 4.1. Compensation should be provided to those local communities hosting a site regardless of whether they willingly accept a new waste disposal facility. Compensation might be economic (tax relief, local investment), environmental (improvement of the area), or in the form of infrastructure, social and public facilities. Local communities will expect careful monitoring and safety requirements, adequate information and also some participation in decision-making.

Intergenerational equity requires that the burdens imposed on future generations from the actions of the present are minimised. The burden is already unavoidable since nuclear waste exists and is accumulating. An argument for deep disposal is that it spatially limits the future problem. But if it encourages the belief that the problem is solved it may justify the expansion of the nuclear industry. In the prevailing uncertainty about the environmental safety of nuclear waste the only way of minimising the burden on future generations is to limit the further creation of wastes. The development of nuclear energy has already been halted for commercial reasons in the United States and the United Kingdom. The problem of nuclear waste provides a further reason for restraining the growth of nuclear energy. Given the economic and environmental problems and political insecurity encountered with other major sources of energy, abandonment of nuclear energy is not an option likely to be entertained in the near future.

6.2 Nuclear, chemical and atmospheric hazards

Here we briefly consider whether the politics of nuclear waste indicate a developing trend in other areas of environmental politics. A comparison can be made with toxic and hazardous wastes, a subject covered in Chapter 6, and with global environmental problems which will be discussed in the last book of this series (*Blackmore & Reddish, 1996*).

1 *Toxic and hazardous wastes.* As with radioactive wastes these are associated with dangerous technologies with low probability/high consequence risks. Disposal of toxic and radioactive wastes creates locally unwanted land uses and there is a need for long term security and surveillance. Fears about local risk and blight, about the dangers to future generations and concern about international trade and dumping have stimulated effective local action and influenced government policies, as we saw in Chapter 7.

Despite these similarities, toxic and hazardous wastes have not generated political movements with the same permanence and power that have emerged in the case of radioactive waste. Toxic and hazardous wastes are not *directly* connected to destructive military purposes. The fear of radioactivity is greater and more pervasive than any other socially induced risk.

2 *Regional and global environmental issues.* These include acid rain, the destruction of the ozone layer and global warming. Like nuclear waste they have achieved considerable political prominence but they are quite different political issues. Regional and global issues, by definition, affect wide areas (perhaps the whole world). While this may cause general anxiety it does not have direct and immediate impact on specific localities, as is the case with nuclear waste. In consequence, regional and global issues do not give rise to local opposition and action; they are handled politically by governments.

These problems tend to raise rather different environmental concerns. They are fundamentally transboundary problems with complex causes and effects that are matters of scientific debate and disagreement. (You might like to refresh your memory on some of the scientific problems and political implications of transboundary pollution by re-reading Chapter 6, Section 4.3). Solution to such problems requires international control and co-operation. National interests may be in conflict with more general international interests. For example, the UK's refusal to join the 30% Club (those countries that agreed to reduce sulphur dioxide emissions by 30% between 1980–93) was strongly influenced by its unwillingness to pay for desulphurisation when most of the benefits were reaped by other countries.

The chances of achieving co-operation in the event of scientific consensus and political agreement on the need for action to prevent global warming are extremely remote. It will require a major reduction in energy consumption in the developed countries and massive compensation to developing countries who forego the exploitation of their resources. The international political arrangements needed to introduce and monitor such changes barely exist, and are also likely to be subverted by national self-interest. These scientific and political problems are the main focus of the next book in the series (*Blackmore & Reddish, 1996*).

By comparison with these problems, nuclear waste seems an easy problem to solve. It is a distinctive political problem which has produced political outcomes that are not paralleled in other environmental problems. Nuclear waste accumulates at specific sites from identifiable industrial processes. It is under the jurisdiction of individual governments. The management of radioactive wastes can be controlled within countries provided that there can be agreement on the methods.

References

BLACKMORE, R. & REDDISH, A. (eds) (1996) *Global Environmental Issues*, Hodder & Stoughton/The Open University, London, second edition (Book Four of this series).

COUNTY COUNCIL COALITION (1987) *The Disposal of Radioactive Waste in Sweden, West Germany and France*, Prepared for the County Councils Coalition by Environmental Resources Ltd.

HMSO (1955) *A Programme of Nuclear Power*, White Paper, Cmnd 9389, HMSO, London.

HMSO (1976) *Nuclear Power and the Environment*, Sixth Report of the Royal Commission on Environmental Pollution, Cmnd 6618, HMSO, London.

HOUSE OF COMMONS ENVIRONMENT COMMITTEE (1986) *Radioactive Waste*, Session 1985–6, First Report, HMSO, London.

NIREX (1983) *The Disposal of Low and Intermediate-level Radioactive Wastes: the Elstow Storage Depot*, A Preliminary Project Statement, NIREX, Harwell.

NIREX (1987) *The Way Forward: a discussion document*, NIREX, Harwell.

NIREX (1989) *Going Forward: the development of a national disposal centre for low and intermediate-level radioactive waste*, NIREX, Harwell.

NIREX (1995) *Evidence to RCF Inquiry*, NIREX, Harwell.

ROTHSTEIN, L. (1995) 'Nothing clean about clean-up', *Bulletin of the Atomic Scientists*, May/June 1995, pp. 34–41.

ROYAL SOCIETY (1994) *NIREX Annual Report*, 1994–95, Royal Society, London.

SARRE, P. & REDDISH, A. (eds) (1996) *Environment and Society*, Hodder & Stoughton/The Open University, London, second edition (Book One of this series).

US DEPARTMENT OF ENERGY (1995) *Closing the Circle on the Splitting of the Atom: the environmental legacy of nuclear weapons production in the United States and what the Department of Energy is doing about it*, and *Estimating the Cold War Mortgage: the baseline environmental management report*, US Dept of Energy, Office of Environmental Management, Washington DC. Available to the public from the US Dept of Commerce, National Technical Information Service.

WILSON, C. L., (1975) 'Nuclear energy: what went wrong?', *Bulletin of the Atomic Scientists*, June, pp. 13–17.

Further reading

AUBREY, C. (1991). *Meltdown*, Collins and Brown, London.

BERKHOUT, F. L. (1991) *Radioactive Waste: politics and technology*, Routledge, London.

BLOWERS, A., LOWRY, D. & SOLOMON, B. (1990) *The International Politics of Nuclear Waste*, Macmillan, London.

CAUFIELD, C. (1990) *Multiple Exposures: chronicles of the radiation age*, Penguin, Harmondsworth.

D'ARCY, S. & EDWARDS, R. (1995) *Still Fighting for Gemma*, Bloomsbury Press, London.

HALL, T. (1986) *Nuclear Politics: the history of nuclear power in Britain*, Penguin, Harmondsworth.

IAEA (1992) *The Safety of Nuclear Power: strategy for the future* (Conference Symposium), International Atomic Energy Agency, Vienna.

KAKU, M. & TRAINER, J. (eds.) (1982) *Nuclear Power: both sides*, W.W. Norton, London.

McSORLEY, J. (1990) *Living in the Shadow*, Pan Books, London.

PATTERSON, W. C. (1983) *Nuclear Power*, second edition, Pelican, Harmondsworth.

WILLIAMS, R. (1980) *The Nuclear Power Decisions*, Croom Helm, London.

ZONABEND, F. (1993) *The Nuclear Peninsular*, Cambridge University Press, Cambridge.

Answers to Activities

Activity 1

For The argument for deep disposal would take the following lines. These are extremely dangerous materials and must be removed from the accessible environment. Deep disposal in engineered repositories in remote areas in stable geological formations will prevent escape of radionuclides into environmental pathways over the necessary periods of time it will take for them to decay to harmless levels. Leaving such dangerous materials as spent fuel scattered over many surface storage sites will require permanent surveillance, will be a hazard to local populations and the workforce and will be vulnerable to terrorist action and the dangers of nuclear theft or war. It would be irresponsible of the present generation to impose such a problem of waste management on subsequent generations. All human activity carries some risks and those from nuclear energy must be weighed against the benefits to present and future generations of a secure energy supply. Similarly the potential dangers of nuclear warfare must be weighed against the benefits of peacekeeping through a nuclear deterrent force.

Against Conversely, it might be argued that over the timescales involved climatic and geological stability is unlikely, and earth movements and the penetration by water of even the most stable formations are a distinct possibility. Political and social changes are inevitable and the permanent monitoring of a hidden danger might become neglected. With surface storage the dangers will remain evident and changes to the storage strategy may be made in good time if necessary. Future generations would be imperilled by the out-of-sight-out-of-mind approach implied in deep disposal. Since the risks from nuclear waste are so large it would be better to minimise waste arisings by ceasing production of nuclear energy now and invest in energy efficiency or more environmentally benign energy sources, and to phase out nuclear weapons production and nuclear-powered submarines.

Activity 2

About 87% of radiation exposure is from natural sources, mostly from cosmic rays, gamma rays from rocks and soils, radon and thoron seeping into buildings, and from particles absorbed internally from eating, drinking and breathing. About 0.1% comes from nuclear discharges.

In Section 4.2 of *Sarre & Reddish, 1996*, Chapter 1, it was pointed out that there was great variability in exposure to radioactivity and that the form of exposure varied according to the type of ionising radiation involved. The precise relationship between exposure and harmful effects is little understood. Average statistics convey no information about the *concentration of exposure* ('hot spots') at specific places. The amounts of radioactivity from the nuclear industry may be small but they are concentrated in specific locations and there is the permanent risk of accidental release.

Activity 3

One obvious answer is that there was relatively little waste produced in the early years and it could be kept in store near the sites of production. But there are also less obvious political reasons encouraging lenient safety standards and primitive management systems. The lack of concern was induced and reinforced by the characteristic features of the nuclear industry evident at this time. It was a military programme, developed in the utmost secrecy, thus obviating independent oversight and regulation. In such conditions radioactive waste management was a non-problem, possibly a result of considerable ignorance about the dangers of radioactivity. It is also conceivable that the problem was recognised but deliberately suppressed: a case of *negative decision-making*, whereby those with power simply defer a decision or decide not to act, comfortable in the knowledge that inaction cannot be challenged.

Activity 4

On the face of it, by accepting the need for a comparative assessment, the government was endorsing a rational strategy. It might also reasonably assume that the opposition would be fragmented, focusing on individual sites. In due course, the relief at the three sites not selected would ensure the isolation of the community at the chosen site. Had the government not capitulated, it is conceivable that the collaboration between the four areas would have foundered once an individual site was selected.

In fact, any hope of a divide and rule strategy was prevented by the united opposition maintained by the four communities. The decision to nominate four sites multiplied the opposition, giving it a national dimension cutting across conventional political boundaries. The Coalition and BOND maintained a united front based on a technical case against shallow burial rather than against individual sites. They were able to put political pressure on the government while providing a coherent alternative which was eventually accepted.

Acknowledgements

Grateful acknowledgement is made to the following sources for permission to reproduce material in this book:

Covers

Front cover, clockwise from top right: David Parker/Science Photo Library; Information Service of the European Community; Roy Lawrance; Martin Bond/The Environmental Picture Library; Janice Robertson; Mike Levers/The Open University; Inco Limited; Information Service of the European Community; Information Service of the European Community; Chris Steele/Perkins/Magnum; *centre*: Janice Robertson; *back cover*: Information Service of the European Community.

Colour plate section

Plate 1: English China Clays Group; *Plate 2*: H. Girardet/The Environmental Picture Library; *Plate 3*: Steven C. Wilson/ENTHEOS; *Plate 4*: British Coal Opencast/Peter N. Grimshaw; *Plate 5*: Cheshire County Council; *Plate 6*: British Coal Opencast/Peter N. Grimshaw; *Plate 7*: Steven C. Wilson/ENTHEOS; *Plate 8*: Steven C. Wilson/ENTHEOS; *Plate 9*: Vanessa Miles/The Environmental Picture Library; *Plate 10*: P. Glendell/The Environmental Picture Library; *Plate 11*: Forestry Research Institute, The Chinese Academy of Forestry/Steve Newman/Fountain Renewable Resources; *Plate 12*: D. Taylor; *Plate 13*: R. Everett; *Plate 14*: D. Olivier.

Figures

Figures 1.1(a), 1.1(b), 1.3, 1.11, 1.13: The BP Statistical Review of World Energy 1995, British Petroleum plc; *Figures 1.9 and 1.12:* Krenz, J. H. (1980), *Energy: From Opulence to Sufficiency*, Praeger; *Figure 1.10:* from *Energy Technologies for UK* 1979 **1**, Department of Energy Paper No. 39, © Crown Copyright. Reproduced with the permission of the Controller of Her Majesty's Stationery Office; *Figure 2.8:* Aerofilms; *Figure 2.9, 2.12 and 2.14(b):* Park, C. P. (1987), *Acid Rain: Rhetoric and Reality*, Methuen & Co.; *Figure 2.10:* Manners, G. (1981), *Coal in Britain*, Allen & Unwin; *Figure 2.11:* Commission on Energy and the Environment (1981), *Coal and the Environment*, © Crown Copyright. Reproduced with the permission of the Controller of Her Majesty's Stationery Office; *Figure 2.15:* Vanessa Miles/Environmental Picture Library; *Figure 3.2:* Saxhof, B., Technical University of Denmark; *Figures 3.4, 3.16 and 3.17: Energy Paper* **60**, Renewable Energy Advisory Group, © Crown Copyright. Reproduced with the permission of the Controller of Her Majesty's Stationery Office; *Figure 3.7:* JET Joint Undertaking in *Culham Laboratory Annual Report 1988 (CLM-AR26)*, (1989); *Figure 3.9:* copyright © James L. Ruhle and Associates; *Figure 3.11:* Wind Energy Group; *Figure 3.12(b):* S. E. A./Lancashire Polytechnic; *Figure 3.13:* Baker, A. C. 'The development of functions relating to cost and performance of tidal power schemes and their application to small scale sites', Symposium on Tidal Power, Thomas Telford Ltd, I. C. E., 1986; *Figure 3.19:* CEGB; *Figure 3.20: Developing Wind Energy for the UK*, (1990), Friends of the Earth; *Figure 4.1:* British Coal; *Figure 4.2:* Inco Limited; *Figure 4.3:* John Blunden; *Figure 4.4:* Gilliland, P. (1979), *Atlas of Earth Resources*, Mitchell Beazley; *Figure 4.6: New Scientist*, 9 December 1982, p. 649; *Figure 4.7: Peruvian Times; Figure 4.8:* West Air Photography; *Figure 4.10:* Singer, D. A. (1977), *Journal Resources Policy*, **3**(2) June, pp. 127–33; *Figure 5.1:* Aerofilms; *Figure 5.2:* Inco Limited; *Figure 5.3: United Kingdom Minerals Yearbook 1994*, © 1995 British Geological Survey; *Figure 5.4:* Copyright © Commission on Mining and the Environment/Photo supplied by Rio Tinto Zinc Ltd; *Figure 5.5:* Aerofilms; *Figure 5.6:* Robert Del Tredici copyright © photograph; *Figure 5.7, 5.8 and 5.9:* Conroy, N., Hawley, K., Keller, W. and LaFrance, C. (Ontario Ministry of the Environment) (1975) Influence of the atmosphere on lakes in the Sudbury area from source (unpublished paper); *Figure 5.10:* Nriagu, J. O. (1984), *Environmental Impacts of Smelters*, John Wiley & Sons; *Figure 5.11:* Troug (1951), *Mineral Nutrition of Plants*, University of Wisconsin Press; *Figure 5.12 (left):* Cotswold Water Park; *Figure 5.12 (right):* RMC Group plc; *Figure 5.13:* Inco Limited; *Figures 6.1–6.8: Industry Technology: Visual History of Modern Britain*, (1965), Studio Vista, a division of Cassell plc; *Figures 6.9, 6.11 and 6.13:* Manchester Central Library, Local Studies Unit; *Figure 6.10:* Guild Hall Library, Corporation of London. Photo: Geremy Butler; *Figure 6.14:* from *Canvey: An Investigation of Potential Hazards from Operations in the Canvey Island/Thurrock Area*, (1978),

© Crown Copyright. Reproduced with the permission of the Controller of Her Majesty's Stationery Office; *Figure 6.15:* Parker Q. C., R. J. (1975), *The Flixborough Disaster: Report of the Court of Enquiry*, Department of Education, © Crown Copyright. Reproduced with the permission of the Controller of Her Majesty's Stationery Office; *Figure 6.17:* Health and Safety Executive (1992), *The Tolerability of Risks from all Nuclear Power Stations*, © Crown Copyright. Reproduced with the permission of the Controller of Her Majesty's Stationery Office; *Figures 6.19 and 6.20:* Allott, K. (1994), *Integrated Pollution Control* Mayer, M. (ed.), pp. 56 and 58, Environmental Data Services Ltd; *Figure 6.23:* Rowe, W. D. (1977), 'A hierarchy of risk assessment terminology', *The Anatomy of Risk*, p. 45, John Wiley & Sons; *Figure 6.24:* Perry, A. H. (1973), *Environmental Hazards in the British Isles*, Allen & Unwin Ltd; *Figures 7.1 and 7.4:* Blowers, A., Lowry, D. and Solomon, D. (eds) (1991), *The International Politics of Nuclear Waste*, Macmillan Press, UK, St Martin's Press, New York; *Figure 7.3: The Hindustan Times*, New Delhi, India; *Figure 7.5:* 'Nuclear dustbins for the centuries', *The Guardian*, 7 January 1974, reproduced with the permission of Professor Bryce-Smith on behalf of all signatories; *Figure 7.6:* Bedfordshire County Council; *Figure 7.7: Evening Gazette*, Teesside; *Figure 7.8:* copyright © Lincolnshire Echo; *Figure 7.9:* Vass, J. 'Man who hopes to see N-dump under his garden', *Aberdeen Press and Journal*, 31 July 1990.

Photographs
Pages 80–81: Peter Thornton/Lakeland Photographic; *Page 110:* photographed from *Revue Archaeologique* 1899, i, p. 329, plate xii; *Pages 280–281:* Echo Newspapers Ltd; *Page 303:* David Austin/*New Scientist*.

Tables
Table 2.2: Skea, J. 'Acid Rain and Urban Atmospheric Pollution: Europe', paper presented to the *4th International Energy Conference: Environmental Challenge: The Energy Response*, The Royal Institute, December 1989; *Table 3.2:* Shea, C. P. 'Renewable Energy: Today's Contribution, Tomorrow's Promise' from Worldwatch Paper 81, Worldwatch Institute, January 1988; *Table 3.3:* Harrison, L. (1992), 'Europe gets clean away', *Windpower Monthly*, September 1992, Torgny Moeller; *Table 4.1:* Gilliland, P. (1979), *Atlas of Earth Resources*, Mitchell Beazley; *Table 4.2:* Bell, E. from Frosch, R. A. and Gallopoulos, N. E. (1989), 'Strategies for manufacturing', *Scientific American*, **261**, September; *Table 5.3:* adapted from proceedings of an international symposium on waste material sponsored by the Réunion Internationale des laboratoires d'Essais et de Recherches sur les Matériaux et les Constructions (RILEM), published in *Materials and Structures*, **12**(70), 1979.

Text
Box 3.1: Shawbond, M. (1995), 'Russia's pollution capital', *Financial Times*, 2 September 1995; *Box 4.3:* adapted from Gilliland, P. (1979), *Atlas of Earth Resources*, Mitchell Beazley; *Box 5.2:* Winterhalder, K. *Proceedings Third Annual Meeting of Canadian Land Reclamation Association*, Laurentian University, 1978; *Box 5.3:* Lertola, J. from Frosch, R. A. and Gallopoulos, N. E. (1989), 'Strategies for Manufacturing, *Scientific American*, **261**, September.

Index

The Open University U206 *Environment* Course Team
Mary Bell, Staff Tutor, Cambridge
Roger Blackmore, Staff Tutor, London
Andrew Blowers, Professor of Planning
John Blunden, Reader in Geography
Stuart Brown, Professor of Philosophy
Michael Gillman, Lecturer in Biology
Petr Jehlicka, Lecturer in Environment
Pat Jess, Staff Tutor, Northern Ireland
Alistair Morgan, Lecturer in Educational Technology
Eleanor Morris, Producer, BBC
Suresh Nesaratnam, Lecturer in Engineering Mechanics
Alan Reddish, Course Team Chair
Philip Sarre, Senior Lecturer in Geography
Varrie Scott, Course Manager
Jonathan Silvertown, Senior Lecturer in Biology
Sandra Smith, Lecturer in Earth Science
Kiki Warr, Lecturer in Chemistry

Consultants
Christine Blackmore, Lecturer in Environment and Development, The Open University
Ian Bowler, Senior Lecturer in Geography, University of Leicester
David Elliott, Senior Lecturer in Design, The Open University
Allan Findlay, Professor of Geography, University of Dundee
Patricia Garside, Professor of Environmental Health and Housing, University of Salford
Owen Greene, Lecturer in Peace Studies, University of Bradford
David Grigg, Professor of Geography, University of Sheffield
Cliff Hague, Professor, School of Planning and Housing, Edinburgh College of Art/Heriot-Watt University
David Lowry, Research Fellow in Energy and Environment Research Unit, The Open University
Christopher Miller, Senior Lecturer in Health, University of Salford
David Olivier, Energy Consultant
Alan Prior, School of Planning and Housing, Edinburgh College of Art/Heriot-Watt University
Jeremy Raemaekers, School of Planning and Housing, Edinburgh College of Art/Heriot-Watt University
Marcus Rand, Energy and Environment Research Unit, The Open University
Shelagh Ross, Staff Tutor, The Open University, Cambridge
Ian Simmons, Professor of Geography, University of Durham
Annie Taylor, Faculty of Social Sciences, University of Southampton
Philip Woodhouse, Research Fellow, Faculty of Technology, The Open University

External Assessor
Professor Sir Frederick Holliday, CBE, FRSE, former Vice Chancellor and Warden, University of Durham

Tutor Panel
Liz Chidley, Part-time Tutor, The Open University, Cambridge
Judy Clark, Part-time Tutor, The Open University, East Grinstead
Margaret Cruikshank, Part-time Tutor, The Open University, Northern Ireland
Helen Jones, Part-time Tutor, The Open University, Manchester

Course Production at The Open University
Melanie Bayley, Editor
David Calderwood, Project Controller
Margaret Charters, Secretary
Lene Connolly, Print Buying Controller
Margaret Dickens, Print Buying Co-ordinator
Harry Dodd, Print Production Controller
Janis Gilbert, Graphic Artist
Rob Lyon, Graphic Designer
Michèle Marsh, Secretary
Janice Robertson, Editor
Paul Smith, Liaison Librarian
Doreen Warwick, Secretary
Kathy Wilson, Production Assistant, BBC